Critical and Clinical Cartographies

New Materialisms
Series editors: Iris van der Tuin and Rosi Braidotti

New Materialisms asks how materiality permits representation, actualises ethical subjectivities and innovates the political. The series will provide a discursive hub and an institutional home to this vibrant emerging field and open it up to a wider readership.

Editorial Advisory board
Marie-Luise Angerer, Karen Barad, Corinna Bath, Barbara Bolt, Felicity Colman, Manuel DeLanda, Richard Grusin, Vicki Kirby, Gregg Lambert, Nina Lykke, Brian Massumi, Henk Oosterling, Arun Saldanha.

Books available
What if Culture was Nature all Along?
Edited by Vicki Kirby
Critical and Clinical Cartographies: Architecture, Robotics, Medicine, Philosophy
Edited by Andrej Radman and Heidi Sohn

Books forthcoming
Architectural Materialisms: Non-Human Creativity
Edited by Maria Voyatzaki

Critical and Clinical Cartographies

Architecture, Robotics, Medicine, Philosophy

Edited by Andrej Radman and
Heidi Sohn

EDINBURGH
University Press

We publish academic books and journals in our selected subject areas across the humanities and social sciences, combining cutting-edge scholarship with high editorial and production values to produce academic works of lasting importance. For more information visit our website: edinburghuniversitypress.com

© editorial matter and organisation Andrej Radman and Heidi Sohn, 2017
© the chapters their several authors, 2017

Edinburgh University Press Ltd
The Tun – Holyrood Road, 12(2f) Jackson's Entry, Edinburgh EH8 8PJ

Typeset in 11/13 Adobe Sabon by
IDSUK (DataConnection) Ltd

A CIP record for this book is available from the British Library

ISBN 978 1 4744 2111 9 (hardback)
ISBN 978 1 4744 2112 6 (webready PDF)
ISBN 978 1 4744 2113 3 (epub)

The right of Andrej Radman and Heidi Sohn to be identified as the editors of this work has been asserted in accordance with the Copyright, Designs and Patents Act 1988, and the Copyright and Related Rights Regulations 2003 (SI No. 2498).

Contents

Acknowledgements vii

The Four Domains of the Plane of Consistency 1
   Andrej Radman and Heidi Sohn

Introduction: A Research into Human–Machine Technologies –
Architecture's Dream of a Bio Future 21
   Arie Graafland

Part I: Architecture

1. Urban Correlationism: A Matter of Access 61
   Stavros Kousoulas

2. Housing Biopolitics and Care 80
   Peg Rawes

3. Amorphous Continua 101
   Chris L. Smith

Part II: Robotics

4. Robots Don't Care: Why Bots Won't Reboot Architecture 123
   Christian Girard

5. The Convivial ART of *Vortical* Thinking 143
   Keith Evan Green

6. Emotive Embodiments 168
   Kas Oosterhuis

Part III: Medicine

7. Ecologies of Corporeal Space 187
   Katharina D. Martin

8. Swimming in the Joint 205
   Rachel Prentice

9. Key-Hole Surgery: Minimally Invasive Technology 221
   Jenny Dankelman

Part IV: Philosophy

10. Elasticity and Plasticity: Anthropo-design and
    the Crisis of Repetition 243
    Sjoerd van Tuinen

11. Automata, Man-machines and Embodiment:
    Deflating or Inflating Life? 269
    Charles T. Wolfe

12. Generative Futures: On Affirmative Ethics 288
    Rosi Braidotti

Notes on Contributors 309
Index 313

Acknowledgements

This book is the result of the international conference Critical and Clinical Cartographies, which was organised by the Architecture Theory section, in collaboration with Hyperbody, of the Faculty of Architecture of the Delft University of Technology (TU Delft), and held in the Armamentarium in Delft in November 2014. The editors wish to thank Arie Graafland for suggesting the core theme for the conference and for generously sharing his research with us. Without him, the conference and this book would never have been possible. We would also like to extend our gratitude to the organising and scientific committees for their insights and feedback in the preparation of the conference and this book, and in particular to the staff of the Architecture Theory section – Patrick Healy, Stavros Kousoulas and Gregory Bracken – for their tireless efforts and enduring support. To the Department of Architecture of the Faculty of Architecture of the TU Delft we owe the generous administrative support they have offered us throughout.

We are especially indebted to Carol Macdonald of Edinburgh University Press for her receptiveness and enthusiasm for our book project, and for her valuable editorial comments and assistance, and to Heleen Schröder, for the magnificent copyediting input and professionalism in preparing this volume for publication.

We extend our special thanks to all the authors in this book for their generous contributions and cooperation. And last, but not least, we wish to thank our students, to whom we dedicate this book.

The Four Domains of the Plane of Consistency

Andrej Radman and Heidi Sohn

> This is Major Tom to Ground Control.
> I'm stepping through the door.
> And I'm floating in a most peculiar way.
> And the stars look very different today.
> (D. Bowie, 'Space Oddity', 1969)

3C Glossary

It is our privilege to introduce the book that was triggered by the conference on Critical and Clinical Cartographies (or 3C for short) that took place at Delft University of Technology, the Netherlands, in November 2014.[1]

First, let us briefly take you through a glossary to situate this neo-materialist project, starting with the three Cs. The first two – *critical* and *clinical* – are directly appropriated from Gilles Deleuze's *Essays Critical and Clinical*.[2] Already in 1988 Deleuze revealed his long-standing plan to write a series of studies on writers as great *symptomatologists* under the overarching label 'Critique et Clinique'.[3] The collection of essays was published five years later.[4] In his English translation from 1997, Daniel W. Smith offered a rare commentary on this aspect of Deleuze's thought.[5] It was only in 2010 that the journal *Deleuze Studies* devoted an issue to the topic of *Deleuze and the Symptom*, edited by Aidan Tynan, with the subtitle *On the Practice and Paradox of Health*.[6] Please note the two new additions to our glossary, namely, the *symptom* and *health*. There is a certain divergence in Deleuze's conception of the clinical, Tynan explains. If in the *symptomatological* register the symptom is diagnostic and relates to the creation of new clinical entities, in the *schizoanalytical* mode it is therapeutic with an injunction to produce. Schizoanalytical practice is attuned to desiring-production and as such repudiates the philosophical and psychoanalytic tie to a hidden meaning to *interpret*.[7] In other words, health already implies practice, insofar as signs imply ways of living, forms of life.

As Deleuze observed, it was this very ability to shift between the two *non*-mutually exclusive perspectives – *illness* and *health* – that constituted the Nietzschean concept of '*great health*'.[8] Let us recall that Deleuze was a student of the author of *The Normal and the Pathological*, Georges Canguilhem.[9] According to Michael Hagner, the radicality of Canguilhem's book lies in the *historical* framing of the human and its autocorrecting functions. As we shall see below, the concept of health determines standards and in turn depends on the context of their determination (double bind).[10] Paradoxically, the blurring of the boundary between the normal and the pathological, Smith surmises, allows for *poor* health to become the very condition of 'great health'. Frailty turns out to be that which forces genuinely creative thought upon us:

> The question that links [art] and life, in both its ontological [what-there-is] and its ethical [how-to-live] aspects, is the question of health. This does not mean that an author necessarily enjoys robust health; on the contrary, artists, like philosophers, often have frail health, a weak constitution, a fragile personal life (Spinoza's frailty, D. H. Lawrence's hemoptysis, Nietzsche's migraines, Deleuze's own respiratory ailments). This frailty, however, does not simply stem from their illnesses or neuroses, says Deleuze, but from having seen or felt something in life that is too great for them, something unbearable 'that has put on them the quiet mark of death.'[11]

The 'great health' requires us to think the critique as a process – productive of the 'real' rather than the Kantian a priori form of 'possible' experience – and to thus rethink the relation between bodies and signs. Symptomatology is thus best defined as the study of signs. As Goethe would have it, this is a *genetic* rather than a *generic* approach.[12] Privileging a method of morpho*genesis* or individuation over mere generic *conditioning* was one of the main motivations behind Deleuze's synthesis of literary criticism and clinical diagnosis. As we will argue in the subsequent part, synthesis – that denotes a constructivist practice in general – is not to be taken as analysis in reverse. Anne Sauvagnargues elucidates further that 'a synthesis, for Deleuze, is not a return to the One, but a disjunctive differentiation that proceeds by bifurcations and transformations, and not by fusion and identity of the same'.[13]

We must confess that, like our colleague Chris L. Smith in his 'Architectures, Critical and Clinical', we too succumb to the following temptation.[14] Whenever Deleuze speaks of the *writer*, we cannot but think of the *architect*, and when Deleuze speaks of *literature* and of its 'revolutionary force', we cannot but substitute it with *architecture*.[15] Let us immediately test it out: 'The architect as such is not a patient but a physician [*care*-taker], the physician of himself and of the world. The world is a set of symptoms whose illness merges with man. Architecture then appears as the enterprise of health.'[16]

It is neither the materiality of the sensuous body nor the immateriality of the signs that render meaning, but a space of reciprocal determination, real yet incorporeal. Deleuze calls this intensive space or *spatium* the 'body without organs' (*corps sans organs*). BwO is not the body (being) but the very process of de-re-territorialisation, that is, *embodiment*. Embodiment is thus defined as the reconversion of a stratified system of expression into a 'pre-individual' field of emergence (becoming). Here comes *the* neo-materialist formula, a prescribed *technology* if you will: 'To make the body a power which is not reducible to the organism, to make thought a power which is not reducible to consciousness.'[17] This is the injunction of immanence.[18]

Finally, what of the third C that stands for *cartography* 'rather than classification', or cartographies, to be precise?[19] Our journey through the glossary has been an enactment of one of the many cartographies to come in the pages that follow, always drawn in a non-totalising manner. As Brian Massumi recently put it, reality is 'both emergent and constructed, as presupposed as it is produced, given for the making and made a given'.[20]

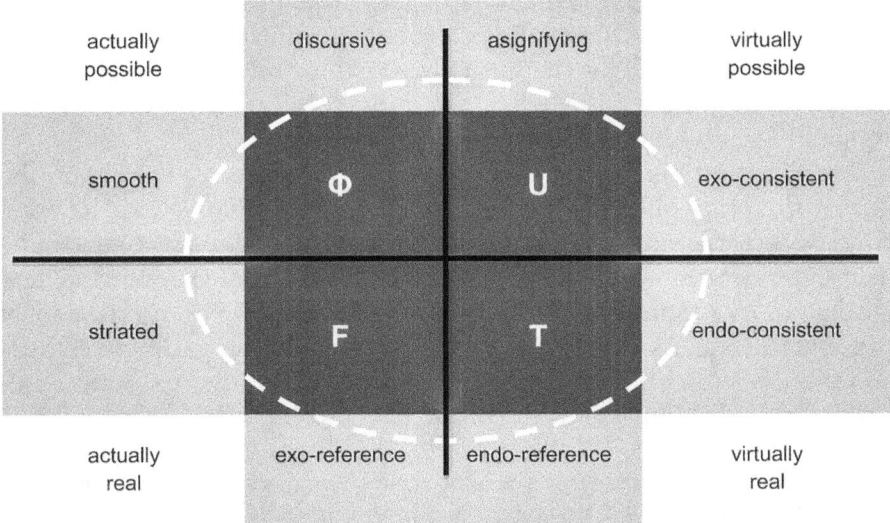

Figure 1 Axes of discursivity/reference and de-territorialisation/consistency, based on Félix Guattari, *Schizoanalytic Cartographies*, trans. Andrew Goffey (London: Bloomsbury, [1989] 2013).
Source: author.

Four Ontological Functors

The book is divided into four parts. Each quadrant consists of three chapters (that is, literary machines). Let us tentatively propose that those four parts are the four ontological functors, as in Guattarian metamodeling.[21] Guattari calls them functors to indicate their transformative effects on the assemblage whose overall dynamics they initiate and sustain:

> The Schizoanalytic Cartographies map out the existential and social parameters within which a desire comes both to problematize itself in thought and to release its otherness in expression – thereby helping to create a new context and to launch a new cycle of transformations . . . Existence itself – or the event that is existence – can be understood as a continuous temporal permutation linking and transforming these four poles; while the condition of domination consists in any attempt to freeze the cycle into a structure of fixed relations, or to guide it along a predetermined and repetitive path.[22]

We start from the middle (*milieu*), that is, from T, the Existential Territory of Architecture. That is where existential apprehension (*umwelt*) starts from anyway.[23] Architecture is virtually real, that is, pathic. The quadrant furthest away from it is Φ, the Machinic Phylum of Robotics, ever proliferating rhizomatically (new technologies always exceed their original premises and uses).[24] Guattari insists that there are no direct tensors between the diagonally placed quadrants of T and Φ, nor between F and U.[25] Φ is actually possible and connects to T only via the domains of F and U. U is the virtually possible Philosophical Universes of Reference (Value). F is Medicine in our case, as an actually real social domain of material and energetic Flows (Fig. 1).

Architectural Territories and Philosophical Universes belong on the side of the quasi-subjective *giving*, while Medical Flows and Robotic Phyla occupy the side of the quasi-objective *given*. When it comes to the cut (schiz) between the virtual (T – U) and the actual (F – Φ) side, it is crucial not to succumb to the fallacies perpetuated by both rationalism and empiricism. As Beth Lord puts it succinctly, the gap between the two should neither be annihilated as in *rationalism* (the real is rational) nor should it be widened as in *empiricism*:

> Dogmatism fills the gap between a determinable object and its conceptual determination either by arguing for the a priori complete determination of the former by the latter . . . or by showing that they necessarily collapse into a new indeterminate unity. But empiricism leaves the determinable object and its conceptual determination separate, such that the determinable–determinant relation is external to the thing to be determined.[26]

THE FOUR DOMAINS OF THE PLANE OF CONSISTENCY

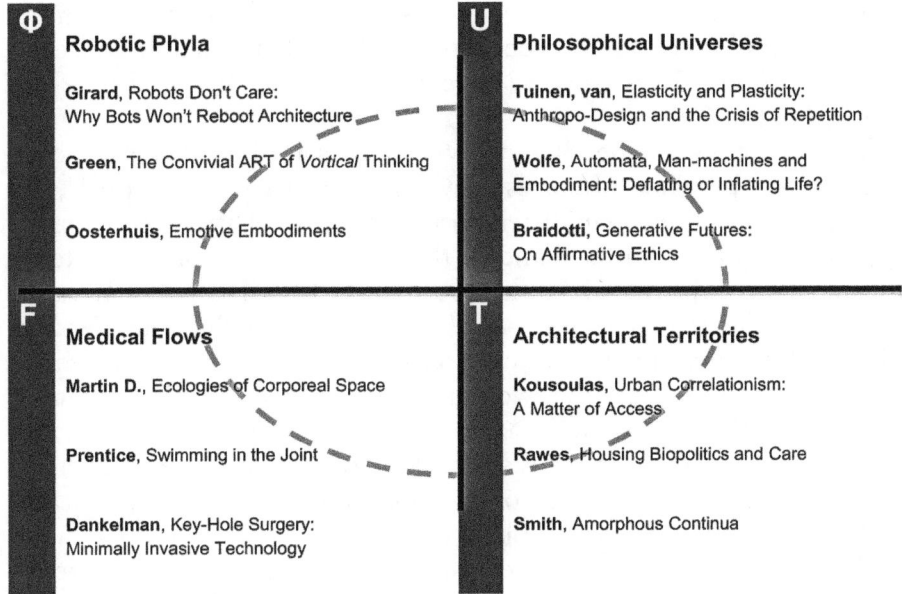

Figure 2 The four parts of the 3C book as ontological functors, each populated by three chapters, modelled after Félix Guattari, *Schizoanalytic Cartographies*, trans. Andrew Goffey (London: Bloomsbury, [1989] 2013).
Source: author.

Lord concludes that a more profound transcendental philosophy must posit the crucial relation – determinant/determinable/undetermined – in such a way that the determination of a thing is never broader than the thing itself. This, in a nutshell, is the 'plastic principle' as explained by Deleuze in his book on Nietzsche. Hence, neo-materialist cartography is meant to render visible a condition 'that changes itself with the conditioned and determines itself in each case along with what it determines'.[27]

Critical and Clinical Cartographies kicks off with an introduction by Arie Graafland who effectively (or should we say counter-effectuatively) positions architectural existential apprehension (T) in relation to the flows of matter and energy (F) on the one hand, and the incorporeal Universes of reference (U) on the other (Fig. 2). The latter are made up of values and non-discursive references that 'escape from the energetic, legal, evolutionary, and existential coordinates' of the other three domains, including the abstract machines that preside over objective laws and changes (Φ).[28] The abstract machinic phyla comprise evolution and blueprints, plans, diagrams, rules and regulations in the cybernetic sense of control mechanisms.[29] To avoid any confusion yet another mini-glossary is in order – that of Graafland's major concepts: F = *Technology*; U = *Pedagogy* (Education); Φ = the *Digital* (Turn); and finally, T = the *Ground*.

The mapping in the 3C involves Guattari's four heterogeneous domains on the plane of consistency: the *critical* half of T and U, the virtual cutouts of existential territories and constellations of ritornellos and the *clinical* half of F and Φ, the actual complexions of material and energetic flows and rhizomes of abstract ideas. The neo-materialist metamodelling remains *ethico-aesthetic* and decidedly non-scientific, where 'ethics' is related to the act of selection and 'aesthetics' to creative productivity.[30] These are the two main Darwinian mechanisms that are coupled together: the *production* of continuous variation and *selection* as discrimination that introduces some kind of bias into the proceedings.

In the concluding chapter of *A Thousand Plateaus* we learn that abstract machines striate the plane of immanence (by way of symmetry-breaking) to produce concrete rules.[31] Rules are practical or, better said, pragmatic, injunctions.[32] This is to say that the source of any normativity must not come from the outside. Deviations are therefore not to be taken as abnormal but as normally *anomalous*.[33] Rule-bonding needs to operate at the level of stratification, at its own terms. Truth and Falsity are not values that exist outside the constitutive problematic fields that give them sense. Ethics, framed in this way, is a problem of *power* and not duty. 'Transgressing the law does not make any sense', explains Sauvagnargues, 'because the law does not exist as an external and transcendent moral imperative that would be possible for anyone to follow or transgress. If there is a law, it regulates real behaviors.'[34] Rather than relying upon the transcendent *logos*, new materialism shifts the emphasis to the 'natural law' of *nomos*.[35] Clinical criticality, in other words, knows no other than the *immanent* criteria of evaluation. It escapes the overcode.[36] To paraphrase Steven Shaviro (paraphrasing Brecht), let us start from the 'bad new things' instead of the 'good old ones'.[37]

If we read Guattari it becomes obvious why Leonardo da Vinci's flying machines never took off in spite of his ingenuity.[38] To cut the long story short, they never did because the project got stuck on the right-hand side of the fourfold matrix in Figure 1. There was not sufficient flux (F) for lift-off at the time, literally speaking. The so-called 'imagination' (voluntarism) does not suffice. It is often said that imagination is your only limit. This is simply wrong, as Benjamin Bretton rightly cautioned in his TED Talk against TED talks:

> If we really want transformation, we have to slog through the hard stuff (history, economics, philosophy, art, ambiguities, contradictions). Bracketing it off to the side to focus just on technology, or just on innovation, actually *prevents* transformation.
>
> Instead of dumbing down the future, we need to raise the level of general understanding to the level of complexity of the systems in which we are embedded and which are embedded in us. This is not about 'personal stories of inspiration', it's about the difficult and uncertain work of demystification and reconceptualisation: the hard stuff that really changes how we think.[39]

The quasi-subjective *pathic* and the quasi-objective *ontic* thrive on each other's energetics. That is why the success of our neo-materialist project depends on prolific disciplinary trespassing and transversality. In the prodigious words of Guattari:

> If one does not want to fall into a childish naturalism opposing nature and culture, infrastructure to superstructure, if one really wants to describe how historic mutations operate, it seems to me necessary to develop expanded concepts of the machine that account for what it is in all its aspects. There are visible synchronic dimensions, but also diachronic virtual dimensions: a machine is something that situates itself at the limit of a series of anterior machines and which throws out the evolutionary phylum [Φ] for machines to come; it is thus a material and semiotic assemblage which has the virtue of traversing, not only time and space, but also extremely diverse levels of existence concerning as much the brain as biology, sentiments, collective investments.[40]

3C is both about 'singularising ritornellisation' *and* re-territorialisation. Simply put, it is against the normative idea of homeostasis. Simpler still, genuine transformation may occur only if the movement between the four domains (T–U–Φ–F), clockwise and anticlockwise, is not arrested (black hole phenomenon).[41] Once again a caveat is necessary: there cannot be any T–Φ and U–F shortcuts. Adjacency itself is crucial, as it engenders the adjacent and not the other way around. This is how Guattari explains resingularisation in his seminal *Chaosmosis*: 'Grafts of transference operate in this way, not issuing from ready-made dimensions of subjectivity crystallised into structural complexes, but from a creation which itself indicates a kind of aesthetic paradigm.'[42] This, in turn, implies a *processual* exploitation of event-centred 'singularities'. Some will find it objectionable to endow architecture with subjectivity, but this is how Simone Brott addresses the issue on the first page of her *Architecture for a Free Subjectivity*:

> Subjectivity is, for Deleuze, not a person, but a power given to immanent forces to act and to produce effects in the world. In short, it is the field of what I call *subjectivization*, meaning the potential for and event of matter

becoming subject, and the multiple ways for this to take place. Deleuze, in fact, tends not to use the word *subjectivity*, speaking instead of 'affects' – the capacity to affect and be affected – and 'pre-personal singularities,' meaning those irreducible qualities or powers that act independently of any particular person.[43]

One must appreciate Guattari's distinction between two modes of destratification, the quasi-objective (F–Φ) in the *parastrata* and the quasi-subjective (T–U) on the *epistrata*.[44] A good example is the Apollo Program.[45] While Leonardo's flying machines never took off due to inadequate *flux* (F), the Apollo Program forcefully landed because of the political and social change in the *phylum*. Everything hinges on ideas in their nascent state (U), but they have to be extended to the *collective* desire. In other words, you need the constituency as a *given* (Φ). You also need the technology as a *given* (F). Both in turn feed the *giving* side of the territory (T) and the values (U).

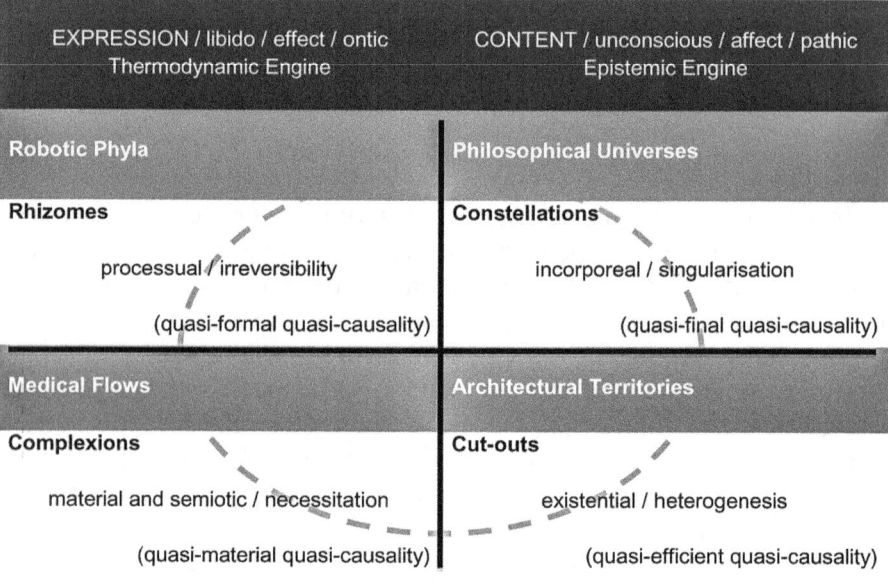

Figure 3 Axes of Expression/Thermodynamic Engine and Content/Epistemic Engine, based on Félix Guattari, *Schizoanalytic Cartographies*, trans. Andrew Goffey (London: Bloomsbury, [1989] 2013) and Peter N. Kugler and Robert Shaw, 'Symmetry and Symmetry-Breaking in Thermodynamic and Epistemic Engines: A Coupling of First and Second Laws', in *Synergetics of Cognition*, ed. Hermann Haken and Michael Stadler (Heidelberg: Springer-Verlag Berlin, 1990), pp. 296–331. *Source: author.*

The neo-materialist approach employed in this book diverges from conventional methods of architectural analysis in that it derives its rationale from a theoretical and philosophical tradition, which bears important connections to architecture as a material practice *both* corporeal and incorporeal.[46] Under this minor tradition, the understanding of 'matter', 'objects' and 'bodies', as well as 'process', 'emergence', and 'change', will necessarily depart from common sense, common logic and common knowledge of these concepts, and their often inaccurate or inappropriate use in architecture. That is why Sanford Kwinter can confidently predicate longevity of architecture on its elimination.[47] By elimination he does not mean annihilation but abstraction. Abstract in Deleuze is synonymous with destratified, that is, unformed matter (U) and unstructured expression (Φ). What matters are the intensities (pun intended). This is the principle of the exteriority of relations. However counterintuitive this may seem, the real relation is external to its *relata*. Sauvagnargues makes this point in her *Deleuze and Art* (2013) repeatedly:

> The body without organs is useful when thinking about the corporeality and morphogenesis of bodies without tying them to an external unifying principle, such as the soul, form, or the unity of organism, but by being located at the level of matter that is not yet informed and is on the plane of forces.[48]

Abstraction, in other words, is the immanent mapping of singularities that hold the assemblage in the absence of a (transcendent) tie. This is the N-1 operation par excellence: it is not about subsuming all kinds of things under a (too-baggy) concept, but about relating each concept to the variables that determine its mutation (think about the rates of change, differentials and not formal or dialectical difference). A concept defined in this way is not even a concept, but a multiplicity always proliferating, rhizomatically.[49] As Rosi Braidotti would say, don't reason, rhizome instead! Janell Watson explains the role of Guattari's schizoanalysis in opening up new possibilities for the libido and the unconscious by choosing the deterritorialised:

> Locating the libido in [Φ], it allows for a process-related energy which pushes dynamic relations into the far-from-equilibrium conditions necessary for transformation and creation [top left quadrant]. This is the domain of the actual possible, discursive yet deterritorialized . . . Schizoanalysis likewise rescues the unconscious from the [T], elevating it to the equally deterritorialized incorporeal [U]. This constitutes the unconscious as 'the set [ensemble] of lines of alterity, virtual possibilities, and unheard-of becomings' (CS 44). The relation between the libido and the unconscious can then become one of processes of singularization, predicated on two-way exchanges between the two sides of the graph.[50]

We draw inexhaustible inspiration from the Guattarian mapping that works impeccably for architecture in its *irreducibility*: the body (T) – auto-affective intensities (U) – innovations (Φ) – institutions (F)[51] (Fig. 3). Again, *giving* means the virtual, and the *given* means the actual. There is isomorphism between the two unequal 'halves' (content is just too big for expression), but no resemblance. The concept of autonomy makes us laugh and we consequently dismiss the cult of the artefact as implausible. Architectural design is action at a distance (aka collective enunciation) in a profound sense.[52] Richard Sennett concurs:

> Architecture forms a special case in relation to the ideal of integrity, for it comes into being in ways paintings, sculptures, and poems do not. The making of a piece of urban architecture is a messy process, involving an army of specialist designers and technicians at war with opposing armies of government officials, bankers and clients.[53]

As keen new materialists we believe that the reality as such is produced through stratification (individuation), be it on the physico-chemical, biological or socio-political (allomorphic) plateau. We learn from Deleuze and Guattari that, apart from the production of variation and selection (connective and disjunctive syntheses), there is a remainder (conjunctive synthesis), a whole that exists alongside its parts.[54] Irreducibility and/or irreversibility is an effect of interacting parts that produce an emergent distributed whole, which in turn constrains the interacting parts (*natura naturans*). The triad of the production production/recording/consumption as real (not dialectical) syntheses resonates strongly with the economy of giving/receiving/returning as theorised by Marcel Mauss in his seminal work titled *The Gift*.[55] The 'gift economy' is a mode of exchange where valuables are given rather than sold without any promise of immediate or future reward. The three stages of exchange have been rendered as the Three Graces.

From the architectural point of view, the concept is most thoroughly taken up by Lars Spuybroek.[56] Spuybroek develops the idea that beauty evolved as an aesthetic concept from gift exchange, which in ancient Greek culture centred on the notion of charis, usually translated as 'grace', though it also means favour, gratitude, pleasure and beauty. The gift cycle embodied by the Three Graces (giving/receiving/returning) is common to all cultures. In contrast to the market economy with its explicit exchange of goods or services for money or other commodity (that is, rational choice theory), gift exchange is governed by social norms (as in *nomos*) and metastable customs. As Massumi put it in his recent book *The Power at the End of the Economy*, freedom is not chosen [or bought], but invented:

This political-intuitive invention is a necessity of life. There is a need to escape the presuppositions of the field of relation into which we are collectively braced. Not by opposing them with an alternative utopian universe where individual choice is finally enabled and allowed free rein (as if such a thing were conceivable, given the web of interdependencies that are part of the warp and woof of life). Not by rationalizing the entire field of life through the good graces of a tribunal of judgment acting from on high (were such a thing possible, even if it were desirable). Rather, by immanently event-converting this rabbit-holed neoliberal world in which we churn. By inventively gesturing neoliberal life toward a whole-field change of state.[57]

The 3C's methodological 'transgressions' seek to provoke and unsettle the conventional systems of classification as hermeneutic vehicles in order to observe processes of emergence, understand the forces that trigger them and map the transformations.[58] As Michel Foucault anticipated in 'Society Must Be Defended', these phenomena are, essentially, aleatory events:

What we are dealing with in this new technology of power is not exactly society (or at least not the social body, as defined by the jurists), nor is it the individual-as-body. It is a new body, a multiple body, a body with so many heads that, while they might not be infinite in number, cannot necessarily be counted. Biopolitics deals with the population, with the population as political problem, as a problem that is at once scientific and political, as a biological problem and as power's problem.[59]

Our book closes on the Spinozian cry: we don't know what a body can do. It is hoped that it is much more than wishful thinking combined with our 'triadomania' (symptom?) that makes us relate the three stages of exchange, personified as the three sister-goddesses – Aglaia/Euphrosyne/Thalia – with the three stages of determination – determinant/determinable/undetermined.[60] In opposition to any foundationalism (that is, fundamentalism), the necessity of *foundation* exists only for *determinable* (yet undetermined) ground, not the final complete determination that remains ipso facto only ever reciprocally determined. Reciprocal determination means that the concrete rules of assemblages effectuate/develop the abstract machines that are, in turn, enveloped in the strata. No wonder that the crucial chapter in *A Thousand Plateaus* 'The Geology of Morals' opens with an image of a lobster: 'God is a Lobster, or a double pincer, a double bind'.[61]

The diagram is neither random nor deterministic and the nature of the 'circle' T–U–Φ–F is to remain radically open. Neo-materialist cartography is not to be mistaken for tracing a fully constituted reality, but mapping the potential for (auto)catalysis, a function of (continuous) variation that is causal in the (quasi)formal not efficient sense. To start from the *milieu* (ethology) is to disregard both the fixation on origins and teleology. In other words, heuristics knows no entailment.[62] Heuristics as a material inference is not an analytical device, but a synthetic operator.[63] Finally, to start from the middle is to zoom in on the ontopowerful dashes between T–F–Φ–U, the locus of the non-subjective and impersonal non-organic vitality.[64] In Braidotti's chapter, which is an epilogue of sorts, we come to realise that we cannot know what kind of 'surplus', re-territorialisation, or new types of subjectivation, we can expect. After all, if effects were reducible to their causes, novelty would be impossible. We can only affirm that whoever engages in the neo-materialist pragmatics of cartography cannot count on the ready-made audience, but on a people to come.[65] The Plane of Consistency is not *given* as in *archi*tectonics. It has to be *made*, over and over again. The pertinent question is not 'why is there something, rather than nothing?' We ought to ask instead, 'why is there difference rather than identity?'

The stars are very different today.

Notes

1. Radman and Kousoulas, eds, *3C: International Conference Proceedings*.
2. Deleuze, 'Foreword', in *Masochism*, p. 14. 'The critical (in the literary sense) and the clinical (in the medical sense) may be destined to enter into a new relationship of mutual learning.'
3. In a conversation with Raymond Bellour and Francois Ewald for *Magazine Litteraire*, 257, September 1988, as reprinted in Deleuze, *Negotiations* (1995), pp. 135–55 (142).
4. In 1993, two years before his death.
5. Smith, 'A Life of Pure Immanence'.
6. Tynan, 'Deleuze and the Symptom'.
7. Schizoanalytical practice repudiates a secret essence to unlock and a repressed trauma to read and realign with the oedipal axis. By contrast, life is seen as an impersonal and non-organic power.
8. 'M as in Malady/Illness' in *Gilles Deleuze's ABC Primer, with Claire Parnet* (directed by Pierre-André Boutang, 1996). Overview prepared by Charles J. Stivale, Romance Languages & Literatures, Wayne State University <http://www.langlab.wayne.edu/Cstivale/D-G/ABC1.html> (accessed 1 February 2016). Deleuze took part in the eight hours of interviews with Claire Parnet that constitute the film project called the *Gilles Deleuze's ABC Primer*, including a discussion of the subject of illness. For Deleuze, illness is not an enemy or something that gives the feeling of death, but, rather, something that gives a feeling of life, not in the sense of 'once I'm cured, I'll finally start living', but rather in the sense of 'illness itself sharpens a sense of life'. Crucially, it is not that one gets tuned in to

one's own life, but into *a* life. Moreover, it is 'impersonal life' that binds thought and excessive affect. Drawing on D. H. Lawrence and Spinoza, who had seen 'something so enormous, so overwhelming that it was too much for them', Deleuze suggests that a very degree of fragility may be what forces genuinely creative thought upon us: 'One cannot think if one isn't already in a domain that exceeds one's strength to some extent, that makes one fragile.' While there is clearly ill health that has a disabling effect, there are degrees of fragility that may enable.

9. Canguilhem, *Normal and Pathological*. In his introduction to the book Michel Foucault argued that Canguilhem developed a philosophy of error, of concept and of life (Rationalism) against a philosophy of sense, of subjects and of experience (Phenomenology and Existentialism). On the (in)commensurability of the 'two philosophies' in the French context see Tom Eyers, 'Living Structures'.
10. See Hagner, 'Normal and Pathological Humanity, Michael Hagner on Canguilhem'.
11. Smith, 'A Life of Pure Immanence', p. xv. Cf. Deleuze and Guattari, *What Is Philosophy?*, p. 172.
12. Cassirer, *Rousseau, Kant, Goethe*, pp. 69, 93. 'To put it briefly and clearly, Goethe completed the transition from the previous generic view to the modern genetic view of organic nature . . . While Kant looks for synthetic principles, for the highest principles of human knowledge, Goethe is looking for the productive principles of creative nature.'
13. Sauvagnargues, *Deleuze and Art*, p. 6.
14. Smith, 'Architectures, Critical and Clinical', p. 233. 'Distilling a history *for* architecture that posits itself symptomatologically is (in one sense) far more difficult than isolating the therapeutic and diagnostic positioning of architecture. Such an architecture would concern itself neither with uncovering underlying or foundational logical structures nor with posing solutions to commonly stated problems. This is because symptomatology is at once more concrete-real and more abstract-real. It is concrete-real in its attention to the actualities of material existence and the temporality of events. It is abstract-real in that any attention paid to the concrete-reality of the world necessarily involves an indulgence in the rich complexities, intensities and contingencies of life. An architecture posited as symptomatology might engage with the immediacy of the present by exploring and experimenting within the world and its "symptoms". This architecture would express new ways of thinking about life and experiment with novel ways of living' (emphasis in the original).
15. Deleuze and Guattari, *Anti-Oedipus*, p. 106.
16. Deleuze, *Essays Critical and Clinical*, p. 3. In the original version: 'The writer as such is not a patient but a physician, the physician of himself and of the world. The world is a set of symptoms whose illness merges with man. Literature then appears as the enterprise of health.'

17. Deleuze and Parnet, *Dialogues*, p. 124. See also Deleuze, 'The Interpretation of Utterances', p. 92.
18. Jobst, 'Gilles Deleuze and the Missing Architecture'.
19. Dolphijn and van der Tuin, eds, *New Materialism: Interviews & Cartographies*, p. 111. 'Thus "neo-materialism" emerges as a method, a conceptual frame and a political stand, which refuses the linguistic paradigm, stressing instead the concrete yet complex materiality of bodies immersed in social relations of power' (from 'Interview with Rosi Braidotti', p. 21).
20. Massumi, *Ontopower: War, Powers, and the State of Perception*, p. 37.
21. Guattari, *Schizoanalytic Cartographies*. For an architectural take on 'asignifying' see: Radman and Hauptmann, 'Asignifying Semiotics as Proto-Theory of Singularity'.
22. Holmes, 'Guattari's Schizoanalytic Cartographies'.
23. In everyday German *umwelt* means 'surroundings' or 'environment' but through the work of the Baltic German biologist Jakob von Uexküll (1864–1944) the term has acquired more specific *semiotic* meanings as the ecological niche as an animal perceives it. Environmental cues could only have an effect on the animal if the combination of stimuli was specific to the respective living being, which means that different species experience the world differently, that is, they have different *umwelts*. See Uexküll, 'A Stroll through the Worlds of Animals and Men'. See also: Buchanan, *Onto-ethologies: The Animal Environments of Uexküll, Heidegger, Merleau-Ponty, and Deleuze*.
24. Shaviro, *No Speed Limit: Three Essays on Accelerationism*. 'There is no preexisting "possibility space" for any ... technology. The development and deployment of a technology generates its own affordances and constraints, which themselves may differ under different economic and social conditions. Technological development will always have a speculative (and nonutilitarian) dimension.' Cf. Graafland and Sohn, 'Technology, Science and Virtuality'.
25. Querrin, *Diagrammes schizoanalytiques*.
26. Lord, 'Deleuze and Kant'.
27. Deleuze, *Nietzsche and Philosophy*, p. 50.
28. Guattari, *Schizoanalytic Cartographies*, p. 52.
29. Watson, *Guattari's Diagrammatic Thought*, p. 99.
30. Watson, *Guattari's Diagrammatic Thought*, p. 97.
31. Deleuze and Guattari, *A Thousand Plateaus*.
32. Brassier, *Concrete Rules and Abstract Machines*.
33. Sauvagnargues, *Deleuze and Art*, p. 25.
34. Sauvagnargues, *Deleuze and Art*, p. 31.
35. Sellars, 'Nomadic Wisdom', p. 71. 'Deleuze and Guattari use the term in its earliest form; *nomos* as pasture or steppe ... The distinction that Deleuze and Guattari want to use is that between the carefully controlled city and the unregulated expanse of the steppe. For them, *nomos* "stands

in opposition to the law or the *polis*, as the backcountry, a mountainside, or the vague expanse around a city".'
36. Holmes, *Escape the Overcode*.
37. Shaviro, *No Speed Limit*.
38. The trilogy: Guattari, *Three Ecologies*; *Schizoanalytic Cartographies*; *Chaosmosis*.
39. Bretton, *We Need to Talk about TED*; emphasis in the original. (TED stands for Technology, Entertainment and Design.)
40. Guattari, *Guattari Reader*, p. 126.
41. Apart from the 'black hole phenomenon', that is, inseparability of the subject from its milieu, there is also a 'lock in phenomenon' as explained by Brian Holmes with reference to the primacy that the VHS video format won over the Beta system in spite of its technical inferiority and thanks to the emerging porn industry market. See Holmes, *Guattari's Cartographies*.
42. Guattari, *Chaosmosis*, p. 7.
43. Brott, *Architecture for a Free Subjectivity*, p. 1; emphasis in the original.
44. Deleuze and Guattari, *A Thousand Plateaus*, p. 53. 'Forms relate to codes and processes of coding and decoding in the parastrata; substances, being formed matters, relate to territorialities and movements of deterritorialization and reterritorialization on the epistrata. In truth, the epistrata are just as inseparable from the movements that constitute them as the parastrata are from their processes.'
45. Guattari, *Guattari Reader*, p. 126. The Apollo programme, also known as Project Apollo, was the third United States human spaceflight programme carried out by the National Aeronautics and Space Administration (NASA), which accomplished landing the first humans on the Moon from 1969 to 1972.
46. For such an inclusive understanding of architecture (in conjunction with political economy, geography and spatial planning) see Volume 5 of Delft School of Design series on architecture and urbanism. Sohn, Kaminer and Robles-Durán, eds, *Urban Asymmetries*. *Urban Asymmetries* aims to disprove some of the prevailing disciplinary discourses in architecture and urbanism that see the city as 'a given' rather than as an evolving sociohistoric phenomenon, and intends to challenge the ubiquitous understanding of architecture as devoid of any social transformative power.
47. Kwinter, 'Four Arguments for the Elimination of Architecture (long Live Architecture)', in *Requiem*, p. 92. 'We should redirect research from the screen-based simulations that have predominated in the last fifteen years toward the considerably greater intelligence that is already impressed into matter.'
48. Sauvagnargues, *Deleuze and Art*, p. 57.
49. Deleuze, 'On A Thousand Plateaus', p. 31.
50. Watson, *Guattari's Diagrammatic Thought*, p. 121. Cf. Guattari, *Schizoanalytic Cartographies*, p. 44.

51. Guattari, *Schizoanalytic Cartographies*. Cf. Kugler and Shaw, 'Symmetry and Symmetry-Breaking'. We ought to stop treating systems as isolated first (structure) and as interacting second (agency). The ecological psychologists Kugler and Shaw propose a different strategy based on the nonlinear coupling of the laws. The transversal coupling is irreversible across *different* scales (symmetry-breaking) and reversible across the *same* scale (symmetry preserving) while 'thermodynamic engines' are exoreferential and 'epistemic engines' are endoreferential.
52. Evans, *The Projective Cast*, p. 363.
53. Sennett, 'Technology of Unity', p. 563. The 'controversy' over the latest Pritzker laureate, the Keynesian Alejandro Aravena, who was reproached for the alleged regressive attitude by the champion of Parametricism the Hayekian Patrick Schumacher, is symptomatic of the lock-in phenomenon. Deleuze and Guattari anticipated the impasse in their first volume of *Capitalism and Schizophrenia* by reference to the dangers of 'neoarchaism' and 'ex-futurism', respectively: 'They are continually behind or ahead of themselves'. <http://architizer.com/blog/patrik-vs-pritzker/> (accessed 2 February 2016).
54. Deleuze and Guattari, *Anti-Oedipus*, p. 338.
55. Mauss, *The Gift*. Cf. Hénaff, *Price of Truth*.
56. Spuybroek, 'Charis and Radiance'.
57. Massumi, *Power at the End of the Economy*, pp. 54–5.
58. Kwinter is a precursor to the neo-materialist approach that frees architecture from the shackles of object-fetishism and the opposite – process-fetishism – alike. See Kwinter, 'Beaubourg or The Planes of Immanence', in *Requiem*, pp. 14–27. 'The first chapter of Gilles Deleuze and Félix Guattari's *Mille Plateaux* [Rhizome] was published at this time as a separate book, which introduced the concept of the "plan d'immanence" into the general culture [*sic*], a concept that would make domain distinctions in cultural practice forever . . . It endowed thought itself with a new role: its task now would be to disengage structures from the real material world and to set them in promiscuous motion tracking their trajectories and migrations from one state of contact and rearrangement to another. The "abstract" and the "concrete" from now on would have lives of their own, participating in a perpetual ballroom dance where partners are exchanged promiscuously *according to design*' (p. 23; emphasis in the original).
59. Foucault, 'Society Must Be Defended', p. 66.
60. Spuybroek, 'Charis and Radiance', p. 123. 'When we transfer this model into figures personifying these acts, we immediately recognize them as the Three Charites, as they were known in ancient Greece, or, in their Roman guise, the Three Graces, the first goddess (Aglaia) embodying giving, the second (Euphrosyne) receiving and the third (Thalia) representing the return.'

61. Deleuze and Guattari, *A Thousand Plateaus*, p. 40.
62. On the concepts of entailment and 'impredicativity' see Radman and Boumeester, 'The Impredicative City', in *Deleuze and the City*.
63. Negarestani, 'Frontiers of Manipulation'.
64. DeLanda, 'Nonorganic Life'.
65. Deleuze and Guattari, *What is Philosophy?*, p. 108. 'Art and philosophy converge at this point: the constitution of an earth and a people that are lacking as the correlate of creation.'

Bibliography

Bowie, David, 'Space Oddity', *David Bowie* (LP album), Philips (label), UK 1969.

Brassier, Ray, *Concrete Rules and Abstract Machines*, introduction by Henry Somers Hall (2015) <https://soundcloud.com/henry-somers-hall/5-ray-brassier-concrete-rules-and-abstract-machines> (accessed 1 February 2016).

Bretton, Benjamin, *We Need to Talk About TED* (2013) <http://www.theguardian.com/commentisfree/2013/dec/30/we-need-to-talk-about-ted> (accessed 1 February 2016).

Brott, Simone, *Architecture for a Free Subjectivity* (Surrey: Ashgate Publishing, 2011).

Buchanan, Brett, *Onto-ethologies: The Animal Environments of Uexküll, Heidegger, Merleau-Ponty, and Deleuze* (Albany: SUNY, 2008).

Canguilhem, Georges, *The Normal and the Pathological* (New York: Zone Books, [1966] 1991).

Cassirer, Ernst, *Rousseau, Kant, Goethe: Two Essays*, trans. James Gutmann, Paul Oskar Kristeller and John Herman Randall, JR (Princeton: Princeton University Press, [1945] 1970).

DeLanda, Manuel, 'Nonorganic Life', in *Incorporations*, ed. Jonathan Crary and Sanford Kwinter (New York: Zone 6, 1992), pp. 129–67.

Deleuze, Gilles, 'Foreword', in *Masochism* (New York: Zone Books, 1991), pp. 9–14.

Deleuze, Gilles, 'On a Thousand Plateaus', in *Negotiations, 1972–1990* (New York: Columbia University Press, [1990] 1995), pp. 25–34.

Deleuze, Gilles, *Essays Critical and Clinical*, trans. Michael A. Greco and Daniel. W. Smith (Minneapolis: Minnesota University Press, [1993] 1997).

Deleuze, Gilles, *Nietzsche and Philosophy*, trans. Hugh Tomlinson (New York: Columbia University Press, [1962] 2006).

Deleuze, Gilles, 'The Interpretation of Utterances', in *Two Regimes of Madness: Texts and Interviews 1975–1995* (Los Angeles: Semiotext(e), 2006), pp. 89–112.

Deleuze, Gilles and Félix Guattari, *Anti-Oedipus*, trans. Robert Hurley, Seem Mark and Helen R. Lane (Minneapolis: Minnesota University Press [1972] 1983).

Deleuze, Gilles and Félix Guattari, *What is Philosophy?* (New York: Columbia University Press, [1991] 1994).
Deleuze, Gilles and Félix Guattari, *A Thousand Plateaus*, trans. Brian Massumi (London and New York: Continuum [1980] 2004).
Deleuze, Gilles and Claire Parnet, *Dialogues* (New York: Columbia University Press, [1977] 1987).
Deleuze, Gilles and Claire Parnet, 'M as in Malady/Illness', in *Gilles Deleuze's ABC Primer* (Directed by Pierre-André Boutang, 1996). Overview prepared by Charles J. Stivale, Romance Languages & Literatures, Wayne State University <http://www.langlab.wayne.edu/Cstivale/D-G/ABC1.html> (accessed 1 February 2016).
Dolphijn, Rick and Iris van der Tuin, 'Interview with Rosi Braidotti', in *New Materialism: Interviews & Cartographies*, ed. Rick Dolphijn and Iris van der Tuin (Open Humanities Press, 2012), chapter 1 <http://hdl.handle.net/2027/spo.11515701.0001.001> (accessed 1 February 2016).
Dolphijn, Rick and Iris van der Tuin, eds, *New Materialism: Interviews & Cartographies* (Open Humanities Press, 2012) <http://hdl.handle.net/2027/spo.11515701.0001.001> (accessed 1 February 2016).
Evans, Robin, *The Projective Cast: Architecture and Its Three Geometries* (Cambridge, MA: MIT Press, 1995).
Eyers Tom, 'Living Structures: Canguilhem, Deleuze, and the Problem of Life', InterCcECT: The Inter Chicago Circle for Experimental Critical Theory (2012) <http://vimeo.com/100439287> (accessed 1 February 2016).
Foucault, Michel, 'Society Must Be Defended', in *Biopolitics: A Reader*, ed. Timothy Campbell and Adam Sitze (Durham, NC and London: Duke University Press, 2013), pp. 61–81.
Graafland, Arie and Heidi Sohn, 'Introduction: Technology, Science and Virtuality', in *The SAGE Handbook of Architectural Theory*, ed. C. Greig Crysler, Stephen Cairns and Hilde Heynen (London, Thousand Oaks, New Delhi, Singapore: SAGE Publications, 2012), pp. 467–83.
Guattari, Félix, *Chaosmosis: An Ethico-aesthetic Paradigm*, trans. Paul Bains and Julian Pefanis (Bloomington: Indiana University Press, [1992] 1995).
Guattari, Félix, *The Guattari Reader*, ed. Gary Genosko (Cambridge, MA: Blackwell Publishers, 1996).
Guattari, Félix, *The Three Ecologies*, trans. Ian Pindar and Paul Sutton (London: Continuum, [1989] 2008).
Guattari, Félix, *Schizoanalytic Cartographies*, trans. Andrew Goffey (London: Bloomsbury, [1989] 2013).
Hagner, Michael, 'Normal and Pathological Humanity, Michael Hagner on Canguilhem', interviewed by Caroline A. Jones in *Thresholds* 42. S. (2014), pp. 100–7.
Hénaff, Marcel, *The Price of Truth* (Redwood City, CA: Stanford University Press, 2010).
Holmes, Brian, *Escape the Overcode* (2009) <https://brianholmes.wordpress.com/2009/01/19/book-materials/> (accessed 1 February 2016).

Holmes, Brian, 'Guattari's Schizoanalytic Cartographies: or, the Pathic Core at the Heart of Cybernetics', in *Continental Drift* (2009) <https://brianholmes.wordpress.com/2009/02/27/guattaris-schizoanalytic-cartographies/> (accessed 1 February 2016).
Holmes, Brian, *Guattari's Cartographies: Territory, Subjectivity, Existence*, EGS lecture (2011) <https://www.youtube.com/watch?v=X0ocpPYpies> (accessed 1 February 2016).
Jobst, Marco, 'Gilles Deleuze and the Missing Architecture', in *Deleuze Studies* 8.2, ed. Ian Buchanan (Edinburgh: Edinburgh University Press, 2014), pp. 157–72.
Keskeys, Paul, *Patrik vs. Pritzker: Schumacher Reignites the Debate over Political Correctness in Architecture*, <http://architizer.com/blog/patrik-vs-pritzker/> (accessed 2 February 2016).
Kugler, Peter N. and Robert Shaw, 'Symmetry and Symmetry-Breaking in Thermodynamic and Epistemic Engines: A Coupling of First and Second Laws', in *Synergetics of Cognition*, ed. Hermann Haken and Michael Stadler (Heidelberg: Springer-Verlag Berlin, 1990), pp. 296–331.
Kwinter, Sanford, 'Beaubourg or The Planes of Immanence', in *Requiem for the City at the End of the Millenium* (Barcelona: Actar, 2011), pp. 14–27.
Kwinter, Sanford, 'Four Arguments for the Elimination of Architecture (Long Live Architecture)', in *Requiem for the City at the End of the Millenium* (Barcelona: Actar, 2011), pp. 89–93.
Kwinter, Sanford, *Requiem: For the City at the End of the Millennium* (Barcelona: Actar, 2011).
Lord, Beth, 'Deleuze and Kant', in *The Cambridge Companion to Deleuze*, ed. Daniel W. Smith and Henry Somers-Hall (Cambridge: Cambridge University Press, 2012), pp. 82–102.
Massumi, Brian, *The Power at the End of the Economy* (Durham, NC and London: Duke University Press, 2014).
Massumi, Brian, *Ontopower: War, Powers, and the State of Perception* (Durham, NC and London: Duke University Press, 2015).
Mauss, Marcel, *The Gift: Forms and Functions of Exchange in Archaic Societies*, trans. Ian Cunnison (London: Cohen and West, [1954] 1966).
Negarestani, Reza, 'Frontiers of Manipulation', *Speculations on Anonymous Materials* symposium (2014) <http://www.youtube.com/watch?v=Fg0lMebGt9I/> (accessed 2 February 2016).
Querrin, Anne, *Diagrammes schizoanalytiques* (2008) <http://www.dailymotion.com/video/x5w6ew_diagrammes-schizoanalytiques-2-5_news> (accessed 1 February 2016).
Radman, Andrej and Marc Boumeester, 'The Impredicative City: or What Can a Boston Square Do?' in *Deleuze and the City*, ed. Hélène Frichot, Catharina Gabrielsson and Jonathan Metzger (Edinburgh: Edinburgh University Press, 2016), pp. 46–63.
Radman, Andrej and Deborah Hauptmann, 'Asignifying Semiotics as Proto-Theory of Singularity: Drawing is Not Writing and Architecture Does Not

Speak', in *Asignifying Semiotics: Or How to Paint Pink on Pink*, ed. Deborah Hauptmann and Andrej Radman, *Footprint*, Vol. 8/1, Issue 14 (Delft: Architecture Theory Chair in partnership with Stichting Footprint and Techne Press, 2014), pp. 1–12.

Radman, Andrej and Stavros Kousoulas, eds, *3C: International Conference Proceedings* (Delft: Architecture Theory Chair in partnership with Jap Sam Books, 2015).

Sauvagnargues, Anne, *Deleuze and Art*, trans. Samantha Bankston (London: Bloomsbury, 2013).

Sellars, John, 'Nomadic Wisdom: Herodotus and the Scythians', in *Nomadic Trajectories*, ed. John Sellars and Dawn Walker in *Pli* (Vol. 7, 1998), pp. 69–82.

Sennett, Richard, 'The Technology of Unity', in *Olafur Eliasson: Surroundings Surrounded: Essays on Space and Science*, ed. Peter Weibel (Karlsruhe: ZKM, 2000), pp. 556–65.

Shaviro, Steven, *No Speed Limit: Three Essays on Accelerationism* (Minneapolis: Minnesota University Press, 2015).

Smith, Chris L., 'Architectures, Critical and Clinical', in *Deleuze and Architecture*, ed. Hélène Frichot and Stephen Loo (Edinburgh: Edinburgh University Press, 2013), pp. 230–44.

Smith, Daniel W., '"A Life of Pure Immanence": Deleuze's "Critique et Clinique" Project', introduction to Gilles Deleuze, *Essays Critical and Clinical* (Minneapolis: Minnesota University Press, [1993] 1997), pp. xi–liii.

Sohn, Heidi, Tahl Kaminer and Miguel Robles-Durán, eds, *Urban Asymmetries: Studies and Projects on Neoliberal Urbanization* (Rotterdam: 010 Publishers, 2011).

Spuybroek, Lars, 'Charis and Radiance', in *Giving and Taking: Antidotes to a Culture of Greed*, ed. Joke Brouwer and Sjoerd van Tuinen (Rotterdam: V2_Publishing, 2014), pp. 119–50.

Tynan, Aidan, ed., 'Deleuze and the Symptom: On the Practice and Paradox of Health', *Deleuze Studies*, Vol. 4, Issue 2 (2010), pp. 153–60.

Uexküll von, Jakob, 'A Stroll through the Worlds of Animals and Men: A Picture Book of Invisible Worlds', in *Instinctive Behavior: The Development of a Modern Concept*, ed. and trans. Claire H. Schiller (New York: International Universities Press, Inc., 1957), pp. 5–80.

Watson, Janell, *Guattari's Diagrammatic Thought: Writing between Lacan and Deleuze* (London and New York: Continuum, 2009).

Introduction: A Research into Human–Machine Technologies – Architecture's Dream of a Bio Future

Arie Graafland

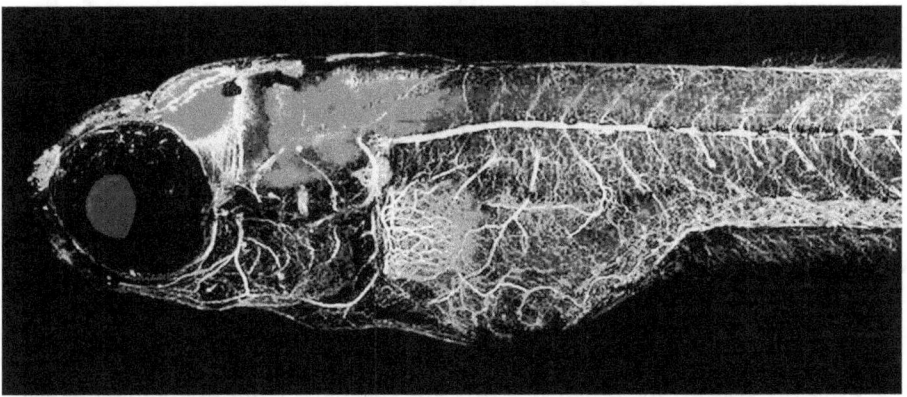

Figure 4 Zebrafish (Danio rerio) is akin to our ecology. Zebrafish is known for its regenerative abilities and has been modified to produce several transgenic strains. It is a tool in the etiology of human genetic diseases like cancer, infectious diseases, kidney failure and diabetes. Zebrafish is akin to Donna Haraway's OncoMouse™. OncoMouse™ is patented. It is a site for the operation of a transplanted, human, tumour-producing gene – an oncogene – that produces breast cancer in women.

Embedded Anthropology

This introduction deals with three related topics.[1] The first topic describes our field of interest: contemporary medical care and human environmental aspects of human–machine interfaces. The second topic deals with the field of architecture and clinic, in particular the design of a dialysis centre. A lot goes wrong here in the design, and to be able to understand why, we need to address a third topic, which deals with education: education in medicine, and education in architecture. What we will be looking into is the relation between the human body as an organism and the machine technologies we use in medical care and architectural design. This is also about thresholds between man and machine that are shifting and crossing borders at the moment. Medical knowledge has advanced rapidly over the past century, progressing at a nearly unimaginable speed today. A lot is at stake here, human health of course, but also money and investment. Medical instruments like magnetic resonance imaging (MRI) scans and dialysis machinery ask for significant and long-term investments. Pharmaceutical industries produce new drugs, and these operate on markets and with products concerning risky investments, research that involves the latest technologies. Their research departments are under constant pressure to come up with better drugs.[2] Diversification and specialisation are needed to survive in a competitive market. The invested money will have to flow back, profits have to be made and patients need secure machineries and drugs, not experiments. This development in the medical sciences relates to more theoretical discourses on 'man and nature' in the humanities at large. It also means different disciplines trying to communicate with one another. Interdisciplinary work is easily stated but not so easily done. The leading question throughout this introduction is the relation of a normative construction of the 'materiality' of the human body, in other words its physical and emotional experiences, and contemporary digital techniques in human–machine interfaces in medical sciences and architectural design.

This introduction is not a study on medical sociology, neither is it a design discourse on advanced technologies. We will enter a field of recently developed 'discourses' and 'practices', which are transgressing their traditional boundaries. Architects are involved in designing buildings. They produce drawings in which 'man' is supposed to get a comfortable life. We will particularly look into the 'interiors' of buildings, assessing their level of 'comfort', which is mainly the domain of interior design and industrial design. Here, already two distinct pedagogical institutions are involved. In addition, we have to go over disciplinary boundaries using 'design' in a more general way to cover both practices.

What we experience today in the humanities is a keen interest in overlapping, or rather, shifting borders of the profession. Annemarie Mol writes that importing texts from other fields tends to be a good way to say 'new' things.[3] These 'new' things do not come from clear-cut disciplines, but from interdisciplinary, slightly undisciplined fields.

In other words, we are particularly interested in the scholars who question current disciplinary boundaries, and are successfully transgressing them.

We could envision our field of interest as a provisionally constructed landscape, a cartography, that you would have to travel through to understand.[4] There is no picture from above, no charts, summaries or shortcuts.[5] We cannot even explain our field 'in a few words'. It is more of an itinerary, guiding you through different territories, but it is not the complete picture. We will be looking into activities, machines, practices and events, and not only the different discourses that are dominant in the different landscapes. Discourses are not as stable as one may think, certainly not in architecture, but to a lesser extent also in medicine there are varying discourses and practices.

We find the relation between the development of human–machine interfaces in medical practices and environmental and architectural design in Katherine Hayles's example of the positron emission topography (PET) scan. Hayles distinguishes between our bodies in the full context of life, and the body as an object for medical practice. Embodiment is the contextualised body. She describes how embodiment is converted into a *body* through imaging technologies that create a normalised construct as a result of averaged data. In contrast to the body, *embodiment* is contextual, enmeshed within the specifics of place, time, physiology and culture, which together compose enactment. Hayles compares it with the digitisation of books, and notices that we lose something in the process. In molecular biology our body is understood as an expression of genetic information and as physical structure. In the literary corpus it is at once a physical object and a space of representation, simultaneously a body and a message. When we look at bodies and books as pure information we lose the resistant materiality that has always marked our experience of living creatures.[6] Hayles's comments also relate to the topic in current parametricist architecture, a technique in architectural design. This resistant materiality is also fading away in architectural education; the vectors in contemporary drawing and representation programmes have no other resistance than data and can easily end up as 'terminal velocities' as Stan Allen claims.[7] We are entering Donna Haraway's world of technoscience: no longer a physical place but the cyberspace of the Internet addressed in the title of her book *Modest_Witness@Second_Millennium*.[8] This also implies a specific idea of medical care. Is care other to technology, Annemarie Mol asks? Is the first humane and friendly, while the second strategic and dependant on rationality alone? The care Mol refers to, is not opposed to, but includes technology.⁹ We may want to compare her *'Body Multiple'* to *'Embodiment'*, which never coincides exactly with 'the body' as fixed entity.

The second topic describes current technologies in the research into human–machine interfaces. Here the focus is on the 'instruments', the 'machinery' and how they relate to the body and the experiences that come with them. Here, we need to establish the distinction between embodiment and body as physical structure. We will address these issues in dialysis clinics. We will show that most architecture is mainly interested in a physical body, although it claims to be directed to embodiment. As a consequence, we need discussion regarding the possible improvements of medical environments. To be able to do that we also need to step back for a moment from the clinic and address more philosophical issues in the humanities about our bodies and the machines we use to cure them. The notion of 'care' will come back, as related to the machinery and experiences they evoke.

The third topic deals with education: education in medicine and education in architecture. Although the disciplines are very different they seem to have something in common in teaching. During the 1980s new forms of education came into being. There was more focus on smaller groups of students and more attention paid to developing practical knowledge in an autonomous setting and students had to do a lot themselves; 'problem-based learning' became a new buzz phrase, both in Western Europe and in the United States. We will discuss examples from Harvard Medical School, and the Netherlands, in particular design education. This part is more design related. Architects and industrial designers may be able to design better environments for care conditions in medical centres. What they will need is transgression of their current practices.

The Changing of the Stage Set

Let us first look into the world of 'man and nature', a recurring phrase in philosophy and the humanities. Byron Good has explored the idea that a view of scientific language as largely transparent to the natural world – a kind of 'mirror of nature' that has been an important line of argument in philosophy since the Enlightenment – has deep affinities with biomedicine's 'folk epistemology' and holds a special attraction for medical behavioural sciences.[10] Good offers many examples in his book, and especially where these medical discourses fail in understanding how narratives about illness are formed by 'folk' languages. He formulates the disjunction between disease as an object or condition of a physical body, as it is popularly (and medically) conceived, and the disease as a presence in a life or in a social world. He qualifies, quite extraordinary, the analysis of illness as 'aesthetic object'. Good is not implying that illness is a thing of beauty, but 'aesthetic object' that determines how we analyse this disjunction between disease as condition of a physical body and in the narratives of our daily life world. The book contains a reference to Elaine Scarry's *The Body in Pain: The Making and Unmaking of the World*, a study into torture and language, most certainly not referring to our daily lives in the West, but to other regions of the world.[11] Her book is based on the horrifying stories recorded by Amnesty International's files on torture and the impossibility to express severe physical pain. Acute pain resists language; it destroys the world we live in. Experience here is singular; the rest is wiped out. Katherine Hayles has argued that feelings constitute a window through which the mind looks into the body. Feelings and emotions are the body murmuring to the mind; in other words, the body provides a ground reference for the mind. According to Antonio Damasio, feelings are not a collection of thoughts with certain themes consonant with a feeling label, such as thoughts of situations of loss in the case of sadness.[12] Feelings arise from any set of homeostatic reactions, not just from emotions proper. They translate the ongoing life state into the 'language' of the mind. For Damasio there are distinctive 'body ways' resulting from different homeostatic reactions, from simple to complex. Feelings are perceptions; the necessary support for their perception occurs in the brain's *body maps*. Damasio's *maps* refer to parts of the body and states of the body. In torture, there is no murmuring left. Body maps fail to function under intense pain, shattering the world of perception and language. The 'aesthetic object' is meant to open up the plurality in the narratives and practices that form the daily life world: of both joy and sorrow, as Spinoza would have it.

Annemarie Mol underlines the importance of different practices in her study on atherosclerosis. If practices are foregrounded there is no longer a single passive object in the middle waiting to be seen from the point of view of a seemingly endless series of perspectives. Instead, objects come into being – and disappear – with the practices in which they are manipulated. Since the objects of manipulation tend to differ from one practice to another, reality multiplies. We find the same primacy of practice in Rachel Prentice, who has studied surgical learning.[13] She quotes a surgeon: 'surgery is a body-contact sport, there is no question about it. You can't be a good armchair surgeon.' It shows how practitioners come to embody biomedical knowledge and values. Prentice argues that medical embodiment goes beyond the acquisition of skills to include the development of perceptions, affects, judgments and ethics that occur through bodily practice in a clinical milieu.[14] Like Mol and Good, she builds on anthropology and science studies. The body, the patient, the disease, the doctor, the technician, technology; all of these are more than one, more than singular.[15] Arthur Kroker calls this phenomenon 'body drift', referring to the fact that we no longer inhabit *a* body in any meaningful sense of the term but rather occupy a multiplicity of bodies: imaginary, sexualised, disciplined, gendered, labouring, technologically augmented bodies. Moreover, the codes governing behaviour across this multiplicity of bodies have no real stability but are themselves adrift: random, fluctuating, changing.[16] Although the medical sciences have progressed enormously, it does not mean that bioscience in itself is a stable univocal concept, unchanged due to its 'scientific' character. Donna Haraway uses the example of the immune system as 'elaborate icon' to show how 'the immune system is a plan for meaningful action to construct and maintain the boundaries for what may count as self and other in the crucial realms of the normal and the pathological'.[17] Haraway qualifies the immune system as both an iconic and mythic object in high-technology culture and a subject of research and clinical practice. She uses Edward Golub's 1987 *Immunology: A Synthesis* to explain the shifting connotations and developments in bioscience in Gershon's four illustrations, dated 1968, 1974, 1977 and 1982, to document the changing 'immunological orchestra'.[18] The orchestra is a wonderful picture of what she calls the mythic and technical dimensions of the immune system; all illustrations are about co-operation and control. Her – at times ironic and sometimes hilarious – description of the Generator of Diversity (GOD), depicts the orchestra of T and B cells and macrophages as they march about the body, an image of a technical-mythic system of molecular biology. The lymphocytes all look like Casper the ghost with the appropriate distinguishing nuclear morphologies drawn in the centre of their shapeless bodies.[19] The shifting connotations show the immune system as 'postmodern pastiche of multiple centres and peripheries'; the actors are there, but the stage set has changed due to a better understanding of what is going on in the body. GOD is no longer in front of the immune orchestra. A special cell, and T suppressor cell, has taken over the role of the conductor. And by 1977 there is no longer a single conductor. The interesting implication of this 'postmodern pastiche' is that science itself is not 'outside' our world, but part of it. Byron Good, for his part, writes that after years of teaching and carrying out research in medical settings, he is more convinced than ever that the language of medicine is hardly a simple mirror of the empirical world. It is a rich cultural language, linked to a highly specialised version of reality and system of social relations, and when employed in medical care, it joins deep moral concerns with its more obvious technical functions.[20]

These conceptual fields in the language of medicine evolve similarly to material culture, in part because concept and artefact engage each other in continuous feedback loops.[21] Katherine Hayles shows us the feedback loops in her analysis of chaos theory as both the subject of scientific enquiry, and the crossroads where various paths within culture converge. It is certainly not easy to draw the line between legitimate and illegitimate extrapolations from chaos theory.[22] In her understanding, chaos theory, like meteorology, epidemiology, irreversible thermodynamics and nonlinear dynamics, in addition to pertaining to scientific discourse, also serves the function of describing and understanding complex behaviours, defeating the conventional methods of formalising a system through mathematics.[23] But, as Kroker explains, Hayles's writing has a wider 'vista':

> [A]t stake is not only a critical reconsideration of the relationship of literature and science but something of greater cultural significance. It's as if in the intellectual persona of a humanities professor interested in the impact of the 'regime of computation' on culture, society, and politics, her thought reveals a broader ontological vista: dissipative structures, bifurcation, recursive symmetry, strange attractors, nonlinear dynamics, life that exists 'far from equilibrium'.[24]

Haraway's analysis fits in with this idea of science and culture: the immune system becomes a historically specific terrain, upon which many diverting forces and elements interact. The immune system is both an iconic mythic object in high-technology culture and a subject of research and clinical practice. Good's analysis approaches Haraway's. While she deals with the changes and relational qualities of bio-sciences and power, Good compares different cultural settings in which different outcomes will appear. Perhaps even more importantly, given the rich cultural frames for conceiving human suffering in many of the societies we study, holding up our own biological language of illness and care as norm seems profoundly inadequate.[25] The language of clinical medicine is a highly technical language of the bio-sciences, grounded in a natural science view of the relation between language, biology and experience. In the words of Good:

[T]he primary tasks of clinical medicine are thus diagnosis – that is, the interpretation of the patient's symptoms by relating them to their functional and structural sources in the body and to underlying disease entities – and rational treatment aimed at intervention in the disease mechanisms . . . Medical knowledge, in this paradigm, is constituted through its depiction of empirical biological reality, . . . diseases are biological, universal, and ultimately transcend social and cultural context.[26]

Good argues that this conception of language and knowledge referred to the 'empiricist theory of medical language' and, as such, serves poorly for cross-cultural research and even for studies of American science and medicine. Those who employ it are led to formulate problems in terms of belief and behaviour, and often reproduce our common-sense views of the individual and society. Good's central goal is to develop an alternative way of thinking about medicine and medical knowledge – in a way comparable to Annemarie Mol's theorising of medicine's ontological politics, a politics that has to do with the way in which problems are framed, bodies are shaped and lives are pushed and pulled into one shape or another.[27] Rosi Braidotti follows the Spinozist concept of 'monistic philosophy of becomings' to consider matter, including the specific slice of it that is human embodiment, as intelligent and self-organising. In her writings matter is not dialectically opposed to culture, nor to technological mediation, but continuous with them.[28] She differs from Mol and Good by radically questioning posthuman subjectivity. The core of it is the generative vitality of life of the non-human, which is intelligent and self-organising: *zoe*. The notion of agency changes: non-human things are figured less as social constructions, and more as actors, as vital materialisation.[29] Agency loses its traditional reference, and is distributed across assemblages (*agencement*) that are ontologically heterogeneous and have a finite life span. Intentionality becomes less definitive of outcomes. For Braidotti, the *zoe*-centred embodied subject is shot through with relational linkages of the contaminating/viral kind that interconnect it to a variety of others, starting from the environmental or eco-others and include the technological apparatus.[30] Braidotti joins Mol in affirmative body politics; critique goes along with creativity in action. She is, however, resistant to what she calls Latour's 'new segregation of knowledge' that is produced along the dividing lines of the 'two cultures': the Humanities and the Sciences. This also produces a different scheme of emancipation and a non-dialectical politics of human liberation. Her post-anthropocentric position, which includes science and technology studies, new media and digital culture, implies that political agency need not be critical in the negative, oppositional sense and thus may not be aimed solely or primarily at the production of counter-subjectivities. Braidotti's subjectivity is rather an autopoietic process of self-styling, involving complex and continuous negotiations with dominant norms and values and hence also multiple forms of accountability.[31]

We do not doubt that the biological sciences have made astounding advances in understanding human physiology, but we are no longer prepared to view the history of medicine as a straightforward recording of the continuous discovery of the facts of nature.[32] There is no straightforward evolutionary development, but we are instead experiencing a *leap forward*. Our present order, Haraway reminds us, may very well be in the region of historical hyperspace of *technoscience*. Technoscience indicates a time–space modality that is extravagant, that overshoots passages through naked or unmarked history.[33] Technoscience exceeds the distinction between science and technology as well as those between nature and society, subjects and objects and the natural and the artefactual that structured the imaginary time called modernity. It is a mutation in historical narrative; it creates blanks in memory and consciousness. Erik Brynjolfsson and Andrew McAfee locate us in 'the second half of the chessboard',[34] which marks a different regime of computing.[35] Brynolfsson and McAfee show the difference in commonly used standard linear spacing graphs. These graphs, however, are not ideal to convey exponential growth. Something similar may be the case with the crisis of scale in digitisation. Antoine Picon discusses the crisis of scale and tectonics in digital design, but another effect of digitisation is a certain loss of memory in the architectural realm. The strong connection between technology and memory found a privileged expression with the tectonic.[36] Picon draws on Laugier's *Essai sur l'Architecture* centred on the link between tectonics and the emergence and development of architecture. Tectonics is related to time, history and memory. The hyperspace of technoscience and our 'learning to race with machines' as Brynjolfsson and McAfee write, seem to be formative for current digital design at the same token.

The examples in the next section are part of technoscience; they deal with the fusion of the human body and their machines, and other prosthetic devices that are life supporting and life saving. While we sustain Haraway's and Hayles's ideas about the permeability of nature and society, subjects and objects, the natural and the artefactual, our examples are closer to Mol's and Prentice's research that shifts away from a more traditional epistemological way of thinking towards a more Latourian approach.[37] Their approach also promotes an ethnographic interest in knowledge practices. Good rests his arguments on the interpretive anthropology of Clifford Geertz among others, and the views of American cultural anthropologist Ruth Benedict on pathology as inseparable from culture, something that continues to challenge empiricist theories about the relation of cultural representation and disease.[38]

Man Machine Technologies

The loss of kidney function is an irreversible, incurable process. Accumulation of toxins will damage other organs leading to heart rhythm disturbances, infections, disorders in the digestive tract or abnormal changes in the nervous system. Patients with chronic renal disease are subject to dialysis treatment. There are two main types of dialysis: *haemodialysis* (HD) and *peritoneal* dialysis (PD). HD is an extracorporeal treatment employing a synthetic membrane as a filter. PD is an intracorporeal treatment employing a natural membrane. PD patients learn a self-treatment and perform treatment at home, while HD patients are treated at dialysis clinics or centres, or ambulatory healthcare facilities. Dialysis rooms are places where the patient gets a treatment that will take four to five hours per session. During that time the patient must remain immobile, lying or sitting on a dialysis chair or bed, bound to a complex machine that purifies up to 400 litres of tap water into dialysis water per treatment.

The design of dialysis clinics requires careful consideration, and is often a core concern of medical care industries. Fresenius Medical Care is a global player in this field, which has established clinics globally. *Dialysis Centres: An Architectural Guide* published in 2012 reunites the expertise of a team of Fresenius Medical Care where an architect, an engineer and a manager lend insights into the design process of dialysis clinics for architects and technicians in many countries across Europe, the Middle East and Africa.[39] The emphasis of the guidebook rests on the design and construction of new dialysis clinics, and in some cases the refurbishment of existing buildings. A guiding factor is based on a functional analysis in architecture: the number of required dialysis 'stations' – basically the dialysis chair and the machinery involved – and the respective floor space, which will to a large degree determine the building costs involved in the design. Activities in the building are grouped in functional areas, a quite normal way of working in architectural practice. A dialysis centre will need dialysis 'stations', medical rooms, patient and staff rooms, support services, water treatment facilities, circulation spaces and the technical facilities like heating, ventilation and cooling. An important issue is surveillance of patients since there has to be a direct line of sight between the dialysis chair and the observation desk.

Most of the items discussed in the guidebook are familiar in most architectural offices: from deciding on the building site and the location of new buildings, questions on whether existing buildings may be refurbished, to minimal details that ensure staff and patient well-being. Considerations of colour, light, materials and furnishing take up an important portion of the guidebook. Other well-anticipated parameters, such as circulations and flow diagrams, are also dealt with in detail. The 'well-designed dialysis room' is in all aspects a medical facility that we know from many modern hospitals. But where is the patient? What is she experiencing in this facility where her blood streams through machineries? Do the bright colours really help her with her leg cramps? Is she looking at the pictures on the wall at all? What is she going to do in these hours? Can she see the gauges on the dialysis machine, see venous pressure, arterial pressure, trans-membrane pressure, the remaining time, her pulse rate and the blood from her body and back? What kind of 'affordances' in J. J. Gibson's sense are at hand here? How do all these 'spheres' of life support hang together? Don't we need a *'spherology'* as in Peter Sloterdijk's work?[40] What kind of 'well-being' do the architects actually provide? Are they not interested in the 'machinery', the 'envelopes'? We do not think so. After all, they are fully 'modern'. They are treating objects unfairly, Latour would say. By treating human life supports as matters of concern, 'we pile concerns over concerns, we fold, we envelop, we embed humans into more elements that have been carefully explicitated, protected, conserved and maintained (immunology being, according to Sloterdijk, the great philosophy of biology)'.[41]

What about the patient's perspective? *View from the Chair* by Thomas V. Carr, and, *Dialysis Advice, A Patient's Point of View* by Kathleen Russell deal with patient-centred experience. These two affordable books share advice on how to overcome treatment in more bearable ways from how to protect yourself from cold drafts and passing time, to deeper concerns such as how to inform yourself of your treatment, knowing your rights and maintaining the right mind-set and mood to endure the treatment better. But, when we put these narratives together, do we have the picture of what it means to undergo dialysis? Do we know what is going on in dialysis clinics? Do the architects know how to design such a building when they have 'both sides'? We doubt it, as these narratives are not complementary. We only have two very different perspectives: on the one hand a fully instrumentalist and functionalist set-up, and on the other personal guidelines for patients who are not really informed and prepared to undergo dialysis. Perspectives, however, are limited and fixed, as they constitute the world as pre-given.[42] Instead, we may see this ecology of a dialysis clinic as 'codetermination' or mutual specification of organism and environment, in the sense of Francisco Varela, who claims that organism and 'environment' are mutually enfolded in multiple ways.[43] According to Félix Guattari, Varela's autopoiesis (defined as organism's self-making) deserves to be rethought in terms of evolutionary, collective entities, which maintain diverse types of relations of alterity, rather than being implacably closed in on themselves.[44] Guattari tries to realise his notion of heteropoiesis, the interconnection between his different ecologies of the social, the political, ethical and aesthetic dimensions, and transversal lines connecting them, thus expanding Varela's idea to the Socius. Enactive here means to emphasise the growing conviction that cognition is not representation of a pregiven world by a pregiven mind but is rather the enactment of a world and a mind on the basis of a history.[45] Varela suggests a change in the nature of reflection from an abstract, disembodied activity to an embodied (mindful) open-ended reflection where body and mind have been brought together. Varela's embodiment also involves cognition even at the level of the cells of the body; structural coupling occurs when autopoietic entities become linked together, which can also happen in a collaborative model. Guattari's expansion relates here and is directly comparable with Braidotti's concept of *zoe*, the materialist and vitalist concept of life itself that flows across all species. *Zoe* – Braidotti's posthuman approach – differs from both Haraway and Hayles, whose work she qualifies as 'high' cyber studies.[46] Braidotti moves on to a post-cyber materialism much in line with Félix Guattari. A better understanding of what is going on in this medical world may lead to what Guattari calls 'putting into circulation tools for transversality'.[47]

Returning to the view from the dialysis chair, we see that this is also part of Haraway's trope of technoscience as something that should not be narrated or engaged only from the point of view of those called scientists and engineers. Technoscience is heterogeneous.[48] We need to foreground the practices as the medical anthropologists Mol, Good and Prentice have shown. The dialysis chair is not in the middle of different perspectives, the body, the patient, the disease, the doctor, the technicians, the technology. Together they form an 'ecology'. We need to know more about this fine 'ecology' and how it relates to Damasio's 'body maps', Gibson's affordances and Good's physical, social and emotional 'context' of the patient.

The optimism in the discussed literature is understandable. The problem with the study on *Dialysis Centres*, however, is the direct 'translation' from a functional and formal layout and set up to an architectural design. There is no intermediate level of translation; function here *is* architectural form. Experience and affect have no place in this deeply positivistic and functionalist paradigm. We have to look closer into architecture's complicated compromise with techniques of representation, as we will explain below. Techniques of representation are central in architectural education, which is almost entirely digital. Architectural pedagogy is the central issue in the next section. The effects of digital techniques are discussed in relation to medical practices, hence, the next part will oscillate between medical and architectural practices with the explicit aim to understand better the latter one.

Education in Medicine and Architecture

Problem-based Learning

More than two decades ago the Faculty of Architecture of the TU Delft experimented with a new teaching curriculum, 'problem-based learning', a model corresponding with a medical teaching model. Initially, this model was developed in McMaster University in Ontario, Canada, and implemented in Maastricht University in the faculty of Health, Medicine and Life Sciences in the Netherlands. By active learning in smaller groups the students acquired their knowledge and skills. The idea is based on how we neurologically acquire knowledge in an active way. The acquired knowledge is always related to the practical condition of the discipline, apparently a perfect model to teach medicine. Rachel Prentice discusses this model in relation to anatomy as an initiation into medicine.[49] The approach allows instructors to build dissection around specific clinical cases or problems. Medical schools have hotly debated the benefits of this situated learning, focusing on the completeness of a highly structured approach by lectures versus the clinical relevance of a problem-based approach. Proponents of the traditional anatomy course argue that problem-based learning is good in theory, but leaves gaps because students are not exposed to the entire body during clinical work. The issue is about whether immersion is immersion in terminology, terms and structures, or immersion in human cadavers.[50]

The Architecture Department at the TU Delft also considered implementing the idea of problem-based learning in their curriculum.

A preparatory course for design instructors was given before the new curriculum was implemented. Instructors compared the human body to the 'architectural body', or better, 'the architectural body of knowledge'. Until then many studios were structured along the type of object, for instance, mass housing, public buildings, villa architectures, interiors, urban design, landscape and so on. During the first three years of education at the TU-Delft, students learn the basics. Not differently than in medicine, in architecture all specialisations are included in the curriculum. But there is no equivalent to anatomy as initiation. In the final two years of education, students are free to elect their graduation projects focusing on architecture, building technologies and urbanism. The students are supervised by a team of disciplinary mentors. Nevertheless, the new teaching curriculum was not only received with resistance, but it has also failed due to the different approaches of the mentors. Like in Prentice's example, some teachers argued for the 'complete body of architecture', while others argued for a more situated knowledge. The discussions could not go any further since the 'different approaches' could not agree with one another on what the 'body of architecture' would look like, and how to 'combine' and work together. These difficulties may have been 'personal' to some extent, but a more serious issue was that the body of architecture itself could not be defined. And worse, no one was really interested in doing so. Some wanted to explore a more 'scientific' way of research and design, sometimes derived from gross cybernetic thinking, sometimes from more sociological, political and urban considerations. A large group defended a more 'architectural', or 'creative' way of working, making a circular definition of the problem. But no concept was found to tackle the 'problem of architecture' and how to teach it. Discussions swept from vocational to scientific, to creative, to computational. The computational turn had not yet been realised in design teaching. In the current situation, this makes it even more complicated. In the end the 'old' system with a new name could prolong its life.

The viability of 'approach' is today a major issue in architectural education. The current situation is even more complex with the introduction of new parametricist technologies. Technology is no longer on the instrumental or 'tool' level, but, on the aesthetic level, designs have to look 'parametrically interesting', as we will discuss later. And it may be that Katherine Hayles, who comes from a different discipline, will provide an explanation. We seem to be in a state of suspension or even a crisis of traditional tectonic assumptions, a situation closely related to the incertitude that affects scale, for it was scale that granted to structure its foundational role. In the same way that Antoine Picon addresses the issue of scale and tectonics, digital culture did not only make the current gigantic projects like Chek Lap Kok Hong Kong International Airport possible, they also create the appropriate cultural context for their reception. It bears the mark of profound incertitude about dimension. On computer screens, forms seem to float without dimension.[51] It also means the weakening or sometimes complete rejection of structural organising rules in design and building production. What we sometimes see is a new poetics based on a ballet of forms yet unheard of, a poetics that has little to do with bearing of loads.

Computation in Design

In current computer technology we see a new conceptual field characterised by its instability and mobility, by its flexibility and propensity to change, the position of architecture within it opening up interesting avenues of enquiry and interpretation.[52] Antoine Picon in his *Architecture and the Sciences* addresses the growing importance of the virtual dimension in contemporary architecture. In current architectural discourse we see an increasing number of images and metaphors from mathematics, physics and molecular biology. He asks whether this is just a new habit, or if there are more profound reasons. Doubtless the use of scientific images and metaphors in architecture is not a recent phenomenon. Science has, in turn, referred to architectural notions throughout history. Referring to Nelson Goodman, Picon argues that architecture, like science, is concerned with making and conceiving of worlds populated with subjects and objects, quite similar to feedback loops. Architecture and science develop along parallel lines, often meeting in their common attempt to shape categories of visual perception. According to Picon we are living in a seamless technological universe where categories borrowed from landscape theory and history may give us a better understanding of our current condition, no longer dependent on structure or system analogies.[53]

Conceptual shifts took place during the development of cybernetics. Contemporary virtual reality may be traced to the Cold War period, when a new phenomenological space was emerging that could be visualised exclusively through the use of screens, maps, diagrams and probabilistic theories of prediction. New visualisations were required and their emergence heralded the destabilisation of form, an important issue in contemporary design. This also has an effect on how we conceptualise the human body in science and architecture. In biomedicine the transformations translate into a body that ceases to be a stable spatial map of normalised functions. It emerges as a highly mobile and unstable field of strategic differences. The body in architectural digital design and image production resembles this concept. Techniques in architectural design are never neutral, as Stan Allen rightfully remarks. The working methods of the architect will always condition the results. Allen mentions that these issues gain urgency at a time when architects work on the computer and almost all architectural representation is filtered through digital media. The success of digital technology (since the 1990s) is a valorisation of a new realism.[54] Whether we talk about Hollywood's special effects or architectural renderings, the success is measured by its ability to seamlessly render 'reality'. But the reality simulated is entirely mediated through the visual conventions of already existing media, primarily cinema and photography. As Christopher Hight reminds us, most of this software was not designed for architectural problems.[55] In an interesting analysis Hight shows that while material logics are embedded into the process of drawing, they resist translation into the conventions of construction. This translation from drawing to building is a central concept of Foreign Office Architects' (FOA) Yokohama Port Terminal. Hight shows that because of the distance to non-uniform rational Basis spline (NURB)-based modelling from conventional orthographic projection and their translation into construction, it is often necessary to move the drawing into software 'closer' to the conventions of construction (for example, from Maya to Auto-CAD). Recent parametric-based software is an attempt to bridge the gap between the practices of architecture and construction. Building Information Management (BIM) is foreclosing the space of translation between drawing and building.[56] In BIM drawings every line or symbol is a 'node' within a networked database of information. The symbols are linked to costs, fabrication and scheduling to be used by other professionals in estimating, planning and construction. A different approach in parametric-based software is Bentley System's Generative Components or those of Gehry Technologies. Here software is developed more out of the practice's experiences in constructing their characteristic designs of complex curvatures. Know-how is accrued over time, Hight writes, across many projects within practice translated into the design of new representational platforms to be used by many others, from steel fabricators to accountants keeping track of costs. According to Hight, the 'hand of the architect' is redistributed into multiple actors, factors, trades and disciplines. But, in the end, there is the building, and 'however transformed, [it] remains the destination of the architect's labor'.[57] Allen's remark that working methods always condition the results is telling: 'the building' has always been a kind of 'pre-select' in the process of design and fabrication. It hovered like a chimera over the software, constantly moving away from 'orthogonal', 'standard' architectures. Here, with the building the aesthetic assessment begins. Architecture is not an autopoietic system as Patrick Schumacher argues in his recent book.[58] A building is defined by codetermination or mutual specification of 'organism' (building) and environment.

In discussing National Aeronautics and Space Administration (NASA) pictures of planet Earth, Haraway talks about the technologies that make visualising possible. Computers, video cameras, satellites, sonography machines, optical fibre technology, television and microcinematography are all involved. However, the technoscientific visual culture signifies *touch*, provoking a yearning for the physical sensuousness of a blue-green Earth.[59] The images are ideologically powerful because they signify the immediately natural and embodied, over and against the supposedly violating, distancing, scopic eye of science and theory. In architectural education and design the situation is no different. Architectural theory is often seen as delaying the process to 'the real'; the image production of computer generated renderings. The digitisation of image has changed the economics of visual production and distribution profoundly. Image has never been more popular than today. More photos are now taken every two minutes than in all of the nineteenth century.[60]

Acceptance of the computer as a visualisation tool in architecture is also market driven. The clients can see what the building will look like before they spend their money.[61] Through a signifying chain that links architectural and artful images, new editing and book design techniques, the competition for best education and interesting curricula and career perspectives, the competitions in architecture, the market, the Grand Projects, the academic journals and architectural magazines, the attraction to the World Class Architecture of the Masters, European and American regulations, enrolment fees, best practices and Star Architects, economic recession and state policies, failing planning bureaus and grass root initiatives, professional fame, scholarship and cut down budgets, exchange programmes and abandoned building sites, visualisation techniques overstress form and appearance.

Ideologically, they are a yearning for embodiment. Visualisation techniques cannot interact with the world in an embodied way; they reproduce a mind without a body. Visualisation techniques are not embodied. They are representational. Time, affect, event and social political agendas in design cannot be addressed through techniques of visualisation. And even more importantly, the marvellous drawings never give an impression of the controversies and the many contradicting stakeholders, as Latour mentions. The space in which these objects seem to move so effortlessly is the most utopian (or rather atopic) of spaces.[62] According to Latour, and it *is* a crucial issue, where are the visualisation tools that allow the contradictory and controversial nature of matters of concern to be represented? Seeing is an *activity*, a balancing of sensations, motor activities and concepts. Visualisation as stable image ignores the fluidity of the eye and peripheral vision, not to mention the rest of the body's senses as we tried to show earlier.[63] Perceiving is a way of acting, Alva Noë writes.[64] Perception is not something that happens to us, or in us. It is something we *do*. The enactive approach means that the perceiver's ability to perceive is constituted (in part) by sensorimotor knowledge, by a practical grasp of the way sensory stimulation varies as the perceiver moves.

Bodies in Medicine and Architecture

In medical teaching the situation is different. Good writes that students in medicine – like geographers – are moving from gross topography to the detail of micro-ecology, covering the whole spectrum of cartography. Within the life world of medicine, the body is newly constituted as a medical body, quite distinct from the bodies with which we interact in everyday life. This is comparable with architectural education where students enter the world of 'world-class architecture' – the canonical architecture – they know from books and especially from the Internet. Most of the time they have not visited the site or the building. The anatomy lab is one critical site where the differences occur. The physical body of architecture is not present in the studio, and as a result there is very little attention to materiality in current architectural education.

Mol shows how vascular surgeons use angiographic pictures as maps during operations. The images help them to decide where to cut and how to cut.[65] Angiograms are part of the diagnosis; they help to decide how and where to operate. But they are not always decisive. A vascular surgeon remarks, 'we treat patients, not pictures'. Architecture students do not have pathology labs; they do not take apart buildings. Practical issues in architectural education are dealt with by construction mentors in particular. To a great extent the construction problems remain abstract problems and solutions abstract solutions. The real difficulties and practicalities manifest later on in the office. There is virtually no 'resistance', both in the material and immaterial sense in architectural education. After a few years the graduates finally are architects, meaning that they are capable of manipulating material practices, negotiating with contractors and clients, handling money flows and solving complex problems at building sites. Study and research mostly end with leaving academia. While medical doctors may find funding from the pharmaceutical industries for new drugs, practicing architects rarely apply for research grants as the building industry conventionally does not fund research in architecture.

Contemporary discourses in architectural and art education no longer deal with materiality but go the opposite way into *dematerialisation*.[66] In medicine the anatomy lab is a ritual space in which the human body is opened to exploration and learning, and in which the subjects of that learning engage in reshaping their experiential world.[67] In medicine there is a *physical* body, a corpse, and in architecture there is a *projected* body. Architects do not make buildings, they produce drawings, as Robin Evans and Stan Allen have so aptly shown us. According to Prentice, the notion of embodied dispositions that guide actions in medical practice resists models of learning that assume that the learner masters a set of rules or plans as in rules-based learning. This explains why the formalised seven steps of 'problem-based learning' could never work in architectural design. The same formalism is apparent in the notion of a general 'theory' in many architecture schools today. Addressing the misconception that there is one overarching architectural theory, Allen claims that architecture is a material practice. In the same way, theory is seen as a plurality of writing practices. Beyond the conventional theory/practice distinction, as Allen has put it, it becomes relevant to distinguish broadly between *practices* that are *primarily hermeneutic* – devoted to interpretation and analysis of representations – and *material practices* like urbanism, ecology, fashion, film making and gardening. The vector of analysis in hermeneutic practices always points toward the past, whereas *material practices* analyse the present in order to project transformations into the future. This will be a useful distinction, as long as we see the actors, agencies and actants as both human and non-human. Material practices today are increasingly digital. The distinctions that Allen and Evans made are useful only to a certain extent, as digitisation has changed the profession profoundly.

Stan Allen's *material practices* are now mostly computer-mediated. 'Hermeneutics' in Allen's sense will not suffice, since architecture theory deals with a lot more, namely philosophy, anthropology, history, sociology, critical theory, cultural studies and semiology. The main issue is not so much the reading of these webs of knowledge, but to show that the established orders in architecture, theory and society are open to change, that the world can be different. Material practices in design are of a different order; they are directed to an open field in order to project transformations into the future. Designing is a performance during which vision maintains a constant interaction between manual, or machine movement, and the resulting inscriptions. And, of course, there are the 'ideas' informing the performance; they are translated from mind to page in a combination of visual and motor activities.[68]

Prentice analyses how medical educators often describe medical learning as the formation of mental models. This cognitivist bias has dominated clinical research and medical decision making since the 1970s. Like in architectural design the computational models for diagnosis and visualisation in medical practice have strengthened the cognitive focus.[69] However, the cognitive approach to learning ignores the embodiment of technique, perception and emotion in the development of a physician's craft. Prentice sees this as the product of the Euro-American traditions that split mind and body. The connection between seeing and doing is neglected. What is also neglected are other senses, including touch, proprioception, sound and smell, which play a certain role in biomedical practice. Prentice quotes an anatomist who explains that he gauges his progress in opening the spine more by sound than by sight. As he chisels through bone, he cannot see when he reaches the spinal column, but the sound changes as his work progresses. Both architectural students and biomedical trainees and practitioners interact with advanced digital technologies. In both cases bodies and practices are reconfigured into new virtual concepts, reassembled into virtual worlds. Both disciplines – medical and architectural – use digital imaging techniques, but the differences here cannot be bigger. Prentice discusses how minimally invasive technologies in medicine foster new relationships between practitioners' and patients' bodies.[70] Minimally invasive surgery is known as key-hole surgery, minimal access surgery or, depending on specialty, arthroscopy, laparoscopy or endoscopy. It means threading a camera into a natural or artificial body and performing work while watching a monitor. The camera is inserted through 'ports' in the body. The comparison with architectural design gets interesting where architects work with digital tools and visualisation techniques. Both surgeon and architect have less kinaesthetic 'feel' for the body when working digitally. Of course there are many obvious differences, and Prentice is very careful in her descriptions and analysis, but one issue is immediately at the forefront. In medical practice, thinking in representational terms captures only one aspect of the work.

Referring to Mol, Prentice argues that the practitioners' hands become the focal point of theorising, moving away from philosophical problems that make knowledge into a mental representation of a reality in the world. Some of the practitioners had the feeling that they no longer worked on patients but on images, that it was more like solving a puzzle. The relationship between surgeons and remote technologies is apparently also dependent on the moment that they were introduced to these techniques. Like architects they had to learn to work with digital tools. But training is not the only issue. One of the surgeons tells her residents that she becomes part of the joint when she does arthroscopy, the apparatus became part of her body. She mentions that her identification of the arthritis she was dealing with was proprioceptive. Another surgeon says he creates a composite bodily understanding of the patient's joint that unites the on-screen visual, kinaesthetic and tactile information coming from instruments, and a patient's body with his surgical experience and anatomical knowledge.

The Institute of Robotics and Mechatronics, German Aerospace Centre (Deutsches Zentrum für Luft- und Raumfahrt) (DLR) is working on telemanipulation in minimally invasive surgery. To overcome the drawbacks of conventional minimally invasive surgery, telepresence and telemanipulation techniques play an important role in the DLR's research. In the case of minimally invasive robotic surgery, the instruments are not directly manipulated anymore. Instead, they are held by specialised robot arms and remotely commanded by the surgeon who sits comfortably at a master console. The surgeon virtually regains direct access to the operating field by having 3D endoscopic sight, force feedback and restored hand–eye coordination. The force feedback is important here; resistance is felt. The surgeon is not limited to seeing but can, via force feedback in the input devices, also feel what he is doing. This correlates to Noë's enactive approach where vision is a mode of exploration of the environment, drawing on implicit understanding of sensorimotor regularities. It involves not only the brain but also the animate body and the world.

North American medical education has faced a drastic reduction of dissection. Its importance was discussed even before computational tools became available. The development since the early 1980s of digital tools for teaching anatomy with the Visible Human Project – a digital database for medical teaching – provided promising alternatives to dissection. This has, according to Prentice, raised the stakes of the debate.[71] Architectural education has seen a similar digital turn. New design tools have been developed and also here the controversies are

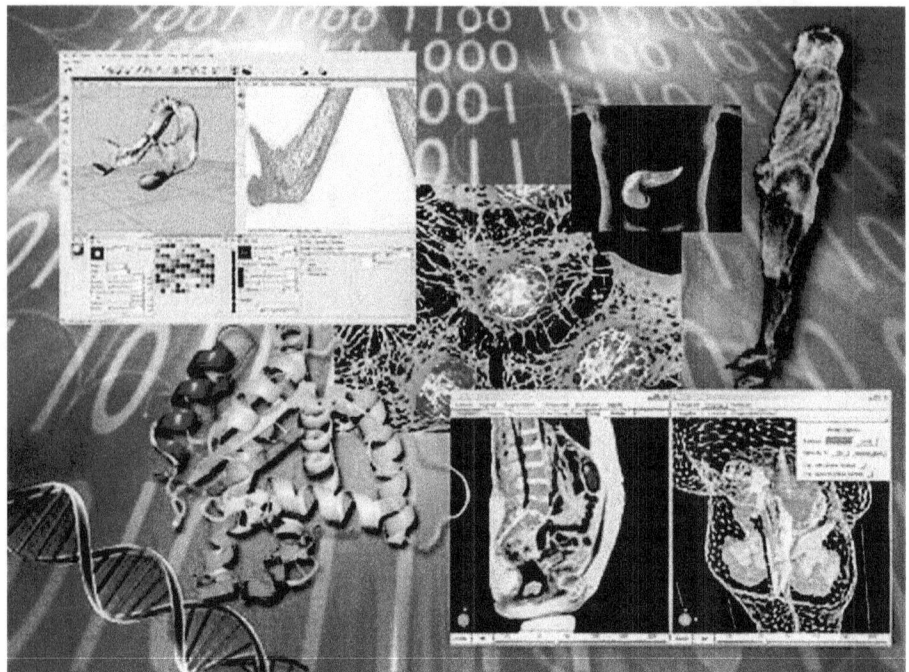

Figure 5 Visible Human Project, University of Michigan.
Source: http://vhp.med.umich.edu/ (accessed 18 October 2016).

manifold, given that architects deal with projection and manufacturing in a computer space that is largely devoid of materiality. Noë suggests that what the problem of vision makes so difficult for computation is that it is framed in an artificially restrictive way.[72] We now need to find out how these computational driven developments have impacted architectural design.

Projection as Flickering Signifiers

Anatomical language forms a platform for medical communication, Prentice writes. Like architectural students, medical students are also expected to understand the spatial relationships of organs and tissues. In architecture, the language has become unstable, however. Concepts and terminology in architecture rapidly change with culture. Over the past few decades we have seen many 'turns' in architecture. This happens in medicine, too, but the changes in architectural discourses follow one another up at ever greater speed since 'culture' changes far more rapidly in architectural discourse. This digital turn has caused information technologies to create '*flickering signifiers*' – a term that Katherine Hayles in *How We became Posthuman* relates to Lacan's '*floating signifiers*' – that are characterised by their tendency toward unexpected metamorphoses, attenuations and dispersions. This, however, does not imply that computational or digital virtual reality is *philosophically* different from the virtual reality of writing, drawing or even thinking. The virtual in creative practices is simultaneously the space of the new, the un-thought and un-realised. And it is precisely here that the real challenge for architecture begins. The 'new' in architecture is certainly not limited to digital techniques as the term 'digital architecture' suggests.[73] This translation is a basic characteristic of every form of creative practice, both architecture and art. It does not mean, however, that architecture theory merely focuses on what already exists. However, when it does, it can be highly speculative about 'the real' and architecture's role in it. Speculation connects both theoretical and material practices. They do not meet on the same structural level, however. Architecture is not a built discourse and it should not be considered a linguistic problem, as we have seen in one of the older 'turns'. This representational idea comes close to the 'feeling' that some surgeons have when working with the Visible Human Project. Dissection in their view comes close to solving a puzzle, 'an exercise that can require considerable spatial orientation skills but far less affective charge'.[74]

Working in and among the world of things – an instrumental practice capable of transforming reality – is not to lose sight of architecture's complicated compromise with techniques of representation as so brilliantly analysed by Robin Evans. Since architects work at a distance from the material reality of their discipline, they necessarily work through the mediation of systems of representation. Allen (in *Practice: Architecture, Technique and Representation*) is right: conventions of representation themselves need to be rethought today. We need new tools. No longer mainly static images, we need to engage time and change, shifting scales and multiple programmes. We are in the phase of 'architecture as interface'.[75] Picon describes the half-century-long history of the architectural use of the computer that has generally privileged form: 'the investigation of shapes in complete contrast with the limited vocabulary of modern architecture. The result has been a proliferation of alternative geometries that are calling for new criteria of evaluation.'[76] The focus on form, however, has prohibited a better understanding of our contemporary informational society of what Picon calls 'cyber-sociability'. 'One of today's most pressing challenges might be how to elaborate a convincing argument regarding the political and social relevance of architectural complexity.'[77] Despite the attention given to Deleuze's *Fold* or Thom's catastrophes theory, relational figures of thought to the Socius are weak or absent. Deleuze's fields and forces are often interpreted as a stylised design principle, not as economic, social and political forces, let alone as vital materialism. And even more important, why this '*Verdopplung der Welt*' (Adorno) is so important in architecture and art remains an open question. When the fluidity and smoothness of multinational capital and information flows are repeated on an architectural level without any intervention we are looking at mainly mimetic responses.

In architectural practice, we need to mediate philosophical thought to concrete forms of human practice and bodily experience. Sociologists and architects deal with a different kind of thinking, a less distanced way of dealing with reality. Referring to David Harvey I wrote elsewhere that architecture theory is never a matter of pure abstraction.[78] My aim is – like Saskia Sassen's – to stay close to the ground. Sassen talks of 'subterranean' trends to call into question familiar categories for organising knowledge about our economies, our societies and our interaction with the biosphere.[79] The specialisation of research, knowledge and interpretation, each with its own methods of protecting boundaries and meaning, does not always help us in our effort of detecting subterranean trends, a comparable way of proceeding I referred to earlier with embedded anthropology. An example of this sociological concreteness may be Luc Boltanski and Eve Chiapello's idea of the 'projective city'. 'Project' temporarily assembles a very disparate group of people, and presents itself as a highly activated section of a network for a period of time that is relatively short:

> Projects make production and accumulation possible in a world which, were it to be purely connexionist, would simply contain flows, where nothing could be stabilised, accumulated or crystallised. Everything would be carried off in an endless stream of ephemeral associations which, given their capacity to put everything in communication, constantly distribute and dissolve whatever gets in them.[80]

Projects have a stabilising effect and make connections irreversible. Boltanski and Chiapello describe it as a temporary pocket of accumulation, creating value. But the projective city is no longer mainly about work and leisure. It is about activity; it surmounts the older categories of the wage-earning classes. Activities generate projects, one integrates into networks, life is conceived as a succession of projects. Frequently named in management literature, projects succeed and take over one another, reconstructing work groups or teams in accordance with certain priorities. In this way the 'projective city' presents itself as a system of constraints placed upon a network world.[81] What matters is never to be short of projects, bereft of an idea, always to have something in mind, a constant desire to do something, no matter what.

Notation beyond Representation

Deleuze's *'affect'* may in a certain way translate well into this stream of very different practices in a network. Affect deals with moods and sensibilities; they influence ethics and politics as much as words do, arguments do and reason does. Although affect is on the architectural agenda, Deleuze and Guattari develop it further into 'material vitality' or 'geo-affect'. Jane Bennett, among many others, follows Deleuze and Guattari in dealing with the affect of technologies, winds, vegetables and minerals.[82] Affect here is directly related to bodily experience; it cannot be simulated by digital technologies. According to Picon, the absence of a body has led to instability of behaviour among participants in online debates. This may be an effect of the 'flickering signifiers' in these online debates. 'Without a body, our opinions tend to fluctuate much more than in "real" life and we become by the same token less reliable.'[83] And it may be more serious than Picon suggests. Francis Fukuyama claims that with the arrival of new technologies we have to consider what he calls 'the great disruption', or the negative trends in society intimately associated with the transition into the information society.[84] Anthony Giddens' 'trust', as well as Fukuyama's *'Trust'*, is at stake here. 'We' have become less reliable.[85] Our 'projects' cannot always be trusted.

Architectural notations always describe a work that is to be realised. They do not trace or represent already existing objects or systems but anticipate new organisations and relationships.[86] Notations may go beyond the visual and engage invisible aspects of architecture, affect being one of them. Allen mentions phenomenological effects of light, shadow and transparency, programme, event and social space. Notations can never be a private language; they are part of a social programme and relate to shared conventions. Architectural drawings are not purely autographic. Architecture itself is marked by its promiscuous mixture of the real and the abstract, he continues, at once a collection of activities characterised by a high degree of abstraction, and at the same time directed toward the production of materials and artefacts that are undeniably real.[87]

Although medicine studies are practical studies, all of its curriculum and practice content comes from the (bio-)sciences. The construction of the natural sciences aspires to universal application. Buildings and plans have a far more unique character, however. The conditions that the architect has to work with come from the outside; programmes are determined beyond the control of the individual architect. Or, as Allen has put it: 'the practice of architecture tends to be messy and inconsistent precisely because it has to negotiate a reality that is itself messy and inconsistent'.[88] In many architecture schools today we find present all of these material practices that Allen is mentioning, all at the same time. This is related to the changing notion of 'culture', and the major changes in this concept over the last thirty years.

The Empire of Sight, Chu's Bio-genetics, Eisenman's Deconstructivism and Koolhaas's Free Fall into Typographic Imagination

According to Terry Eagleton, culture traditionally signified almost the opposite of capitalism.[89] By the 1960s and 1970s, however, culture was also coming to mean film, image, fashion, lifestyle, marketing, advertising and the communication media. Signs and spectacle were spreading throughout social life. There were anxieties in Europe about cultural Americanisation. We seemed to have achieved affluence without fulfilment, which brought cultural or 'quality of life' issues sharply to the fore. Culture in the sense of value, symbol, language, art, tradition and identity was the very air that new social movements like feminism and Black Power breathed.[90] The halcyon days of cultural theory lasted until about 1980, several years after the oil crisis that heralded a global recession, the victory of the radical right and the ebbing of revolutionary hopes. In the 1980s and 1990s it turned into postmodernism. Investment shifted away from industrial manufacture to the service, finance and communication sectors. As big business became cultural, ever more reliant on image, packaging and display, the culture industry became big business. In universities and academies, pleasure, desire, art, language, the media, the body, gender and ethnicity became the dominant topics. A single word to sum all these up would be 'culture'. Today we can add computation, information technology, genetic makeup and biology. Naturally, this had its effect on architecture and architecture schools all over Europe and the United States. The triumphalism in Karl S. Chu's plea for 'xenoarchitecture' is no different from the concept of autonomy in architecture from the 1980s. 'For the first time in humankind,' Chu writes, 'humankind is finally in the possession of the power to change and transform the genetic constitution of biological species, which, without a doubt, has profound implications for the future of life on Earth.'[91] This 'new architecture' ('new' is the preferred currency of many contemporary architects) will be adequate to the demands imposed by computation and the biogenetic revolution. But Chu's architecture is not much different from Eisenman's notations in the Biocenter design (1987). The layout of Eisenman's design represents a deoxyribonucleic acid (DNA) molecule.[92] The most striking characteristic of the plan is the literal graphic copying of the four nucleotides in the design of the building.[93] In Chu we see the same procedure, but now as 'biogenetics'. The bio-machinic mutations of species is nothing else than simulated semiotic material resulting in, quite predictably, 'organic' architecture. Chu's architecture is certainly not 'a living rather than simply animated architecture'.[94] There is nothing to fear, no 'Frankenstein flavour' as Picon says. It is just another effort to reinstall autonomy in architecture. There is nothing alive in architecture, there is no body in architecture, dead or alive, there is no cell structure, no tissue, no DNA molecule. The architects' way of communicating is first and foremost through the drawing while text is less and less important due to the penetration of new communication and representation techniques into the profession, and the wider cultural issues that Eagleton is writing about. Some of the architects are addressing this quite clearly in their current practice. The effects are ambivalent, both positive and negative as Braidotti writes. The proliferation of discourses is both a threat and an opportunity in that it requires methodological innovations, such as a critical genealogical approach that by-passes the mere rhetoric of the crisis.[95] The Office for Metropolitan Architecture (OMA) and its partner research office AMO is an example of this ambivalence. It specialised in 'commercial communication' through different disciplines and clients. Koolhaas's *S,M,L,XL* was co-authored by Bruce Mau, the designer of *Zone*. Next to built work the book is also a 'free fall in the space of the typographic imagination'. Architecture students and critics alike find Rem Koolhaas difficult to situate; on what side of 'culture' and 'critique' is he? The answer to the question of where to situate him is not so much personal as it is about the new connotation of culture after the 1990s. Koolhaas is one of the few architects closely following the commercial trail. This difficulty to 'situate' OMA, and many other offices, also shows the problem that architecture theory is facing. If the Humanities are under attack, as Braidotti rightfully remarks, then this is also affecting philosophy and science studies in the technical universities, and in particular theory and history in architecture faculties. As part of a scientific field, most faculties of architecture fall under technical umbrellas. This, together with funding, the length of the curriculum and the number of students, are vital for the survival of architecture in the academic and educational fields. However, architecture theory can no longer just address the 'master narratives' of Le Corbusier, Mies, Koolhaas or Eisenman. It will have to address techno-science issues in a field impregnated by digitisation, culture studies, posthuman theories and contemporary philosophies on 'life' and 'matter'. We are facing more and more complexity, but 'leaving the Vitruvian frame'[96] in architectural design once and for all is not all that easy. Visualisation can prove woefully misleading, Hugh Aldersey-Williams argues, and scientists have long debated whether it is a useful tool after all.[97] 'Visualisation becomes more treacherous the further you travel away from the human scale.'[98] What we need is not another turn to autonomy, this time computational, but the 'ecological registers' or the three ecologies of Guattari: environmental, social and mental. We need to think 'transversally', to interlace the mechanosphere, the social sphere and the inwardness of subjectivity.[99]

A Write-up as Conclusion

Although writing and publication in architecture has become more important in universities over the last twenty years, it is not in the way Good envisions a way of proceeding like in medicine. Good draws his examples from his participation in conversations with a group of Harvard medical students. His issue is how medicine constructs the 'objects' to which clinicians attend, arguing that medicine formulates the human body and disease in a culturally distinctive fashion.[100] For more than four years Good and his wife Mary-Jo interviewed a cohort of approximately fifty students, selected primarily from the graduating class of 1990. In addition, they interviewed faculty and administrators about the curricular reform, Harvard's New Pathway to General Medicine Education and about their experiences in teaching in both the traditional or class curriculum and the New Pathway. During the first two years learning to see is fundamental. In the later years this is different; students finally enter the world of the hospital and join teams of interns, residents and attending physicians who care for the sick. During this time, students learn to construct sick persons as patients, perceived, analysed and presented as appropriate for medical treatment. Learning to write up a patient correctly is crucial to this process. Good mentions how a student compared the medical learning to reading a book, and writing a character study of one of the characters in the book. He went on to describe the standard categories of a medical interview – chief complaint, history of present illness, review of symptoms, past medical history, family and social history, and physical exam.[101] It is itself a formative practice as Good describes, a practice that shapes talk as much as it reflects it. It is also in line with the first steps of problem-based learning. They don't want to hear the story of the person, another student says. They want to hear the edited version. The edited version is quite often the problem in architectural education. Good's analysis of the impact and meaning of these stories fits the architectural discourse very well. Stories are one means of organising and interpreting experience, of projecting idealised and anticipated experiences like in architecture, a distinctive way of formulating reality and idealised ways of interacting with it. These presentations are not merely a way of depicting reality but a way of constructing it. It is one of a set of closely linked formative practices through which disease is organised and responded to in contemporary American teaching hospitals. Many of the so called 'approaches' in architectural teaching are circular; how the differences in approach should be made 'operational', and more importantly, 'understandable' to others by a 'write-up' is often left open. In other words, when we return to the questions of the teaching programme in architecture schools, what the 'parameters' are in a 'scientific approach', and which ones in 'a creative approach', are mostly open questions. So what's next?

We have to learn 'to race with the machine', as Brynjolfsson and MacAfee write. In architecture we cannot go back to the pre-digital. Earlier we referred to Robin Evans's work, which is pre-digital, but who formulates an important issue about the 'ideas' informing the performance in architectural design, ideas that are translated from mind to page in a combination of visual and sensori-motor activities. Brynjolfsson and MacAfee call this *ideation*, or coming up with new ideas or concepts. There we beat the machines. No one has ever seen a truly creative machine, they write. 'Programs that can write clean prose are amazing achievements, but we've not yet seen one that can figure out what to write about next. We've never seen software that could create good software; so far attempts at this have been abject failures.'[102] In the beginning of this introductory chapter we wrote that in many architecture schools, 'theory' is seen as delaying the design process. Now at the end we want to reverse this idea by claiming that 'sensible theory' is speeding up the process of thought and design. The 'write-up' in architectural education has to be improved greatly.

Notes

1. The text is for the greater part written in a 'we' form; only in a few cases I directly relate to my earlier writings in 'I'. The 'we' in this case are Alfred Jacoby and myself. We are both interested in what in the opening is called 'Embedded Anthropology'; authors like Annemarie Mol, Rachel Prentice and Byron Good have our ongoing attention. The many discussions we had about these medical anthropologists and how they could be of value to the research project of Anhalt University of Applied Science, and Fresenius Medical Care were the starting point of this text. In the end I pushed forward to write the text. The text also builds on an earlier one, 'Science/Technology/Virtuality' published in *The Sage Handbook of Architectural Theory* (2012). The title is almost the same; 'Virtuality' in the *Handbook* text is replaced by 'Education'. I try to explain what virtuality means in architectural education by comparing it to medical education. Both disciplines are working with digital 'tools'. 'Virtuality' relates the two, but to my mind in very different ways.
2. GlaxoKleinSmith, www.gsk.com (accessed April 2014).
3. Mol, *Logic of Care*.
4. Rosi Braidotti describes it as follows: 'Cartographies aim at epistemic and ethical accountability by unveiling the power locations which structure our subject-position. As such, they account for one's locations in terms of both space (geo-political or ecological dimension) and time (historical and genealogical dimension). This stresses the situated structure of critical theory and it implies the partial or limited nature of all claims to knowledge.' Braidotti, *Posthuman*, p. 164.
5. The aerial picture as absolute point of view implies 'the basic wisdom of critical sociologists' as Latour would have it. Latour, *Reassembling the Social*, p. 32. A lot of present social studies are 'case studies'; the time of the grand theories seems to be over. Whether this is 'After Theory' is to be seen. Many Marxist thinkers like Walter Benjamin (*The Arcades Project*) and Theodor Adorno were fully aware of Latour's critique very early on. Theodor Adorno's paratactic mode of presentation in his Hölderlin essay is a strong argument against any form of absoluteness. Parataxis violates discursive logic, even common sense, and it is rough and abrupt, juxtaposing seemingly disparate elements, much like modernistic music. See Adorno's *Noten zur Literatur III*. Oskar Negt and Alexander Kluge's *Geschichte und Eigensinn* (1981) may be another example. *Mille Plateaux* by Gilles Deleuze and Félix Guattari with a similar relational thread was published in 1980. The English translation followed seven years later. And it is at least remarkable that Donna Haraway posits her thinking still as Marxist, be it in line with Latour (and others) on the relational character of humans and non-humans as *socially* active partners.
6. Hayles, *How We became Posthuman*, p. 29.

7. Allen, *Practice*.
8. Haraway, *Modest_Witness*, p. 49.
9. Mol, *Logic of Care*, p. 3.
10. 'Western medicine understands the body as a complex biological machine, while the Zinacanteco see the body as a holistic integrated aspect of the person and social relations. Our treatments are mechanical and impersonal, our healers characterised by distance, coolness, formal relations, and the use of abstract concepts; their curing makes use of emotionally charged symbols, and a treatment relationship is characterised by closeness, shared meaning, warmth, informality, and everyday language. Western curing is aimed exclusively at the mechanical body, while Zinacanteco procedures are directed at social relations and supernatural agents.' Good, *Medicine, Rationality, and Experience*, p. 27.
11. Scarry, *Body in Pain*.
12. Damasio, *Looking for Spinoza*, p. 86.
13. Prentice, *Bodies in Formation*.
14. Ibid. p. 6.
15. Mol, *Body Multiple*, p. 5.
16. Kroker, *Body Drift*, p. 2.
17. Haraway, 'Biopolitics of Postmodern Bodies', p. 204.
18. Plates 3, 4, 5 and 6 of Gershon's *Immunological Orchestra* (courtesy of Edward Golub) found in: Donna Haraway, 'The Contest for Primate Nature: Daughters of the Man-the-Hunter in the Field, 1960–80', in *Simians, Cyborgs and Women: The Reinvention of Nature* (New York: Routledge, 1991), pp. IV–VII (plates 3–6), pp. 81–108; see also Edward Golub, *Immunology: A Synthesis*, ed. Donna Haraway (Michigan: The University of Michigan, 1987).
19. Haraway, 'Biopolitics of Postmodern Bodies', p. 206.
20. Good, *Medicine, Rationality, and Experience*, p. 5.
21. Graafland and Sohn, 'Science/Technology/Virtuality', p. 469.
22. Hayles, *Chaos and Order*, p. 15.
23. Graafland and Sohn, 'Science / Technology / Virtuality', p. 469.
24. Kroker, *Body Drift*, p. 70.
25. Good, *Medicine, Rationality, and Experience*, p. 23.
26. Ibid. p. 8.
27. Mol, *Body Multiple*, p. viii.
28. Braidotti, *Posthuman*, p. 40.
29. Bennett's example is the North American blackout in August 2003. For a vital materialist the grid is better understood 'as a volatile mix of coal, sweat, electromagnetic fields, computer programs, electron streams, profit motives, heat, lifestyles, nuclear fuel, plastic, fantasies of mastery, static, legislation, water, economic theory, wire, and wood – to name just some of the actants'. Bennett, *Vibrant Matter*, pp. 23–5.
30. Braidotti, *Posthuman*, p. 193.

31. Ibid. p. 35.
32. Good, *Medicine, Rationality, and Experience*, p. 22.
33. Haraway, *Modest_Witness*, p. 3.
34. Brynjolfsson and McAffee, *Second Machine Age*, pp. 48–9.
35. 'After thirty-two squares the inventor went through the first half of the chessboard. After thirty-two squares, the emperor had given the inventor about four billion grains of rice. That's a reasonable quantity – about one large field's worth – and the emperor did start to take notice.' Kurzweil, *Age of Spiritual Machines*, p. 36.
36. Picon, *Digital Culture*, p. 134.
37. Bruno Latour's research in the practices of the sciences has led him to a redefinition of the concepts of nature, society and technology. His 'sociology of science' aims at dismantling and discharging both the concepts of nature and society, since both 'monsters', as he calls them, were born in the same season and for the same reason. 'Modern' designates two sets of entirely different practices, which must remain distinct.
38. Good, *Medicine, Rationality, and Experience*, p. 35.
39. Boccato et al., *Dialysis Centres*, p. 10.
40. See Peter Sloterdijk's *Sphären Trilogie*, specifically *Sphären I (Blasen)* and *Sphären II (Globen)* published in German from 1998 to 2004. Volume I, Bubbles, and II, Globes, have been translated into English by Wieland Hoban and published by the MIT Press in 2011 and 2014.
41. Latour, 'Cautious Prometheus?'
42. Varela et al., *Embodied Mind*, p. 202.
43. Varela, 'Early Days of Autopoiesis'.
44. Guattari, *Chaosmosis*, p. 40.
45. Varela et al., *Embodied Mind*, p. 9.
46. Braidotti, *Posthuman*, p. 104.
47. 'The tools for transversality imply another role for intellectuals, no longer erecting themselves as master thinkers or providers of moral lessons, but to work, even in the most extreme solitude, at putting into circulation tools for transversality.' Guattari, *Chaosmosis*, p. 130.
48. Haraway, *Modest_Witness*, p. 50.
49. Prentice, *Bodies in Formation*, p. 83.
50. Ibid.
51. Picon, *Digital Culture*, p. 124.
52. Graafland and Sohn, 'Science/Technology/Virtuality', p. 470.
53. Allen, 'Landform Building'.
54. Allen, *Practice*, p. 74.
55. Hight, 'Manners of Working', p. 419.
56. Ibid. p. 423.
57. Ibid. p. 427.
58. Schumacher, *Autopoiesis of Architecture*.
59. Haraway, *Modest_Witness*, p. 174.

60. Brynjolfsson and McAffee, *Second Machine Age*, p. 126.
61. Allen, *Practice*, p. 75.
62. Latour, 'Cautious Prometheus?', p. 12.
63. Until the period in the sixties architectural form was considered the ultimate result of a process of research, both in its aesthetic form and material substance, like in the Bauhaus tradition in Germany. The Bauhaus is also the heir of the Arts and Crafts movement, a practical-minded philosophy. Both De Stijl and Bauhaus adopted Arts and Craft's anti-industrial programme and applied it to modern industrial design, establishing a radical fusion between fine art and design. This blurring of the boundaries between fine art and design has remained an important characteristic of the arts and visual culture in general in the Netherlands. See: Cramer, 'Interventions, Experimentation, Markets'.
64. Noë, *Action in Perception*.
65. Mol, *Body Multiple*, p. 93.
66. Graafland and Sohn, 'Science/Technology/Virtuality', p. 465.
67. Good, *Medicine, Rationality, and Experience*, p. 72.
68. Evans, *Projective Cast*, p. 368.
69. Prentice, *Bodies in Formation*, p. 13.
70. Ibid. p. 172.
71. Ibid. p. 70.
72. Noë, *Action in Perception*, p. 20.
73. Graafland and Sohn, 'Science/Technology/Virtuality', p. 481.
74. Prentice, *Bodies in Formation*, p. 91.
75. Picon, *Digital Culture*, p. 55.
76. Ibid. p. 62.
77. Ibid. p. 81.
78. Graafland, 'Looking into the Folds', pp. 143–4.
79. Sassen, *Expulsions*, p. 6.
80. Boltanski and Chiapello, *New Spirit of Capitalism*, pp. 104–5.
81. 'Projective city' may be confusing for an architectural audience; Boltanski and Chiapello call it 'projective', but we could also call it the 'project city'. It is not about a design issue as in Stan Allen, but about the continuous stream of activities in different projects.
82. 'The association with passivity still haunts us today, I think, weakening our discernment of the force of things. But it might be only a small step from the creative agency of a vital force to a materiality as itself this creative agent.' Bennett, *Vibrant Matter*, pp. 61–5.
83. Picon, *Digital Culture*, p. 54.
84. Graafland and Sohn, 'Science/Technology/Virtuality', p. 475.
85. Fukuyama, *Trust*.
86. Allen, *Practice*, p. 64.
87. Ibid. p. xvii.
88. Ibid. p. xi.

89. Eagleton, *After Theory*, p. 24.
90. Ibid. p. 25.
91. Chu, 'Metaphysics of Genetic Architecture', p. 423.
92. Graafland, 'Peter Eisenman', p. 108.
93. Ibid.
94. Picon, *Digital Culture*, p. 99.
95. Braidotti, *Posthuman*, p. 144.
96. Ibid. p. 167.
97. Aldersey-Williams, 'Applied Curiosity', p. 52. The example he gives is the bitter argument between physicists Werner Heisenberg and Erwin Schrödinger in the 1920s. Schrödinger's image of a wave in a box describing the behaviour of a small particle in a field of force, such as the negatively charged electron of a hydrogen atom held in orbit around a positively charged proton nucleus, was derided by Heisenberg, who felt that visualisation was invalid for quantum phenomena occurring on a scale below the wavelengths of light.
98. Ibid. p. 53.
99. Bennett, *Vibrant Matter*, p. 114. See also: Guattari, *Three Ecologies*, pp. 41–2.
100. Good, *Medicine, Rationality, and Experience*, p. 65.
101. Ibid. p. 77.
102. Brynjolfsson and McAffee, *Second Machine Age*, p. 191.

Bibliography

Aldersey-Williams, Hugh, 'Applied Curiosity', in *Design and the Elastic Mind*, ed. Paola Antonelli (New York: The Museum of Modern Art, 2008), p. 52.

Allen, Stan, *Practice: Architecture, Technique and Representation* (London: Routledge, 2009).

Allen, Stan, 'Landform Building', keynote lecture delivered at the symposium 'The Geologic Turn: Architecture's New Alliance', Taubman College of Architecture and Urban Planning, University of Michigan, 10–12 February 2012.

Bennett, Jane, *Vibrant Matter: A Political Ecology of Things* (Durham, NC: Duke University Press, 2010).

Boccato, Carlo, Guido Giordana and María Mello de Miguél, *Dialysis Centres: An Architectural Guide* (Lengerich: Pabst Science Publishers, 2012).

Boltanski, Luc and Eve Chiapello, *The New Spirit of Capitalism* (London: Verso, 2007).

Braidotti, Rosi, *The Posthuman* (Cambridge: Polity Press, 2013).

Brynjolfsson, Eric and Andrew McAffee, *The Second Machine Age: Work, Progress, and Prosperity in a Time of Brilliant Technologies* (New York: Norton & Company, 2014).

Carr, Thomas V. *View from the Chair: A Beginners Guide to Life on Dialysis* (Charleston: CreateSpace Independent Publishing Platform, 2011).

Chu, Karl S., 'Metaphysics of Genetic Architecture and Computation', in *Architectural Design*, Vol. 76, Issue 4 (2006) [quoted from Sykes, A. Krista (ed.), *Constructing a New Agenda, Architectural Theory 1993–2009* (New York: Princeton Architectural Press, 2010).

Cramer, Florian, 'Interventions, Experimentation, Markets', in *Reinventing the Art School 21st Century* <http://www.wdka.nl/wp-content/uploads/sites/4/2015/01/WdKA_Re-inventing-the-art-school.pdf> (accessed 14 September 2016).

Damasio, Antonio, *Looking for Spinoza: Joy, Sorrow, and the Feeling Brain* (Orlando: Harcourt Books, 2003).

Gilles, Deleuze and Félix Guattari, *Mille Plateaux* (Paris: Les Editions de Minuit, 1980).

Gilles, Deleuze and Félix Guattari, *A Thousand Plateaus: Capitalism and Schizophrenia*, trans. Brian Massumi (Minneapolis: University of Minnesota Press, 1987).

Eagleton, Terry, *After Theory* (London: Basic Books, 2003).

Evans, Robin, *The Projective Cast: Architecture and Its Three Geometries* (Cambridge, MA: The MIT Press, 2000).

Fukuyama, Francis, *Trust: The Social Virtues and the Creation of Prosperity* (New York: Free Press, 1996).

Golub, Edward, *Immunology: A Synthesis* (Michigan: The University of Michigan, 1987).

Good, Byron, *Medicine, Rationality, and Experience: An Anthropological Perspective* (Cambridge: Cambridge University Press, 1994).

Graafland, Arie, 'Peter Eisenman: Architecture in Absentia', in *Peter Eisenman: Recent Projects*, ed. Arie Graafland (Amsterdam: SUN, 1989), pp. 95–125.

Graafland, Arie, 'Looking into the Folds', in *The Body in Architecture*, ed. Deborah Hauptmann (Rotterdam: 010 Publishers, 2006), pp. 138–57.

Graafland, Arie, and Heidi Sohn, 'Science/Technology/Virtuality', in *The Sage Handbook of Architectural Theory*, ed. Greig Crysler, Stephen Cairns and Hilde Heynen (London: Sage, 2012).

Guattari, Félix, *Chaosmosis: An Ethico-Aesthetic Paradigm* (Indianapolis: Indiana University Press, 1995).

Guattari, Félix, *Three Ecologies* (London: The Athlone Press, 2000).

Haraway, Donna, 'The Biopolitics of Postmodern Bodies: Constitution of Self in Immune System Discourse', in *Simians, Cyborgs, and Women: The Reinvention of Nature*, ed. Donna Haraway (New York: Routledge, 1991), p. 206.

Haraway, Donna, *Modest_Witness@Second_Millenium.FemaleMan©_Meets_Onco Mouse™* (New York: Routledge, 1997).

Hayles, Katherine N., *Chaos and Order: Complex Dynamics in Literature and Science* (Chicago: University of Chicago Press, 1991).

Hayles, Katherine N., *How We became Posthuman: Virtual Bodies in Cybernetics, Literature, and Informatics* (Chicago: University of Chicago Press, 1999).

Hight, Christopher, 'Manners of Working: Fabricating Representation in Digital Based Design', in *The Sage Handbook of Architectural Theory*, ed. Greig Crysler, Stephen Cairns and Hilde Heynen (London: Sage, 2012), pp. 410–29.

Kroker, Arthur, *Body Drift: Butler, Hayles, Haraway* (Minneapolis: Minnesota University Press, 2012).

Kurzweil, Ray, *The Age of Spiritual Machines: When Computers Exceed Human Intelligence* (Harmondsworth: Penguin Books, 1999).

Latour, Bruno, *Reassembling the Social: An Introduction to Actor-Network Theory* (Oxford: Oxford University Press, 2005).

Latour, Bruno, 'A Cautious Prometheus? A Few Steps Toward a Philosophy of Design (with special attention to Peter Sloterdijk)', keynote lecture for the Networks of Design Meeting, Falmouth, Cornwall, 3 September 2008 <http://www.bruno-latour.fr/sites/default/files/112-DESIGN-CORNWALL-GB.pdf> (accessed 14 September 2016).

Mol, Annemarie, *The Body Multiple: Ontology in Medical Practice* (Durham, NC: Duke University Press, 2003).

Mol, Annemarie, *The Logic of Care, Health and the Problem of Patient Choice* (London: Routledge, 2008).

Negt, Oskar and Alexander Kluge, *Geschichte und Eigensinn* (Frankfurt am Main: Zweitausendeins Verlag, 1981).

Noë, Alva, *Action in Perception* (Cambridge, MA: The MIT Press, 2004).

Picon, Antoine, *Digital Culture in Architecture: An Introduction for the Design Professions* (Basel: Birkhäuser, 2010).

Prentice, Rachel, *Bodies in Formation: An Ethnography of Anatomy and Surgery Education* (Durham, NC: Duke University Press, 2012).

Russell, Kathleen, *Dialysis Advice, A Patient's Point of View* (Seattle: Walrus Productions, 2013).

Sassen, Saskia, *Expulsions: Brutality and Complexity in the Global Economy* (Cambridge, MA: Belknap Press/Harvard University Press, 2014).

Scarry, Elaine, *The Body in Pain: The Making and Unmaking of the World* (Oxford: Oxford University Press, 1986).

Schumacher, Patrick, *The Autopoiesis of Architecture: A New Framework for Architecture* (West Sussex: John Wiley & Sons, 2011).

Sloterdijk, Peter, *Sphären I – Blasen, Mikrosphärologie* (Frankfurt am Main: Suhrkamp Verlag, 1998); *Bubbles: Spheres Volume I: Microspherology*, trans. Wieland Hoban (Los Angeles: Semiotext(e), 2011).

Sloterdijk, Peter, *Sphären II – Globen, Makrosphärologie* (Frankfurt am Main: Suhrkamp Verlag, 1999); *Globes: Spheres Volume II: Macrospherology*, trans. Wieland Hoban (Los Angeles: Semiotext(e), 2014).

Sloterdijk, Peter, *Sphären III – Schäume, Plurale Sphärologie* (Frankfurt am Main: Suhrkamp Verlag, 2004).

Theodor, Adorno, *Noten zur Literatur III* (Frankfurt am Main: Suhrkamp Verlag, 1965).

University of Michigan, Visible Human Project, led by Dr Brian Athey, serves The National Library of Medicine's Visible Human data to health science students, clinicians, educators, and researchers <http://vhp.med.umich.edu/> (accessed 18 October 2016).

Varela, Francisco J., Evan Thompson and Eleanor Rosch, *The Embodied Mind: Cognitive Science and Human Experience* (Cambridge, MA: The MIT Press, 1993).

Varela, Francisco J., 'The Early Days of Autopoiesis', in *Emergence and Embodiment: New Essays on Second-Order Systems Theory*, ed. Bruce Clarke and Mark B. N. Hansen (Durham, NC: Duke University Press, 2009), pp. 62–76.

PART I

Architecture

CHAPTER 1

Urban Correlationism: A Matter of Access

Stavros Kousoulas

In his books *The Order of Things* and *The Archaeology of Knowledge*, Michel Foucault introduced the concept of *episteme*, not as the scientific position in each discipline that may be in oppositional relationship with others, but rather as a population of statements that function as the historically produced a priori of a specific discourse.[1] Foucault underlines that 'in any given culture and at any given moment, there is always only one episteme that defines the conditions of possibility of all knowledge, whether expressed in a theory or silently invested in a practice'.[2] While one may disagree with Foucault's stance on the exclusivity of a dominant episteme, one can certainly adopt his definition of episteme as the discursive apparatus that formulates the boundaries of the universally accepted inputs and outputs of a discourse. I join those who consider correlationism as the episteme of contemporary philosophy insofar as this intellectual framing apparatus is limited to a discursive practice.[3] However, I will attempt to show that it is a basic ontological premise and the episteme of not only architectural and urban design but also of theory production.

In order to do so the concept of episteme has to be expanded. For that I will use some of the insights by philosopher of science Isabelle Stengers. The point for Stengers is to move away from the binary of socially constructed versus objectively true scientific knowledge and to highlight that it is the practices of knowledge production that either constrain or pluralise methods of epistemological enquiry.[4] For that purpose, Stengers introduces and develops the concept of an 'ecology of practices'. This radically differentiates her approach from the Foucauldian episteme, while permitting her to remain within the insights of the episteme as a practical – and therefore material – framing apparatus. Through this concept, Stengers explores how scientific practices impinge on, and relate to other practices. In other words, for Stengers the question of how and what sciences discover about the world cannot be separated from how sciences impinge on it.[5] In her own words, the first step towards an ecology of practices is the 'demand that no practice be defined as "like any other", just as no living species is like any other'.[6] An ecology of practices focuses on the precise demands and requirements that each practice formulates in order to perform, and consequently on the obligations that this imposes on those who participate in it. By doing so, she distinguishes 'creative' practices, those of invention and discovery, and the modernist ambition to formulate so-called objective criteria for judging the scientific value of other practices.

Despite providing the methodological ground for a genealogical and non-linear discourse analysis, it becomes evident that Foucault's episteme lacks the practical materiality that bares the potential of complexifying and pluralising such an effort. Even within the same scientific domain, practices of enquiry – from theoretical research methodologies to experimental projections – set up a population of differential spatio-temporal rhythms, defining the discourse frameworks. In other words, one has to speak of epistemes and their material ecologies. As social theorist and philosopher Brian Massumi puts it, scientific progress is not to be linked with any kind of universal truth, but rather with a social technology of 'belonging' among divergent practices and their active practitioners.[7] Architectural discourse is inseparable from material practices. Research methodologies in space, design and form are inseparable from the 'laboratories' where experimental epistemological systems are developed. Urban theory is rather theories of multiple urban 'embodiments'.

For philosopher Karen Barad it is of utter importance to replace the term 'interaction' with 'intra-action', as the former wrongly presumes the existence of independent entities before their relationships. In Barad's ontology, intra-actions include all manifestations of matter that produce 'agential cuts': *relata* within events that emerge through intra-actions and formulate the separability of agency as distributed in matter. Agential cuts stand in for causality, the issue being not one of addition to an extensive list of casual agents, but rather that of subtraction from a pre-existing intensive agency. It is from this point that Barad's ontology becomes extremely relevant to issues of epistemological enquiry. And that is because phenomena are produced by apparatuses, not as a means of gaining access to the world, but rather as 'boundary-drawing practices, specific material (re)configurings of the world – which come to matter'.[8] Architectural theory does not provide us with a better version of a world's observation nor capture an essence hidden beneath it. It does not open a window in the world's surface. Instead, it embodies a specific apparatus involving the intra-action between material and discursive practices of 'cutting' the world through a manifested agency.

It is for their intra-active capacities that epistemes ought to be understood as an ecology of 'material-discursive' practices. Any embodied apparatus is in fact a boundary maker, the 'machinic phylum' of an episteme that does not simply unfold in space and time, but rather produces space and time – the relational becoming of a differential form. It is machinic due to the existence of processes that act on a population of heterogeneous elements and cause them to come together, consolidating into a novel entity. Nonetheless, it is also a phylum due to the existence of a common body plan that through different operations yields a concrete actualization.[9] In Barad's diffractive methodology, heterogeneous elements of a population of theoretical approaches are brought together. Diffractive reading advances the emergence of new concepts, probing for conceptual differences that make a difference. In what follows, I will attempt to perform a diffractive reading, entangling various elements of architectural, urban and spatial theories in an effort to assemble a genealogy. Based on my initial argument that correlationism, namely the privilege to access, is their base, I will affirmatively trace their intra-connections. This, in turn, will enable me to offer a materialist, morphogenetic account of architectural and urban phenomena.

Correlational Groundings

As a philosophical term, correlationism was first coined by philosopher Quentin Meillassoux. In his book *After Finitude* Meillassoux departs from a long-lasting debate among philosophers, that of the primacy between primary and secondary qualities.[10] The distinction has significant effects on both ontological and epistemological levels. It is the very nature of reality that is at stake here. In brief, primary qualities have been considered as the properties of objects that exist independently from the observer, while secondary ones are those that produce sensations in the observer: motion, figure, number, solidity and extension, opposite to colour, sound, taste and smell, just to name a few. While it is evident that we are dealing with issues of subject–object relations, of the knower and the known, there is another issue emerging out of this dichotomy, namely the issue of *access*.

Once you 'remove the observer ... the world becomes devoid of these sonorous, visual, olfactory, etc., qualities, just as the flame becomes devoid of pain once the finger is removed'.[11] Meillassoux himself admits that there is no reason to question the fact of emergent properties as a result of relational capacities. The sensible is a relation, and not a set of properties inherent and intrinsic to singular agents.[12] What he considers problematic is the very distinction itself: the emphasis on the presence of a subject so as for the secondary qualities to be able to manifest. Therefore, his thesis is twofold:

> [O]n the one hand, we acknowledge that the sensible only exists as a subject's relation to the world; but on the other hand, we maintain that the mathematizable properties of the object are exempt from the constraint of such a relation, and that they are effectively in the object in the way in which I conceive them, whether I am in relation with this object or not.[13]

While I object to his effort to mathematise ontology, I will join his plea for a mind independent existence of the world. It is the mind independent existence of reality that has been traditionally labelled as a realist one.[14] On this ontological commitment, Meillassoux defines correlationism as consisting in the disqualification of the possibility to consider subjective or objective realms independently of one another.[15] To put it in simpler words, correlationism is the ontological stance according to which an object is unable to be conceived as in-itself, in isolation from its relation to a subject observing it.[16] In order to do so, one must assume and adopt a version of nature that is radically separated from 'sentience'; one must fully adopt what Whitehead refers to as the 'bifurcation of nature'.

Let us now turn to architectural practice and discourse and their relationship with issues of access. The aim is not to provide a historical analysis of architectural discourse, but to trace a genealogical development that can be clearly connected with modernity. I join camps with architectural theorist Sanford Kwinter in claiming that the very notion of modernist space is nothing but a footnote in the much broader notion of modernity.[17] Modernity is primarily a philosophical problem and subsequently one of historical periodisation. Like Kwinter, I argue that modernity should not be understood as a synonym of the new or the contemporary, but rather as a virtual possibility – rarely actualised yet persisting – aiming at the transformation of the very conditions of possibility.[18] The problem of modernity is a problem of ontological value. It substitutes the mutual relationship between the virtual and the actual with static Platonic ideals or Aristotelian hylomorphism, or both.

As scholar of cultural studies Lawrence Grossberg argues, it is this distinction between virtuality and possibility that is crucial, considering that most theories of imagination have been theories of possibility.[19]

I will diffractively expose a genealogical line that runs from the first attempt to theoretically reflect on issues of spatial production, their later interpretation and critical transformation in defining the outlines of formal architectural qualities, to the combination of both under the premise of modernist emancipation. It is a brief exposition of the relation between architecture and the issue of access. However, as Deleuze would have it, 'it's not a matter of bringing all sorts of things under a single concept, but rather of relating each concept to the variables that explain its mutations'.[20] It is a genealogy that still remains to be written in its full extent, encompassing many more instants than the ones I will refer to in this essay. For the present, Vitruvius, Alberti and Le Corbusier will be considered as the 'conceptual personae' who highlight issues attached to spatial production while nevertheless expanding beyond it. While intervention in the world may be architecture's priority, the questions that any form of intervention poses (be it on space, on the human body or on any process of creation for that matter) still remain to be fully articulated. The question of access appears under the mode of a problem, forcing one to consider its implications on a population of heterogeneous fields. The 'question-problem complex' needs us to ask not only *what is* access but rather who, which one, how many, how, where, when, in which case and from what point of view.[21] In these minor questions lays the potential of tracing a continuous variation and not paving the passage from opinion to *doxa* to *Urdoxa*.

Access through Function

The *Ten Books of Architecture* is considered the oldest surviving treatise on architecture. In the first volume of *De Architectura* Vitruvius boldly argues that architecture is to be distinguished from random building practices through a concise set of principles: *ordinatio, dispotitio, eurythmia, symmetria, decor* and *distributio*.[22] As many scholars have underlined, the sequence of these basic principles is what positions architecture as the 'first art'. Yet, on closer inspection, this rather classical reading of practices of spatial production may surprise us. One would expect *symmetria* to be the most prominent feature of architectural principles, or even the ultimate goal of the practice, however is only to be found in the fourth position, just before decor and distribution.[23] And that is because, according to Vitruvius, *ordinatio* is what crucially defines architecture. In his words,

> order is the balanced adjustment of the details of the work separately, and, as to the whole, the arrangement of the proportion with a view to a symmetrical result. This is made up of Dimension, which in Greek is called posotes. Now Dimension is the taking of modules from the parts of the work: and the suitable effect of the whole work arising from the several subdivisions of the parts.[24]

Vitruvius' architectural theory can be seen as the effort to bridge the thought of two major, yet opposing, philosophers: Plato and Aristotle. Residing on possibility instead of virtuality, Vitruvius sets the foundations for a typological architectural thinking in which spatial individuation is 'achieved through the creation of classifications and of formal criteria for membership in those classifications'.[25] Vitruvius combines Platonic transcendent essences and Aristotelian immanent natural states to formulate a theory of spatial production. While the references to a structurally rigid list of principles that define architecture are transcendent elements, Vitruvius demands the knowledge and manipulation of materiality: air and wind, earth and water, fire and the sun will guide architecture towards a naturalised typologisation. Symmetry and proportion – as well as all the other principles of architecture – are the *substantial forms* that give reason (*ratio*) to the design, while matter itself is an *accidental form*, to use the Aristotelian division. A building can lose or gain material properties as long as the conceptual principles that constitute its essence remain intact and ordered. The premises of the *hylomorphic scheme* are now within architectural production and theory. Essence as an ideal form pre-exists any of its incarnations as an actual form, and matter is but a malleable receptacle for exterior forms. The inert and passive wilderness (note the contradiction) has to be tamed. It is then that we can speak of the notorious triad: *firmitas, utilitas, venustas*.

The Vitruvian triad is a plea for coherence – a plea for a rationalised order, against contingency, encompassing not only architecture but modernity as a whole.[26] It is, to use a metaphor, a *solace in the prescription*, safely guiding both architectural production and theory.[27] What is essentially faulty in the triplet of the Vitruvian qualities is their self-evident nature. As architectural theorist Jeremy Till notes, 'they should be background beginnings rather than the foreground ends that the Vitruvian dogma suggests'.[28] Based on the self-evidence of these qualities, the *whole* body of architectural knowledge and production can be manipulated. As theorist Indra McEwen underlines, 'bodies were wholes, whose wholeness was, above all, a question of coherence. The agent of coherence – in the body of the world and in all the bodies in it – was ratio.'[29] Once again, Reason attained through Order(*ing*) of a (whole) Body. However, one can never fully exhaust the field of capacities given a body in its milieu, rendering thus Spinoza's quote that 'no one has yet determined what the body can do' even more relevant when related with Vitruvius's conception of architecture.[30] More so, if the process of ordering this body requires its division in primary and secondary qualities activated by a subjective access to their relationship.

Let me now attempt a reading of the triad's elements that highlights their intricate connections with a bifurcated understanding of nature – or, to paraphrase Whitehead, with the *bifurcation of architecture*. The key element in the case of the Vitruvian triad, as well as in its numerous interpretations and adaptations, is that of *utilitas*. Standing as the triggering cause–effect between materiality and aesthetics, *firmitas* and *venustas*, the notions of utility and functionality activate the subjective access to the great architectural outdoors. A functional closed system of predicted needs and end results where the presence of the architect is reified as the 'creator' who bridges the gap between an untamed world (of matter) and an ethereal after (of beauty). Architecture, Vitruvius constantly reminds, is the means to provide solutions to high-order problems. Let me expand further on how one can conceive *firmitas* as a set of primary qualities and *venustas* as secondary ones.

Firmitas, the strive for a built environment that can effectively endure – even eternally as Vitruvius would demand – is the art of mastering materiality; quantifiable matter that can be extensively measured and arranged through design, forming volumes and spaces for desired functions. The proper handling of matter requires a mathematised space with the architect bearing the scientific knowledge not only to distribute it – under Euclidian geometry – but also to explore the inherent, primary qualities of the building material. Next to the architecture that is the cause of awareness, comes the architecture that is apprehended in awareness. Not the quantifiable or the quantitative, but the qualitative – the psychic additions of a perceiving mind. If the former is the architecture that is the cause of awareness in the form of its building material, the latter is the architecture that is formulated via perception; in other words, it is an architecture of secondary qualities: form, ornamentation, the play of light, rhythm and proportions, colour, some of the elements that give a so-called poetic dimension to built space. These are elements that qualify as the set of qualities through which *venustas* is achieved.

A 'bifurcated' architecture, then, is divided between its fundamentals (solid and invisible to the eye, known only to science) and its aesthetics (what the mind adds to make sense of it). Among them, *utilitas*, the functionality of space, separates the architect from the builder and from the artist, and allows her to access and arrange the relationship of objective materiality with subjective experience. It is from this point on, that a few hundred years after Vitruvius, Leon Battista Alberti would attempt to intensify the triad through the addition of a crucial concept, one that would envision formalism – and not only function – as the key to Reason.

The Form of Access

In his book *Momus*, Alberti uses a much more fluid language to describe his most intriguing concept of *lineamenta* not in technical terms but rather in a form of philosophical fiction.[31] Alberti advocates the superiority of visual over philosophical knowledge, where philosophy and architecture seem to be in competition regarding their epistemological validity.[32] Irrespective of who gets the upper hand, it is the primacy of epistemology over ontology, or to be even more precise, the dormant separation of the two that affirms the bifurcation into primary and secondary qualities. Especially in the fourth book of *Momus*, where we come across the dialogues between Charon, the ferryman to the underworld and Gelastus, the shade of the dead philosopher, Alberti discusses extensively how something comes to being – the actions of a creator, the making of a substance, the impelling of form and the changing of one thing into another.[33] As Charon says,

> [O]h, what a fine philosopher who understands the ways of the stars but not those of men! Learn from Charon the ferryman to know thyself. I will tell you what I remember hearing, not from a philosopher – for all your reasoning revolves only around subtleties and verbal quibbles – but from a certain painter. By himself this man saw more while looking at lines than all you philosophers do when you are measuring and investigating the heavens.[34]

The painter, embodied in the gazing eye, does not only know how to look at the world but is also able to masterfully guide its reproduction via representation. The eye is able to know more about the world than any thinking mind, than any reasoned imposition of projected needs and function. Reason is approached through exhaustive observation. *Lineamenta*, a concept central to Alberti's aesthetic theories, stands next to *materia*, forming the *corpus*, the body, of built space. While *lineamenta* is the product of a thinking subject, aware of its environment and struggling to make sense of it, *materia* is the product of nature, dependent on selection and preparation.[35] If for Vitruvius it was the process of defining functions and providing solutions on their projection that affirmed the role of the architect, then for Alberti the gap between primary and secondary architectural qualities can be bridged by the human subject when it becomes able to master the *surface* of the world – both in documenting it and representing it, approximating thus a higher, divine formal order. In his own words:

> [A]ll the intent and purpose of lineaments lies in finding the correct, infallible way of joining and fitting together those lines and angles which define and enclose the surfaces of the building ... Nor do lineaments have anything to do with material, but they are of such nature that we may recognise the same lineaments in several different buildings that share one and the same form ... It is quite possible to project whole forms in the mind without any recourse to the material ... Since that is the case, let lineaments be the precise and correct outline, conceived in the mind, made up of lines and angles, and perfected in the learned intellect and imagination.[36]

For Alberti, matter is to blame for a folly of outrageous proportions – the deviation that humankind undertook from the form that was bestowed on it from the Creator. This theological stance lies at the very core of his aesthetics. It is his constant preoccupation with the disjunction between appearance and reality that is articulated almost everywhere throughout the corpus of his treatises. It is, however, an exclusive disjunction, one that presupposes a fundamental *either...or...* dispositioning reality and its multiple, variable and transformable appearances. Where Alberti positions form, an essential lineament, as the underlying locomotive of spatial studies and production, he does so in a manner that attributes form a solely signifying power. That is why *lineamenta* carry for him not only the meaning of the world (what makes it make sense to us) but also the potential of transforming it via the addition of more form to that closed world system, the addition of meta-signifying levels of an ideal order. This is why Alberti's formalism, understood as a plea to 'formal' form, coupled with Vitruvian functionalism, as a plea to 'functional' function, compose the *abstract machine* that runs through 'modernities'. This is a non-temporal spatial theory that is based on the premise of a possible break with an idealised time. But in order to attain the temporal rupture one has to 'bifurcate' reality. And that is achieved by mastering the manipulation of primary architectural qualities through accessing the secondary ones.

Access as Rupture

From mass to surface to plan; from *volume*, to *lineaments*, to *function*; from the material primary to the intangible secondary and a reminder for each. That is how Le Corbusier makes his plea for a new architecture. The present is a nightmare from which we have to disassociate.[37] Before that, however, one must recall the eternal ideals that constitute the world. One must accept that the Engineer and the Architect stand separated – and necessarily so. The Engineer, Le Corbusier claims, stands high, while the Architect has lost the way. Moreover 'the Engineer inspired by the law of Economy and governed by mathematical calculation, puts us in accord with universal law. He achieves harmony.'[38] On the contrary, the Architect 'by his arrangements of forms, realizes an order which is a pure creation of his spirit'.[39] An immaterial, and sometimes ill spirit, which has to be reminded of the basics in order to be cured. If for Vitruvius function was what gave purpose and meaning to the architect's profession and if for Alberti it was the handling of form, then for Le Corbusier it was both. However, neither function nor form occupied the same gravitational force than with his predecessors. What constituted the radical break with the past was the break itself; form and function, both masterfully manipulated, would assist with that. A new (possible, not contingent) future could come forth. If the futurity envisioned by Le Corbusier is a naked fact, then it is a fact that 'is a medium for ideas only by reason of the *order* that is applied to it'.[40]

Once again, that very same trajectory we have been examining since Vitruvius, places Order as what knits the threads of this world. And order of that kind is achieved through function, through the Plan. Without it 'you have lack of order and wilfulness'.[41] The Plan is the generator.[42] The lineaments, the surface, as the arrangement of volumes leads to the plan while the plan simultaneously assures the value of the form – that it is not 'shapeless, poor, disordered and graceless'.[43] In short, 'where order reigns, well-being begins'.[44] In that sense when Le Corbusier declares that 'to create architecture is to put into order,'[45] one has to consider what exactly he had in mind as disordered. As Till underlines, what Le Corbusier envisioned as the exemplar case of disorder was an organism in a state of sickness – an ill organism, falling far away from its own ideal, the one that the Creator imagined, designed and build to perfection.[46] Similar to Alberti, the disjunction between appearance and reality preoccupies Le Corbusier as well. Since the disjunction is conceptualised once more as an exclusive one, then for the treatment of the illness, for the recuperation of order, only radical surgery can do well. A radical break with the past, a break that would not cure the cancer, but rather remove it once and for all. Once order, the means to achieve and the goal of the temporal cut, has been mastered – through the ingenious mastering of form and function – then the Architect can eventually stream sensations through primary matter, 'then can arise those multifarious sensations, which evoke all that a highly cultivated man may have seen, felt and loved'.[47] Le Corbusier could not have expressed this position more clearly than in his early work *Purism*, where he openly states that

> the goal of art is to put the spectator ... in a state of an elevated order. To conceive, it is first necessary to know what one wishes to do and specify the proposed goal ... Conception is, in effect, an operation of the mind which foreshadows the general look of the art work ... Possessed of a method whose elements are like the words of language, the creator chooses among these words those that he will group together to create a symphony ... One comes logically to the necessity ... of logical choice of themes, and the necessity of their association not by deformation, but by formation.[48]

I will not expand further on the crucial difference between 'formation' and 'deformation', but, following Massumi, I will argue that architecture essentially deals with the latter rather than with the former.[49] It is worth mentioning, however, that by deformation one should consider the capacity of design processes to deal with the production of form topologically. While architecture seen as formation considers form as both the raw material and the end product, its origin and *telos*, architecture understood as deformation shifts focus to the design process itself – no longer a means of arranging order and ordering arrangements but rather the eventuating practice of 'intuiting' topological transformations.[50]

It should be clear by now that Le Corbusier's aesthetic theory manages to combine the most crucial elements of an entire genealogy in order to propose a methodology of attaining an ordered *rupture*. In the very spirit of that rupture, Le Corbusier is to be seen not as the inventor of modernism – that would not only be inaccurate, it would also mask the genealogy I wish to expose – but as a consequence of modernity: a modernity long before him and one that continues long after him. Le Corbusier is a symptom not a cause.[51] Therefore, and despite the paradox, Le Corbusier's definition of architecture is not his; it is the contingently obligatory – not the logically necessary – transformation of a methodology of shaping the world.

> Architecture is the masterly, correct and magnificent play of masses brought together in light. Our eyes are made to see forms in light; light and shade reveal these forms; cubes, cones, spheres, cylinders or pyramids are the great primary forms which light reveals to advantage; the image of these is distinct and tangible within us without ambiguity. It is for this reason that these are beautiful forms, the most beautiful forms. Everybody is agreed to that, the child, the savage and the metaphysician.[52]

In his definition, Le Corbusier refers to something much broader than architecture, namely to the process of creation itself. It is the masterly, correct – learned and magnificent – composition of fixed forms that are initially drawn from an a priori vocabulary, that includes the great primary forms that light afterwards reveals to advantage. The creator is in possession of the supreme combinatorial skills, able to formulate the perfect assemblages of forms; as Massumi puts it, 'creation is an individual expression of the artist at the same time as it accedes to universality'.[53] The correct combination of the elementary forms, as brought into light, can give birth to beauty, something that everyone will agree on: Charon, Momus and Gelastus, the child, the savage and the metaphysician. Indeed, each one has once experienced beauty. That is a given. What is not a given – and demands to be studied – is how each one ended up experiencing beauty. The very existence of a universal state does not qualify to universalise the conditions that made it emerge.

For all three conceptual personae, Vitruvius, Alberti and Le Corbusier, the condition was still much larger than that that it conditioned. They all succumbed to what is essentially a rigid or at best elastic principle, in strict contrast to what Deleuze has called a 'plastic principle' where a condition is not larger than what it conditions. Moreover, plasticity assumes a condition that is determined at the same time that it determines.[54] The 'creative' condition, for all three, transcends the created and it is the access to the condition that mobilises creation. The problem, then, is that of the retroactive hypostatisation or how creation adds anything to its concept.[55] Moreover,

> to the extent that the possible is open to 'realization', it is understood as an image of the real, while the real is supposed to resemble the possible. That is why it is difficult to understand what existence adds to the concept when all it does is double like with like. Such is the defect of the possible: a defect which serves to condemn it as produced after the fact, as retroactively fabricated in the image of what resembles it.[56]

What is called the 'possible' is condemned to be forever governing the structure of the real and it is universalised to cover all experience.[57]

But if the given does not explain but must itself be explained, then the given is to be seen as a result of a (morpho)genetic process, not an origin but that that must be originated. Even if one insists on preserving the term 'essence', it should be preserved on the condition of saying that the essence is an accident, a singular determination, an event.[58] The ontological acknowledgement of the event and the *in-between-ness* that it entails, is what will shift the focus from access to a relational understanding of an amplified – and ontologically flat – agency.

Towards a Non-Bifurcated Architecture

The *more* of nature against the *less* of its concepts is a cornerstone of the empiricist thinking. It is on this that Whitehead bases his concept of nature. For Whitehead, nature is what we are *aware of in* perception.[59] Essentially, what he is formulating is an astonishingly simple question; since one is aware of the relational, complex reality of perception, why should this process – and the nature it takes place in – be reduced to anything less than what they are. Therefore, the reference to awareness is a double one: on the one hand to deal with what one perceives and on the other to remain true to the very process of perception without diminishing it. The effect would be a truly ground-breaking potential: to pluralise perception and its awareness as a post-human, non-anthropocentric, multi-agential ability.

To do so, an ontology of the event that avoids the wavering between subjective and objective, secondary and primary qualities has to be articulated – a *return to experience itself*.[60] It is the return to experience through an ontology of the event that can introduce time, not in the form of a condemned past or a longed-for future, but rather in the form of 'duration'. It is the acknowledgement of time as the duration of the event, which can affirm the non-divisible nature of a post-human perception. The process of perception and the process of its conceptualisation are of different durations, even if they coincide from time to time. To confuse the two is to fall back on the epistemic fallacy. And this retreat, common in architectural theory, eventually necessitates the formulation of givens so as to excuse the *less* of the concept in comparison with the *more* of experience. It is the *less* of the architectural concepts that demands a bifurcated version of experience. The bifurcation of architecture in its primary and secondary qualities not only sets the foundations for the exclusive disjunction between appearance and reality, the infinite list of non-plastic 'either . . . or' principles, but also activates the separation between a subject and an object – a separation that takes the form of a rivalry disputed by the ability of a 'meta-subject', the creative mind, to access the bifurcation – be it on the base of function, form or rupture – and project their reconciliation.

However, if, as Whitehead suggests, the presupposition of a mind stands as a barrier in accounting for what experience in nature is, then this is because mind is presupposed categorically. Mind, for most sciences, is a term, a *relatum*. A radical way out of a bifurcated nature, as media philosopher Steven Shaviro puts it, presupposes the existence not only of a non-correlational mind but of a non-correlational thought. In his own words

> a thought – or consciousness, or sentience, or feeling, or phenomenal experience – that is non-phenomenological insofar as it goes on without establishing relations of intentionality to anything beyond itself and even without establishing any sort of reflexive relation to itself.[61]

This is a thought therefore that is fully material and practical, in the sense of an agential awareness of perception: a double-bind between nature and what constitutes it, the condition and the conditioned. If nature is what we are aware of in perception, then radicalising this awareness as a capacity of everything that constitutes nature, manages to provide an immanent account of the transcendental conditions of experience. If thought is not to be related in any way with rationality, intention or correlation – those being variations of a theme rather than the theme itself – then thought is agency as distributed in material awareness and a mind is a relation, not a *relatum*. No longer differences in kind, but merely in degree.

Hence, what is of importance is to fully and openly acknowledge experience as the eventuating dramatisation of the striving of entities for individuation – a striving contingent but not depended on their agency, in no need of a universal epistemological truth or method of enquiry to translate experience (*more*) to subjective experience (*less*). No entity can ever be fully exhausted in its conceptualisation as it is the remaining non-conceptualised portion that allows us to speak of that entity. Any form of representation, including architectural, rests on the 'non' that is essentially non-representable, non-conceptualisable. When philosopher of science Roy Bhaskar, asks 'what the world must be like for science to be possible?'[62] he is not referring to how we have access to the world, but rather what one ought to presuppose of the world so that any form of scientific practice – and representation – is possible.[63] This question demands a transcendental response, dealing not with Kantian 'possible experience', but with 'real experience' and its real (*plastic*) conditions.

For Bhaskar, it is the persistent existence of what he calls 'intransitive objects' that makes science possible. These are real structures that nevertheless exist independently of human minds and are more than often *out of phase* with actual patterns of events.[64] Representation is founded on the intransitive nature of these populations of objects. It is not only that representation is about what is essentially non-representable; as a condition, representation is possible because of the non-representable. If in any form of representation, an apparatus that reproduces the casual sequence of events within a closed system is deployed, it is the intransitive persistence of an open system that *injects* causality – not in the form of a natural law, but in the form of a contingent virtuality. If experience not only precedes but also conditions any of its representations, then the capacity to affect and be affected is prior to any formal manifestations, dealing not with formation – as Le Corbusier would have it – but with deformation, the cascade of ongoing material transformations. Aesthetics become the key to causality, not as a signifying symbolism of a tamed matter, but as the locomotive of change – change as an aesthetic transformation and causality as wholly aesthetic.[65]

That is why the attempt to represent or project axioms regarding subjective experience, understood as what one perceives in the form of secondary qualities, is condemned not only to be inaccurate, but to never surpass a descriptive state. If *firmitas*, *utilitas* and *venustas* remain self-evident, it is because they intend to contain the more of experience in the less of concepts, and if *lineamenta* remain purely schematic, it is because they attempt to capture a constant process of formal dramatisation as a case of drawn outlines. While undeniably advancing architectural and urban production, most of the theories developed within the discipline seem to ignore a quite simple fact: a horizontally expanded experience is 'the phenomenon whose existence is more certain than the existence of anything else'.[66] A return to experience does not stand as a return to subjective experience, but to experience itself: wild, free and intransitive. It is on this premise that architectural theories should shift focus from the *relata* of the discipline to the experiential relations that precede them. For the sake of their own persistence and individuation, architectural theories ought to embrace a non-correlational architectural mind and thought.

Notes

1. Foucault, *Order of Things*. See also: Foucault, *Archaeology of Knowledge*.
2. Foucault, *Order of Things*, p. 168.
3. Bryant, *Democracy of Objects*, p. 37.
4. Shaviro, 'Cosmopolitics', in *The Pinocchio Theory*.
5. Ibid.
6. Stengers, 'Introductory Notes on an Ecology of Practices', p. 184.
7. Massumi, 'Political Economy of Belonging', pp. 174–89.
8. Barad, *Meeting the Universe Halfway*, p. 140.
9. DeLanda, 'The Machinic Phylum'.
10. Meillassoux, *After Finitude*.
11. Ibid. p. 1.
12. Ibid. p. 2.
13. Ibid. p. 3.
14. DeLanda, *New Philosophy of Society*, p. 1.
15. Meillassoux, *After Finitude*, p. 5.
16. As media philosopher Steven Shaviro notes, there is a long genealogy to be traced here: from Descartes to Locke to Hume, the world has been constantly partitioned in primary and secondary qualities. It is this partitioning that generated what has been called the crisis of Humean scepticism, later resolved by Kant in arguing that all unknown realities out there ought to be organized by the conditions that our minds impose. See: Shaviro, *Universe of Things*.
17. Kwinter, *Architectures of Time*, p. 34.
18. Ibid. p. 35.
19. Grossberg, 'Affect's Future', p. 320.
20. Deleuze, *Negotiations*, p. 31.
21. Deleuze, *Difference and Repetition*, p. 66.
22. 'Now architecture consists of order, which in Greek is called taxis, and of arrangement, which the Greeks name diathesis, and of eurythmia and symmetry and decor and distribution which in Greek is called oeconomia.' See: Vitruvius, *De Architectura*, I, 2, 1.
23. Lefas, 'Fundamental Terms of Vitruvius', ff.179.
24. Vitruvius, *De Architectura*, II, 10, 3.
25. DeLanda, *New Philosophy of Society*, p. 41.
26. Till, 'Architecture and Contingency', p. 120.
27. Groak, *Idea of Building*, p. 54.
28. Till, 'Architecture and Contingency', p. 121.
29. McEwen, *Vitruvius*, p. 17.
30. Spinoza, *Ethics*, E3P2S.
31. Alberti, *Momus*.
32. Pearson, 'Philosophy Defeated', p. 3.
33. Ibid. p. 5.

34. Alberti, *Momus*, pp. 306–9.
35. Pearson, 'Philosophy Defeated', p. 7.
36. Alberti, *De Re Aedificatoria*, p. 7.
37. 'A man of the eighteenth century, plunged suddenly into our civilization, might well have the impression of something akin to a nightmare. A man of the 'nineties, looking at much of modern European painting, might well have the impression of something akin to a nightmare. A man of today, reading this book, may have the impression of something akin to a nightmare.' See: Le Corbusier, *Towards a New Architecture*, p. v.
38. Ibid. p. 1.
39. Ibid.
40. Ibid, p. 26; emphasis added.
41. Ibid, p. 45.
42. Ibid.
43. Ibid, p. 48.
44. Ibid, p. 54.
45. Le Corbusier, *Precisions*, p. 68.
46. Till, 'Architecture and Contingency', p. 124.
47. Le Corbusier, *Towards a New Architecture*, p. 143.
48. Le Corbusier and Ozenfant, 'Purism', pp. 65–7.
49. Massumi, 'Sensing the Virtual'.
50. Ibid.
51. Till, 'Architecture and Contingency', p. 126.
52. Le Corbusier, *Towards a New Architecture*, p. 29.
53. Massumi, 'Sensing the Virtual', p. 19.
54. Deleuze, *Nietzsche and Philosophy*, pp. 50–7.
55. Bryant, *Difference and Givenness*, p. 34.
56. Deleuze, *Difference and Repetition*, p. 212.
57. Bryant, *Difference and Givenness*, p. 34.
58. Deleuze, *Difference and Repetition*, p. 191.
59. Whitehead, *Concept of Nature*, p. 28.
60. Ibid. p. 43.
61. Shaviro, *Universe of Things*, p. 125.
62. Bhaskar, *Realist Theory of Science*, p. 21.
63. Bryant, *Democracy of Objects*, p. 42.
64. Bhaskar, *Realist Theory of Science*, p. 64.
65. Morton, *Realist Magic*, p. 19.
66. Strawson, 'Realistic Monism', p. 3.

Bibliography

Alberti, Leon Battista, *De Re Aedificatoria*, I.1, trans. J. Rykwert, J. Leach and R. Travenor (Cambridge, MA: MIT Press, 1988).

Alberti, Leon Battista, *Momus*, ed. Virginia Brown and Sarah Knight, trans. Sarah Knight (Cambridge, MA: Harvard University Press, 2003).

Barad, Karen, *Meeting the Universe Halfway: Quantum Physics and the Entanglement of Matter and Meaning* (Durham, NC: Duke University Press, 2007).

Bhaskar, Roy, *A Realist Theory of Science* (New York: Routledge, 1998).

Bryant, Levi, *Difference and Giveness* (Evanston: Northwestern University Press, 2008).

Bryant, Levi, *The Democracy of Objects* (Ann Arbor: Open Humanities Press, 2011).

DeLanda, Manuel, 'The Machinic Phylum', in *TechnoMorphica*, ed. Joke Brouwer and Carla Hoekendijk (Rotterdam: V2 Publications, 1997), pp. 76–9.

DeLanda, Manuel, *A New Philosophy of Society* (London: Continuum Books, 2006).

Deleuze, Gilles, *Nietzsche and Philosophy*, trans. Hugh Tomlinson (New York: Columbia University Press, 1983).

Deleuze, Gilles, *Negotiations, 1972–1990*, trans. Martin Joughin (New York: Colombia University Press, 1995).

Deleuze, Gilles, *Difference and Repetition*, trans. Paul Patton (London: Continuum, [1968] 2001).

Foucault, Michel, *The Archaeology of Knowledge & The Discourse on Language*, trans. A. M. Sheridan Smith (New York: Pantheon Books, 1972).

Foucault, Michel, *The Order of Things: An Archeology of the Human Sciences*, trans. Alan Sheridan (New York: Vintage Books, 1994).

Groak, Steven, *The Idea of Building* (London: E & FN Spon, 1992).

Grossberg, Lawrence, 'Affect's Future: Rediscovering the Virtual in the Actual', in *The Affect Theory Reader,* ed. Gregory Seigworth and Melissa Gregg (Durham, NC: Duke University Press, 2010), pp. 309–38.

Kwinter, Sanford, *Architectures of Time* (Cambridge, MA: MIT Press, 2001).

Le Corbusier, *Towards a New Architecture*, trans. Frederick Etchells (London: Architectural Press, [1927] 1986).

Le Corbusier, *Precisions: On the Present State of Architecture and City Planning*, trans. Edith Schreiber Aujame (Cambridge, MA: MIT Press, 1991).

Le Corbusier and Amadee Ozenfant, 'Purism', in *Modern Artists on Art*, ed. Robert L. Herbert (Englewood Cliffs, NJ: Prentice-Hall, 1964), pp. 58–73.

Lefas, Pavlos, 'On the Fundamental Terms of Vitruvius's Architectural Theory', *Bulletin of the Institute of Classical Studies*, Vol. 44, Issue 1 (2000), pp. 179–97.

McEwen, Indra Kagis, *Vitruvius: Writting the Body of Architecture* (Cambridge, MA: MIT Press, 2003).

Massumi, Brian, 'The Political Economy of Belonging and the Logic of Relation', in *Anybody*, ed. Cynthia Davidson (Cambridge, MA: MIT Press, 1997), pp. 174–89.

Massumi, Brian, 'Sensing the Virtual, Building the Insensible', in *Architectural Design (Hypersurface Architecture)*, ed. Stephen Perrella, *Architectural Design* (Profile n. 133), Vol. 68, Issue 5/6 (1998), pp. 16–24.

Meillassoux, Quentin, *After Finitude: An Essay on the Necessity of Contingency* (London: Continuum, 2008).

Morton, Timothy, *Realist Magic* (Ann Arbor: Open Humanities Press, 2013).

Pearson, Caspar, 'Philosophy Defeated: Truth and Vision in Leon Battista Alberti's Momus', *Oxford Art Journal*, Vol. 34, Issue 1 (2011), pp. 1–12.

Shaviro, Steven, *The Universe of Things: On Speculative Realism* (Minneapolis: University of Minnesota Press, 2014).

Shaviro, Steven, 'Cosmopolitics', in *The Pinocchio Theory*, 2005 http://www.shaviro.com/Blog/?p=401 (accessed 5 May 2015).

Spinoza, Benedict de, *Ethics*, trans. E. Curley (London: Penguin Classics, 2005).

Stengers, Isabelle, 'Introductory Notes on an Ecology of Practices', *Cultural Studies Review*, Vol 11, Issue 1 (2005), pp. 183–96.

Strawson, Galen, 'Realistic Monism: Why Physicalism Entails Panpsychism', in *Consciousness and its Place in Nature: Does Physicalism Entail Panpsychism?* ed. Anthony Freeman (Exeter: Imprint Academic, 2006), pp. 3–31.

Till, Jeremy, 'Architecture and Contingency', *Field: A Free Journal for Architecture*, Vol. 1 (2007), pp. 120–35.

Vitruvius, *De Architectura*, volume 1, ed. and trans. Frank Granger (London: W. Heinemann, 1931).

Whitehead, Alfred North, *The Concept of Nature* (Cambridge: The University Press, [1920] 1957).

CHAPTER 2

Housing Biopolitics and Care

Peg Rawes

This chapter examines how Spinoza's seventeenth-century philosophy constitutes a humane form of rational thought – what I call ethical ratio – through which to consider the biopolitical structure of the housing crisis in the UK. In July 2015 the British Government's 'regressive'[1] budget withdrew welfare funding, including housing benefits and 'family credit', which supports low-income families in rental accommodation and low-paid work. Within twenty-four hours of this news, the Government announced two other decisions that undermine designing high-quality and low-carbon housing: first, the relaxation of planning laws that, in effect, incentivises housing developers to further reduce the quality of space allocated to new housing[2] and, second, the withdrawal from a commitment to zero-carbon requirements in new housing.[3] In March 2016 these measures were followed by the Housing and Planning Bill, resulting in even less housing security for tenants in social housing provision, and making parts of the south east in the UK wholly unaffordable for low-income families.[4] These recent political changes to the financial, material and environmental structure of UK housing provision add weight to the worryingly inhumane ratios that structure the sector, which currently ensure maximum financial gain to the developer, versus the right to affordable high-quality housing for both the tenant and owner-occupiers.

Post-war Labour and Conservative UK government's housing policies addressed the lack of good-quality housing with specific consideration for the needs of the lowest socio-economic groups in society. However, since the 1980s Conservative government's policy of 'right-to-buy' that encouraged the sale of council homes, together with reductions in building new homes since the late 1960s, local authority housing provision has dropped, and the housing market has been more explicitly designed as a driver of short-term speculative economic interest (reflected of course in the USA's 2008 sub-prime mortgage crash).

This dysfunction in housing has escalated to disastrous proportions, most recently reaching unsustainable levels of insecurity because of steep rises in unregulated rents, poor-quality house-building schemes by large-scale developers (what Shelter calls an issue of 'little boxes'[5]), together with poor maintenance of council housing stock, and the regeneration of social housing as assets on the international global market:[6] all combining to represent highly competitive supplies of rented and sellable housing in speculative markets. Conspicuously, this is now the norm across society (rather than a concern for the minority), affecting communities ranging from students, young professionals and middle-class families in rented accommodation, and preventing a generation of the under-forties from being able to 'get on to the property ladder'. Journalists and economists, including the Governor of the Bank of England, have commented on the unsustainability: 'Only 43 homes in London are affordable for first-time buyers';[7] the situation represents 'a human disaster';[8] and Mark Carney's observation in May 2014 that the UK housing sector has 'deep, deep structural issues'.[9]

Without doubt then, well-being, or the care of the self, as an individual, and for other social groups or communities, is now thoroughly framed by biopolitical concerns: both because of the distribution of resources, taxation and 'design' of society by the modern state, and from research that has been conducted by charities and researchers involved in supporting those with the lowest levels of social, political and economic equality. Drawing from Foucault's writings on 'technologies of the self', and Spinoza's geometric essay *The Ethics* (1677) I consider how Spinoza's seventeenth-century 'radical enlightenment' definition of an ethical form of rational thought has resonance with architects and professionals who seek to demystify and resist the critically severe and inhumane ratios of inequality that form the current housing crisis in the UK today. Spinoza's thinking therefore previews these current conversations of biopolitical social justice and difference, particularly if his notion of ethical ratio is understood as a kind of *zoe* (that is nonstandard or non-human life), which thinkers, including Félix Guattari and Rosi Braidotti, have promoted in ethical ontologies, especially of those lives designated to be outside the democratic majority.

Inhumane Ratios

Michel Foucault's 1970s biopolitical lectures on the power relations that form modern neoliberal government highlights how societal and political structures shape the formation of the individual, on macro-social and micro-individuated scales. Harnessing the biological self to the political formation of subjectivity his relational analysis shows that 'design' is not just carried out through techniques that are scaled at political, infrastructural, educational and social forms but extend throughout the design of buildings and of cities, in to forms of self-management that the individual enacts on him/herself.

In his late biopolitical work, published as seminar proceedings from Vermont in 1984, Foucault examines the structures of inhumane social, political and economic regimes that design (actually, the term he uses is 'control') society, focusing on notions of self-management by the individual and the formation of the city. In a discussion on the 'art of living' from ancient Greek thought he draws attention to the relationship between the built environment, the self, care and art, in a manner that is also promoted by contemporary design communities and philosophers who are concerned with the right to, and design of, humane forms of life and inhabitation (including Spinoza, Guattari, Braidotti, Haraway and Esposito). Analysing texts from the early seventeenth century, he distinguishes between their early modern technocratic forms of 'care of the self' in contrast to ancient Greek society's notions of care, which are constituted by the 'art' of life: '"*epimelesthai sautou*," or "to take care of yourself," "the concern with self," "to be concerned, to take concern of yourself."'[10] He explains that, for the Greeks, this idea was 'one of the main principles of cities, one of the main rules for social and personal conduct and for the art of life', and observes that under modern moral philosophical conventions 'care' is replaced by 'knowing'.[11]

In the final chapter of this publication, Foucault addresses the historical formation of Western rationalist thought, assessing the manner in which it organises the distribution and forms of political power. He critiques utopian enlightenment claims for democracy by highlighting its technocratic form of governance, especially through 'control', questioning reason's ethics in educational, technological, legal and welfare structures:

> [T]he coexistence in political structures or large destructive mechanisms and institutions oriented toward the care of individual life is . . . one of the central antinomies of our political reason. It is the antinomy of our political rationality which I'd like to consider . . . Even when we vote for or against a government which cuts social security expenses . . . we do these things not only on the ground of universal rules of behaviour but also on the specific ground of a historical rationality. It is this rationality . . . which is one of the main features of the modern political rationality, developed in the seventeenth and eighteenth centuries through the general idea of the 'reason of the state' and also through a very specific set of techniques of government.[12]

Foucault discusses these issues in a seventeenth-century French essay on a 'well-governed state' by Louis Turquet de Mayenne in 'La Monarchie aristo-democratique' (1611), which identifies four 'boards' that manage ('police') society. The first regime looks after positive forms of life, principally education. The second oversees 'negative aspects of life, that is the poor, widows, orphans, the aged, who required help'. This is a system of welfare that is 'concerned also with people who had to be put to work and who could be reluctant to go to work, those whose activities required financial aid, and it had to run a kind of bank for the giving or lending of funds to people in need', and for maintaining public health. Board three manages commercial society, and the final board is directed towards the management of the built and natural environments, public and private property.[13] These constitute a 'political rationality' that operates through 'economical, social, cultural, and technical processes . . . always embodied in institutions and strategies and has its specificity',[14] but the original invention of this 'utopic' reasoning is the extent to which it designs governance through control. In addition, he notes that the failure of these utopic projects is not because of their politics or theory, but because of

> the *type of rationality in which they are rooted*. The main characteristic of our modern rationality in this perspective is neither the constitution of the state . . . nor the rise of the bourgeois individualism . . . The main characteristic . . . is the fact that this integration of the individuals in a community or in a totality results from *a constant correlation between an increasing individualization and the reinforcement of this totality*.[15]

Foucault's analysis of inhumane forms of reason, governance and design of society, including the management of the built environment, provides a platform from which I now move to consider how Spinoza's ideas of geometric ratio suggest a form of early-enlightenment thinking in which political reason is founded upon more distinctly humane and ethical biopowers, enabling the individual and society the capacity to 'artful' care of the self. According to this humane ratio, the individual's biopowers are increased, enabling self-determination or care, but not in subservience to the authority of the state (for Spinoza, of God or Nature). Instead, the individual and society are brought into an equilibrium of human and non-human qualities, providing an interesting historical parallel to questions of housing ethics today.

This alternative form of biopower is a precursor to the affirmative thinking of Guattari and Braidotti and others, who designate legitimate rights to life as *zoe*. It is therefore distinct from Foucault's focus on the inhumane manifestation of reason as control in modern societies. Instead, Spinoza's biopolitics accords with architecture and housing in which ratio promotes humane, ethical and ecological relations.

Dissimilar Substance

Spinoza's intensive analysis of the mental and physical, social and ecological (that is, non-human) powers and patterns that constitute the diversity of nature and the human agency previews Foucault's biopolitical understanding of subjectivity: in particular, Spinoza carefully outlines a complex relational ontology that is defined through dissimilarity rather than similarity. For Spinoza, all entities are singular and self-determining modes, and our powers of expression (our affects) are distinct, yet relational. Subjectivity is therefore composed of ratios of dissimiliarity, indicating that his understanding of the biological accords with non-human *zoe* rather than the stately form of *bios*.

For Spinoza, ratio is an environmental, human and univocal form of material organisation or life. Called 'substance', it is not composed of equal parts or modes, but of the dissimilar powers that reflect the 'truth' of God/Nature. Ratio (or reason), for Spinoza, reflects the dissimilarity of communities in society. Moreover, rather than being defined through a hierarchy of legitimate to illegitimate modes of existence, Spinoza shows that the irreducible natural (and divine) powers of substance also constitute everyday humane biopowers in the imagination and emotions.

This humane form of ratio then designs the relationship between the individual, nature and God, and affirms individuated but relational subjectivity: for example, in Definition 3, substance's powers of self-determination are 'that which is in itself and is conceived through itself; that is, that the conception of which does not require the conception of another thing from which it has to be formed'.[16] Related to nature because of the univocity of substance, the individual's rights to self-determination are always distinct from the universal biopowers of God and nature, however, Spinoza does not then dismiss them as illegitimate powers of reasoning. Instead, he defines a distinct biopower in the individual's right to exist – the conatus – that is also an expression of substance. Spinoza therefore shows how an ethical form of ratio is the basis of an individual's care or design of itself, 'the power of any thing' to be, 'by which it endeavours to persist in its own being'.[17]

He details this humane rational form of existence – our capacity for care or design – in the complex ratios of power that exist between the distinctly human mind and body, the agreement and disagreement between our affects that determine our capacity for joy, sadness and emotional forms of 'reason' (sense-reason), the power of our imagination, as well as in the common biodiversity of the world in nature. Part IV is the most intense axiomatic analysis of these dissimilar ratios (and also begins with a preface on the powers of the affects for design judgement in building a house). Even though Spinoza observes that our emotions generated by our powers of reasoning are more desirable than those that are not (for example, melancholy/depression), he recognises that all are necessary constituents of human experience, and of our capacity for well-being, even if they require correction.[18] Also, not only do the specific ratios of intensity generated between our different emotional states define the way in which we achieve self-determination and, hence, 'freedom', but it is the very relationality or difference between these states that defines the biopower's humane *ratio*nalism: our affects are therefore not just measured quantifiably or subservient to a 'pure reason', but calibrate the diversity of what it means to be human.

In addition, his use of the geometric method in the *Ethics* is redirected from a history of disembodied mathematical proof (and, for some readers, a marker of the disembodied basis for inhumane forms of rationalist thought), into a rational materialisation that accords with ethical approaches to geometric thinking in twentieth-century social housing design. His repurposing of Euclid's classical geometric thinking to generate patterns of dissimiliarity transfers a method more commonly seen as the production of similiarity/harmony, into one of dissimilarity.

Spinoza's inventive alterity in rational thinking is therefore a distinctly ethical form of ratio that resonates with designers who are concerned with the production of diverse relational biopowers (but not in common with recent calls by the avant-garde architectural design community that desires a return to romantic notions of autonomy). Instead, what makes Spinoza's thinking about non-human forms of biopower relevant for today's housing crisis (especially for those communities who are traditionally excluded) is the capacity for an ethical ratio(nalism) to articulate full self-determination, care of the self and the community.

Spinoza's contribution to a history of technologies of self-care also resonates strongly with the social, economic, ethical and environmental forces at play today in advanced capitalist and developed societies. On the one hand, reflecting the pressures of rapid capitalist forms of urban development in the late seventeenth century, yet also written at a time when the codification of enlightenment principles of individual equality as universally agreed legal and social rights were still being debated,[19] Spinoza shows a striking resistance to claims for moral forms of democratic political thought that are also in crisis today (since universal human rights and social justice may still not be deployed equally). Spinoza's thinking about what constitutes humane society and life is of value because it shows that an *ethical* society is one that defines human rights as dissimilar, rather than *universally* equal. For Spinoza, our capacity to care, for oneself, and as a member of society, is based upon understanding that our needs *are differential*. Ratio, for Spinoza, is a concept that, by its nature, is composed of the differences between its various parts. Writing at the time of early 'radical' enlightenment Spinoza does not seek to impose an order of universal self-same equality or measure, but mobilises the complex dissimiliarity of its nature.

Spinoza's *zoe*-power is therefore a seventeenth-century proto-biopolitics that is transversal and ethical. It is an analysis of a humane form of rational thought that is developed through calibrations (that is, ratios) of care within the individual and between entities: a relational ontology that is constituted out of complex self-other relations. As a result, ratio is always a calibration of dissimilarity, of the 'power of any thing' and of the non-human (*zoe*), and so provides a powerful critique for demystifying and understanding the structures of management, care and design in contemporary housing provision ratios.

It also has more in kind (has a care) with modern European social and social modernist housing projects (for example, the Amsterdam School's social housing schemes, Berlin's Hufeisensiedlung and Vienna's Rabenhof estates, and Le Corbusier's Unité d'habitations), highlighting how ratios of care in social, spatial, material and cultural life *have previously* been successfully designed *into housing*. In the UK, housing estates once overlooked and politically maligned in the 1980s and 1990s have now been retrieved as esteemed examples of social housing, not least because the cost of the embodied energy in their concrete now urgently needs to be economically, politically and environmentally valued in a serious commitment to reducing carbon emissions (rather than continuing to waste building resources by demolition). These schemes include: the 1930s Park Hill in Sheffield, the Smithsons' Robin Hood Gardens (1968–72), and innovative local authority architect office designs, such as Alexandra Road, by Neave Brown for Camden Council (1978), Kate Macintosh's Dawson's Heights in Dulwich (1972) and Ted Hollamby's Central Hill in Crystal Palace (1974).[20]

Ratios of inequality are also high on the agenda of researchers and charities, charting the impact of poor health and income equality in the UK, and correlate in severity with the increasing housing crisis.[21] These empirically rigorous social science reports corroborate the arguments about the inhumane biopolitical organisation of society that is observed by political philosophers, arts and design professionals and housing activists; however, politicians have also cynically used their arguments about well-being and economics in their agenda to design 'fitter' societies.[22]

Humane Ratios or Transversal Forms of Care and Design

Returning to the disastrous status of the UK's current housing provision, we can argue that negative biopowers have been normalised into the dominant forms of technocratic housing infrastructure and governance, and in the management and 'design' of the built environment at a macro-scale, but also internalised at a micro-scale within the individual, and in the community. Closing this chapter, however, I discuss two examples of recent research and guidance by architects and planners that show how *zoe* and ratio can be brought together into more positive biopolitical relations, rather than designing in a manner that reinforces the denigration of communities as 'lacking' or 'nonhuman'. In these schemes, biopolitical diversity is affirmed in the architectural design process, rather than 'managed'.[23] Félix Guattari and Rosi Braidotti's writing underpins this argument because each mobilises the 'alterity' identified by Foucault not to delimit subjectivity to normative forms of life. When affirmatively articulated as non-human(ist), new ontological categories of individuation, critique, resistance and artful self-care are possible. These approaches therefore represent ethical ratios of self-care, and care for others: what Braidotti has called 'transversal counter models', not 'confined to negative bonding in terms of sharing the same planetary threats: climate change, environmental crisis or even extinction',[24] and accord with Guattari's 'transversal ecosophy' through which new subjectivities and ecological relations are created.[25]

First, however, it is worth highlighting some of the historical and structural social, political and design biopolitics that define the current crisis. Some of the most toxic ratios of inequality, which were previously features of poor late-nineteenth century housing, have now returned to current housing provision. Studies by Shelter and NatCen, for example, observe alarming increases in overcrowding-related health issues, such as asthma in children and mental health issues in both adults and children[26] – features that nineteenth-century housing philanthropists such as Octavia Hill, Thomas John Barnardo and Joseph Rowntree sought to address, together with Ebenezer Howard's Garden City plans for improving the environmental quality of urban space. In addition, the post-WWI Tudor Walter's report led to regulations on internal and external design standards in the 1919 Town Planning and Housing Act, including improvements to light provision, street layout and reductions in terrace rear elevations.[27] Post-WWII, Sir William Beveridge's 1942 *Report on Social Insurance and Allied Services*, which was to form the basis of the Welfare State, drew attention to the increase in an ageing population and suggested that services concerned with the welfare of the old should include the provision of special housing. In 1949 these recommendations formed part of the post-WWII Housing Act, which identified the need for flexible provision across the community, not just for the traditional working classes, and gave local authorities increased powers to improve housing stock and maintenance.[28]

However, since the 1980s, the housing market has rapidly increased in dysfunction to its highest levels of disequilibrium since the strong political leadership on housing in the 1950s and 1960s that prioritised social and welfare housing provision.[29] As mentioned earlier in the chapter, there are a number of contributing factors, including the decrease in affordable house building by successive governments since the late 1960s and the increased reliance by governments on fiscal policies designed to create short-term economic gain from speculative rental and buying markets. Cheap mortgages and unregulated rental markets have been designed to favour the landlord rather than the long-term well-being or security of the tenant; for example, in 2014 a criminally small 'semi-studio' flat in London (with the most basic furniture, for example, a fridge, shower cubicle and single bed), and below the legal requirements for a studio occupancy, could cost £1,000 rent per month.[30] In addition, since 2013 a 'bedroom tax' has meant that council tenants have their housing benefit reduced if they do not fully occupy all their bedrooms (legislation that has failed to take into account disability and child needs);[31] previously-designated social housing – often in the most deprived areas – is systematically sold to developers who resell the refurbished homes as assets on the international market, and that may well remain unoccupied;[32] Housing Associations and Councils also have to compete against developers and private landlords on the open market so that previously designated social housing is now re-valued beyond the affordability of those who previously lived there; and significant numbers of new homes fall below recommended space and design standards. All signs that the housing market is a seriously toxic biopolitical form of anti-social design: a profoundly inhumane ratio of dissimilarity.

Spatial (geometric) dissimiliarity is a key feature of these inhumane ratios. In the UK – unlike Europe and the US – housing is sold per room rather than square meterage and, in contrast to Germany, France, the Netherlands and the Scandinavian countries, the UK does not have a legal minimum standard for new housing. New homes have been getting smaller. In 2011 the RIBA's *Case for Space* report observed that the average UK house is 85m^2 with 5.2 rooms, an average of 16.3m^2 per room, but more recently dropping to just 76m^2.[33] In addition, a one-bed flat was 4m^2 short of its required space (the equivalent to a table and sofa); a three-person house was short of a whole one-person bedroom (8m^2) – all well below recommended space allocations.[34] In London in 2010 only 18 per cent of housing stock met recommended space standards.[35] In comparison, Irish new homes were 15 per cent larger (87.5m^2), Dutch 53 per cent larger (115.5m^2) and Danish 80 per cent larger (137m^2).[36]

Positive efforts to design humane space ratios into housing did continue up to the late 1970s, however, partly due to the 1961 Government-commissioned Parker Morris Report, *Homes for Today and Tomorrow*, which established space standards as key criteria for housing design through the 1960s and 1970s, and still regarded as important by designers today.[37] But the standards were only made compulsory for all local authority housing in 1969, and their subsequent implementation during the 1970s also had mixed results. Issues that combined to remove support for the standards so that they were abolished in 1980 included: the implementation of the standards to the bare minimum, and cuts to the public sector that undermined local authority investment, together with the use of new, but unreliable industrialised concrete slab technologies, which resulted in strong public and political criticism, and calls for the demolition of large-scale housing schemes built during this era.[38] More recently, Parker Morris's legacy for improving housing design has been reappraised within discussions of flexible use[39] and recent guidance on space standards. Planner Matthew Carmona has noted the shift from the era of Parker Morris which resulted in restrictive design decisions that too heavily relied upon spatial concerns, to an evaluation of space as one of a number of design quality indicators, within a broader understanding of the diversity of families' needs today. He notes that this broader set of ratios also reflects the celebrated mid-twentieth-century social housing projects, such as Park Hill in Sheffield and Bevin Court, Islington, London, where flexibility and a shared communal understanding of home and the built environment were prioritised.[40]

Other significant examples of innovative housing schemes that also met the standards includes the Smithsons' Robin Hood Gardens in Poplar, London (1968–71). This scheme was notable for its design of two large concrete slab blocks on either side of an open landscaped mound – that the Smithsons' called 'landcastles' – located within a densely built inner-city area between a tunnel and a dual carriageway. However, their Brutalist design contributed to a mixed reputation and fierce criticism of its utopic 'streets in the sky' walkways, which together with stairwells and circulation routes, were seen as sources of anti-social behaviour.[41] These issues, combined with periods of poor maintenance by the Local Authority (although improved during the late 2000s[42]) and the increased pressure on cheap real estate that has come with Canary Wharf's consolidation into an extension of the City, have added to its 'historic' reputation as a 'failed' estate. Most recently, the estate twice failed to get listed status (first in 2009 and again, in the summer of 2015), despite high-profile campaigns by architectural professionals. Its status remains polarised: Tower Hamlets Council is now demolishing it so that a new high-density and largely private housing development can be built that will meet the City's aspirational property interests. On the other side of the debate, the estate continues to be championed by the architectural, planning and heritage professions, who are concerned about the environmental waste, social and cultural destruction that will come with its demolition and removal of its tenants.

Within this context, Sarah Wigglesworth Architects' 2011 research project for the Twentieth Century Society is an example of a sustainable social housing design that creatively updates Robin Hood

Gardens' original Parker Morris space standards in order to better meet the needs of the community that occupied it.[43] Figures 6 and 7 show designs from the practice's proposal, which address spatial flexibility, environmental improvements to the building's thermal performance, and its South East Asian tenants' housing needs for large extended families. Figure 6 shows how the blocks' overcrowded flats and maisonettes could be modified to enable increasing their occupancy up to families of eight. In order not to lose the integrity of the original design, these changes are made economically feasible by removing only non-structural walls and retaining all existing party walls, staircases and circulation access.[44] In addition, the density of the blocks is increased by the addition of lightweight duplex units on the roof, with winter gardens, and additional units on top of the currently poorly managed single-storey garage blocks. Second, the firm showed that an inexpensive approach to retrofitting insulation panels into flats and the walkways, together with fitting double-glazed windows, could reduce household energy costs by 30 per cent annually (see Fig. 7).[45]

The Parker Morris space standards also underpin the aims of the 2010 London Design Guide.[46] One of the report's authors, Alex Ely of Mae Architects, and a contributor to policy documents for CABE and the RIBA, has observed that architects' support for space standards is situated within the profession's commitment to promoting high-quality flexible space design and use, rather than the previously misjudged belief that space is an end in itself.[47] Hence, Parker Morris's findings have been updated to provide 'future proof' housing for subsequent generations, and towards enabling better renovation and environmental performance. Seven criteria outline the Guide's aims: understanding space as an integral quality of urban design and the context in which housing is situated; designing for the diversity in the population's needs; maintaining the quality of design throughout exterior to interior spaces, entrances and shared, communal space; ensuring that internal space standards have high flexibility and a concomitant quality of plan, including storage space; designing to maximise privacy and psychic well-being, including noise mitigation, light and height standards; maximising environmental performance and carbon emission mitigation; and committing to responsible and good governance of the design process.

Like Wigglesworth, Ely and other architects known for their award-winning housing designs, such as Peter Barber Architects, there is a shift

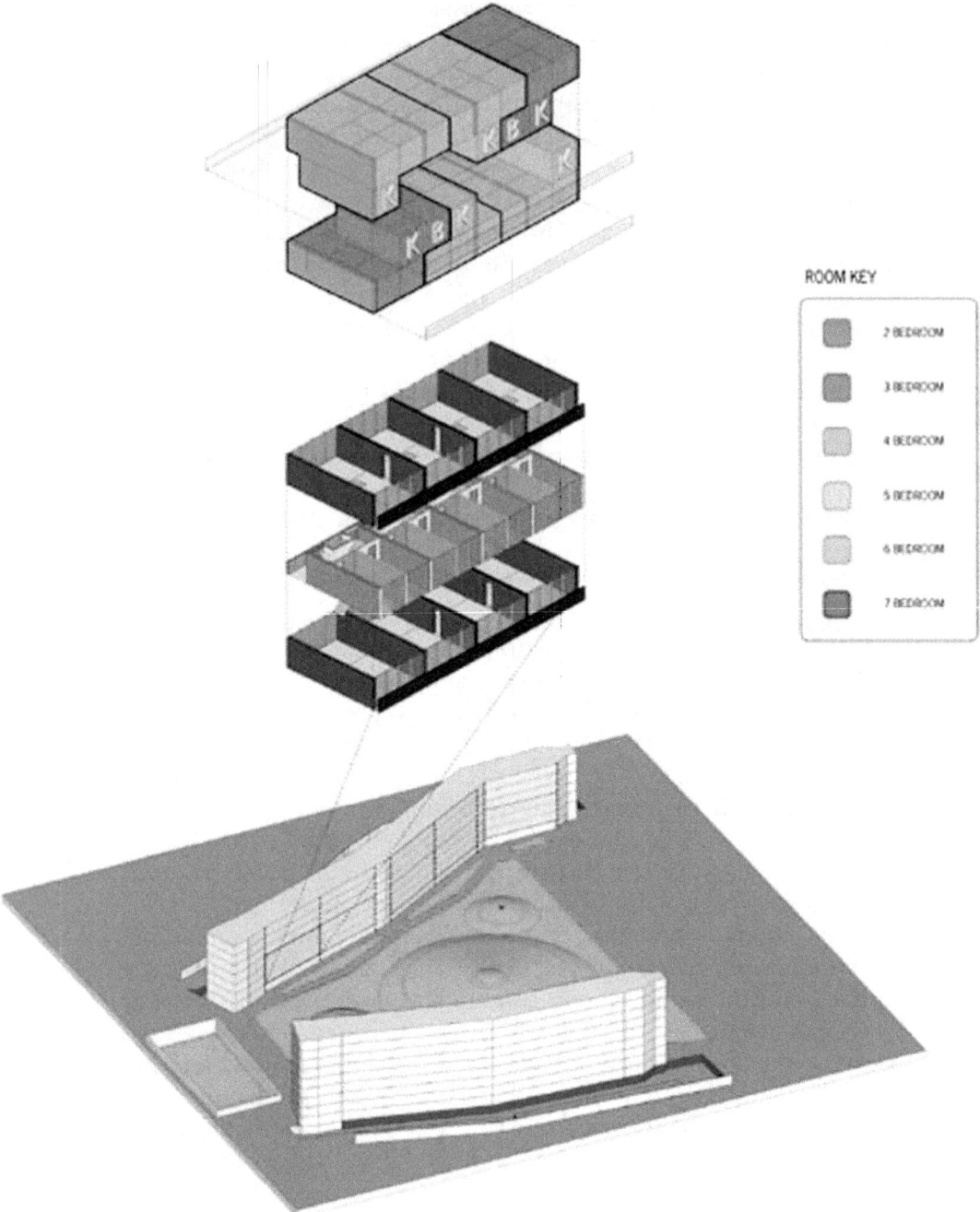

Figure 6 Robin Hood Gardens: Axonometric and bedroom permutations (double page spread)
Source: Sarah Wigglesworth Architects, 2011.

HOUSING BIOPOLITICS AND CARE

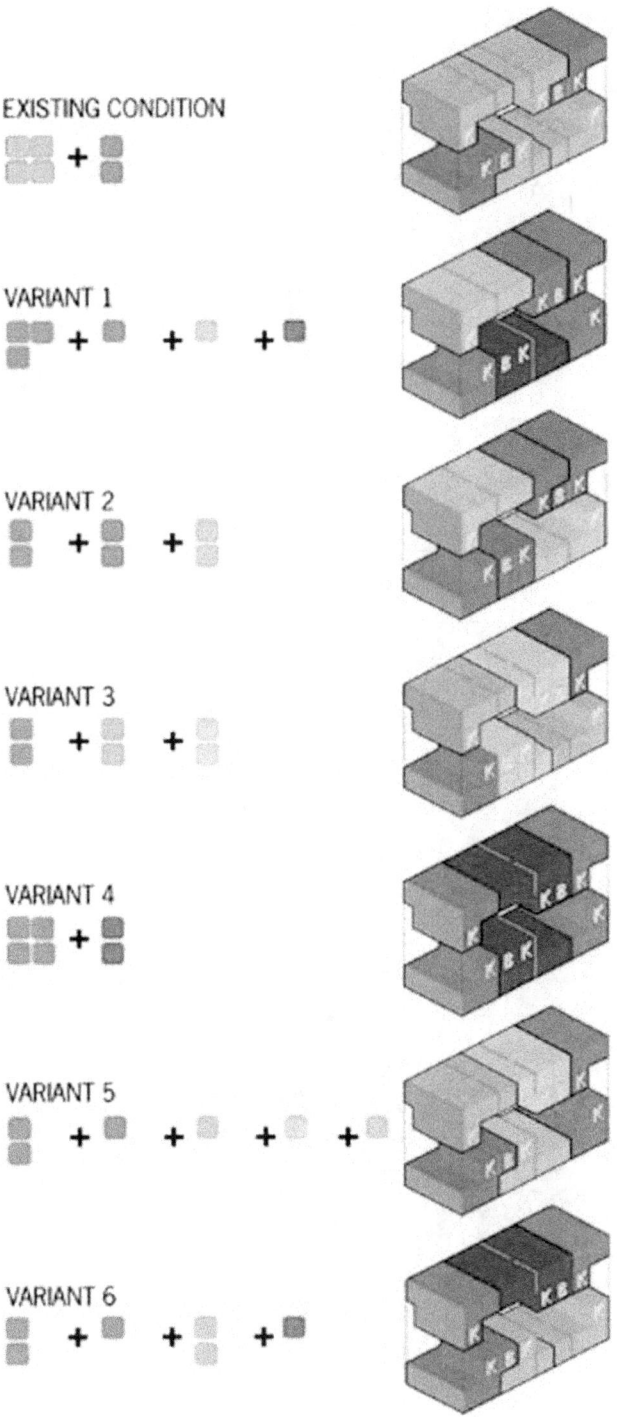

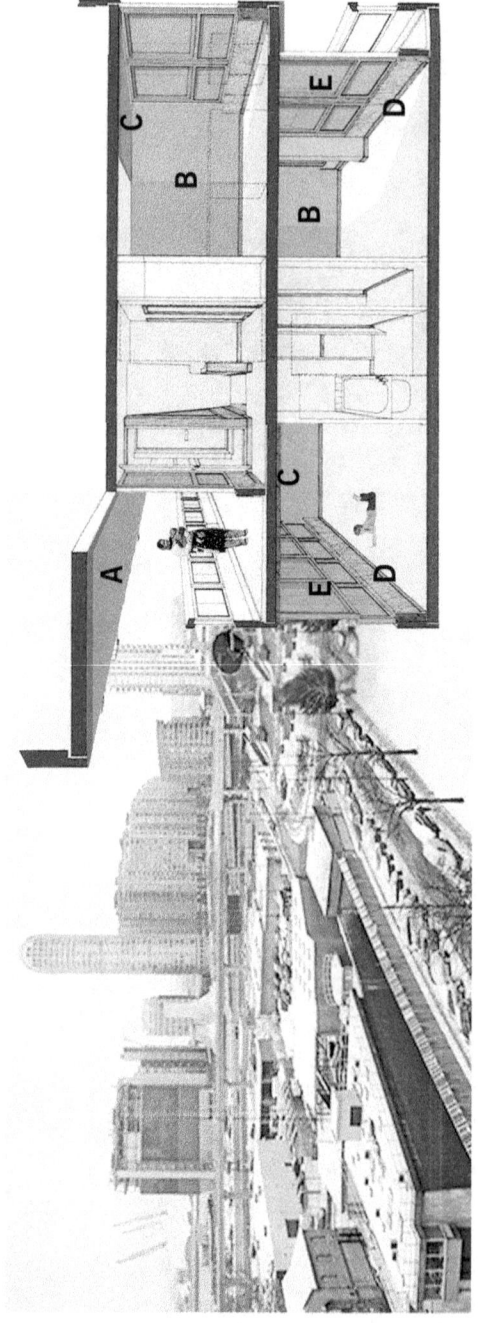

A. insulation to soffit of walkway for flat above
B. 50mm insulation layer to party walls to minimise thermal bridging
C. insulation to soffit beneath balcony
D. Thickened external Walls below windows with 90mm internal insulation layer wrapped down face and across slab to minimise thermal bridging
E. New double glazing installation

Figure 7 Robin Hood Gardens: Thermal section.
Source: Sarah Wigglesworth Architects, 2011.

in emphasis from minimum space standards as an isolated criterion of quality, to an emphasis on ratios of space that enable the flexible and diverse needs of today's households (for example, the multi-use home where pram, bike, wheelchair access and storage quality are indicators of cross-generational need):[48] flexible spatial housing design therefore recognises the importance of soft ratios and the different patterns of daily and familial habits, from the young, the disabled, to the old.[49]

These soft ratios reflect not just the geometric spatial ratio but the human ratio across the various patterns of inhabitation in the home. Overall, however, while these are positive examples of the architectural and built environment professions working together to improve housing quality, the overriding issue that prevents their creative and robust implementation is the continued lack of political engagement in housing design as economically viable *and* ethical investment in long-term forms of care. Depressingly, while the Government's 2014 Housing Standards Consultation[50] drew substantially from research by architects, planners and housing professionals about how these values *are integral* to high-quality design *and* social well-being, the standards have only been implemented as recommendations for local authorities, rather than as national legal requirements. In addition, the recent Government's Starter Homes Design guide retreats entirely from this expertise, focusing instead on unimaginative and conservative forms of exterior 'dressing' that have no commitment to improving the internal design or space quality.[51] The need for a comprehensive political revision to this damaging mismanagement of our homes and cities therefore remains the most severe obstacle to the sustainability and quality of the UK's housing. This chapter has attempted to show that within the historical and contemporary architectural and housing professions, 'care of the self' has been, and remains, a politically, aesthetically and economically powerful constituent in design. Spinoza's thinking about these ethical biopowers resonates with this discussion by offering an unusually humane form of ratio that is much needed now, and will be for generations to come.

Notes

1. Paul Johnson of the Institute for Fiscal Studies, London, quoted in Giles, 'Higher Wage Will Not Compensate for Lost Tax Credit Cuts', *The Financial Times*, 9 July 2015.
2. Wainwright, 'Osborne's Planning Reforms Risk Slum House Building', *The Guardian*, 10 July 2015.

3. Oldfield, 'UK Scraps Zero Carbon Home Target', *The Guardian*, 10 July 2015.
4. See, for example, a recent data-map that shows how the withdrawal of these finances will result in low-income families being effectively prohibited from large parts of the UK, especially cities, and entirely from the South East. Butler and Arnett, 'Lower Benefit Caps Exclude Poor Families', *The Guardian*, 20 July 2015.
5. Shelter, *Little Boxes, Fewer Homes: Setting Housing Space Standards Will Get More Homes Built*, Shelter, April 2013.
6. Wainwright, 'Revealed: How Developers Exploit Flawed Planning System to Minimize Affordable Housing', *The Guardian*, 25 June 2015.
7. Elledge, 'Only 43 Homes in London are Affordable for First-time Buyers', *The Guardian*, 30 April 2015.
8. Moore, 'Britain's Housing Crisis is a Human Disaster', *The Observer*, 14 March 2015.
9. Inman, 'Mark Carney', *The Guardian*, 19 May 2014.
10. Martin et al., *Technologies of the Self*, p. 19.
11. Ibid. p. 19.
12. Ibid. pp. 147–8.
13. Ibid. pp. 154–5.
14. Ibid. p. 161.
15. Ibid. pp. 161–2; emphasis added.
16. Spinoza, *Ethics*, p. 31.
17. Ibid. p. 108.
18. Ibid. Part IV, Proposition 61, and Part V, Proposition 7.
19. See Israel, *Radical Enlightenment*.
20. See, for example, Cordell and Walsh, *Utopia London* and Colquhoun, *RIBA Book of British Housing*.
21. See, for example: Martin, 'Why the UK Leads the Way on Inequality', *New Economics Foundation*, 9 January 2015; Joseph Rowntree Foundation, *Housing and Poverty Blog*; and Dorling, *All That is Solid*. Prior to these recent studies, epidemiological and economic research into inequality by Wilkinson and Pickett brought to light inhumane ratios as factors in well-being in *The Spirit Level*. Also see my research with Beth Lord on these issues in the AHRC-funded project, *Equalities of Wellbeing*.
22. David Cameron's speech on 'Big Society' in February 2011 was seen to be a cynical attempt to address well-being and equality agendas, especially in light of Wilkinson and Pickett's research.
23. Also see Rawes et al., eds, *Poetic Biopolitics*.
24. Braidotti, *Posthuman*, p. 103.
25. Guattari, *Three Ecologies*.
26. See, for example, RIBA, *Case for Space*, p. 13; NatCen and Shelter, *People Living in Bad Housing* ; and Harker, *Chance of a Life Time*.
27. See, for example, Swenarton, *Homes Fit for Heroes*.

28. Survey of London, 'Public Housing in Poplar'.
29. See, for example, Pepper, 'Three Ages of Post-war Housing'.
30. Gilman, 'Charming but Compact', *The Daily Mail*, 30 August 2014.
31. Dugan, 'Exclusive', *The Independent*, 13 September 2013.
32. Osborne, 'Hundreds of Flats in Canary Wharf Development Sell in Less than Five Hours', *The Guardian*, 13 July 2015. Fifty per cent of these 'off-plan' sales were made to overseas buyers, and marketed through an event in London and Hong Kong.
33. RIBA, *Case for Space*, p. 10.
34. Ibid. p. 5.
35. Mayor of London, *Housing Design Standards*, p. 10.
36. RIBA, *Case for Space*, p. 5.
37. Parker Morris Report, *Homes for Today and Tomorrow*.
38. See, for example, Ronan Point, East London, which was demolished in 1968 after its structural failure. Hulme Crescent (1972–91) in Manchester and, most recently, the Heygate (1974–2014) in Elephant and Castle, have all faced strong criticism for being 'failed' estates.
39. Drury, 'Parker Morris – Holy Grail or Wholly Misguided?'
40. Carmona, *Space Standards*, pp. 3–4.
41. Alan Powers notes how the estate was seen as a social and design failure in the 1970s. Powers, ed., *Robin Hood Gardens*, p. 20.
42. Robinson, 'Fit for Purpose', p. 128.
43. See Sarah Wigglesworth, 'Robin Hood Gardens', in which she proposed a social, economic and sustainable approach. Also see the project brief and designs at <http://www.swarch.co.uk/projects/robin-hood-gardens/info/> (accessed 20 July 2015).
44. Wigglesworth, 'Robin Hood Gardens', *The Architects' Journal*, 31 January 2011.
45. Ibid.
46. Mayor of London, *London Housing Design Guide (Interim Edition)*, August 2010, p. 6, and Mayor of London, *Housing Design Standards*, July 2010, pp. 14–15.
47. Ely, 'The Case for Space', *Equalities of Wellbeing*, April 2014.
48. Garvie, 'Little Boxes, Fewer Homes', *Equalities of Wellbeing*, April 2014.
49. See also Anwan et al., *Spatial Agency*.
50. Department for Communities and Local Government, *Building Regulations: Housing Standards Review*, 14 March 2014.
51. Wainwright, 'Tories New Housing Design Guide Backs Tiny Unliveable Homes', *The Guardian*, 2 April 2015.

Bibliography

Anwan, Nishat, Tatjana Schneider and Jeremy Till, *Spatial Agency: Other Ways of Doing Architecture* (London: Routledge, 2011).
Braidotti, Rosi, *The Posthuman* (Cambridge: Polity, 2013).

Butler, Patrick, and George Arnett, 'Lower Benefit Caps Exclude Poor Families', *The Guardian*, 20 July 2015 <http://www.theguardian.com/society/2015/jul/20/lower-benefit-caps-exclude-poor-families-make-cities-unaffordable> (accessed 20 July 2015).

Carmona, Matthew, *Space Standards: The Benefits* (London: UCL, Report for CABE, 2010).

Colquhoun, Ian, *RIBA Book of British Housing: 1900 to the Present Day*, second edition (Oxford: Elsevier, 2008).

Cordell, Tom and Niamh Walsh, *Utopia London* (London: Cordell and Walsh, 2010).

Department for Communities and Local Government, *Building Regulations: Housing Standards Review*, 14 March 2014 <https://www.gov.uk/government/uploads/system/uploads/attachment_data/file/291796/140313_Building_Regulations.pdf> (accessed 14 July 2015).

Dorling, Danny, *All That is Solid: How the Great Housing Disaster Defines Our Times, and What We Can Do about It* (London: Allen Lane, 2014).

Drury, Andrew, 'Parker Morris – Holy Grail or Wholly Misguided?' *Town and Country Planning*, Vol. 77, Issue 10 (August 2008), pp. 403–5.

Dugan, Emily, 'Exclusive: 5000 People are Now Facing Eviction after Bedroom Tax', *The Independent*, 13 September 2013 <http://www.independent.co.uk/news/uk/politics/exclusive-50000-people-are-now-facing-eviction-after-bedroom-tax-8825074.html> (accessed 1 September 2014).

Elledge, Jonn, 'Only 43 Homes in London are Affordable for First-time Buyers. So Who's to Blame?' *The Guardian*, 30 April 2015 <http://www.theguardian.com/commentisfree/2015/apr/30/housing-crisis-first-time-buyer-affordable-homes-london> (accessed 1 May 2015).

Ely, Alex, 'The Case for Space', *Equalities of Wellbeing*, UCL London, April 2014, <http://www.equalitiesofwellbeing.co.uk/publications-from-equalities-of-wellbeing-housing-workshop/> (accessed 14 July 2015).

Garvie, Deborah, Shelter, 'Little Boxes, Fewer Homes – Why Housing Space Standards Will Get More Homes Built', *Equalities of Wellbeing*, April 2014 <http://www.equalitiesofwellbeing.co.uk/publications-from-equalities-of-wellbeing-housing-workshop/> (accessed 14 July 2015).

Giles, Chris, 'Higher Wage Will Not Compensate for Lost Tax Credit Cuts', *The Financial Times*, 9 July 2015 <http://www.ft.com/cms/s/0/87b00b52-264e-11e5-9c4e-a775d2b173ca.html#axzz3hO2EXMET> (accessed 14 July 2015).

Gilman, Ollie, 'Charming but Compact', *The Daily Mail*, 30 August 2014 <http://www.dailymail.co.uk/news/article-2738462/Charming-compact-pay-1-000-month-live-one-tiny-semi-studios.html> (accessed 1 September 2014).

Guattari, Félix, *The Three Ecologies*, trans. Ian Pindar and Paul Sutton (London: Athlone Press, 2000).

Harker, Lisa, *Chance of a Life Time: The Impact of Bad Housing on Children's Lives* (London: Shelter, 2006).

Inman, Phillip, 'Mark Carney: Rising House Prices Pose Biggest Risk to Recovery', *The Guardian*, 19 May 2014 <http://www.theguardian.com/business/2014/may/18/mark-carney-house-prices-risk-economy-bank-of-england> (accessed at 30 June 2014).

Israel, Jonathan, *Radical Enlightenment: Philosophy and the Making of Modernity 1650–1750* (Oxford: Oxford University Press, 2001).

Joseph Rowntree Foundation, *Housing and Poverty Blog*, July 2015 <http://www.jrf.org.uk/topic/housing-and-poverty> (accessed 14 July 2015).

Martin, Alice, 'Why the UK Leads the Way on Inequality', *New Economics Foundation*, 9 January 2015 <http://www.neweconomics.org/blog/entry/why-the-uk-leads-the-way-on-inequality> (accessed 14 July 2015).

Martin, Luther, Huck Gutman and Patrick Hutton, *Technologies of the Self: A Seminar with Michel Foucault* (Amherst, MA: University of Massachusetts Press, 1988).

Mayor of London, *Housing Design Standards*, Evidence Summary (London: July 2010).

Mayor of London, Design for London, *London Housing Design Guide (Interim Edition)*, London Development Agency, August 2010.

Moore, Rowan, 'Britain's Housing Crisis is a Human Disaster. Here are 10 Ways to Solve It', *The Observer*, 14 March 2015 <http://www.theguardian.com/society/2015/mar/14/britain-housing-crisis-10-ways-solve-rowan-moore-general-election> (accessed 14 March 2015).

National Centre for Social Research (NatCen) and Shelter, *People Living in Bad Housing – Numbers and Health Impacts* (London: Shelter, 2013).

Oldfield, Philip, 'UK Scraps Zero Carbon Home Target', *The Guardian*, 10 July 2015 <http://www.theguardian.com/environment/2015/jul/10/uk-scraps-zero-carbon-home-target> (accessed 14 July 2015).

Osborne, Hilary, 'Hundreds of Flats in Canary Wharf Development Sell in Less than Five Hours', *The Guardian*, 13 July 2015 <http://www.theguardian.com/money/2015/jul/13/canary-wharf-flats-maine-tower-overseas-buyers> (accessed 14 July 2015).

Parker Morris Report, *Homes for Today and Tomorrow* (Great Britain Ministry of Housing and Local Government, December 1961).

Pepper, Simon, 'Three Ages of Post-war Housing', *Equalities of Wellbeing*, London: UCL, Housing and Design Seminar, 30 January 2015 <http://www.equalitiesofwellbeing.co.uk/podcasts-from-housing-and-design-seminar/podcast> (accessed 14 July 2015).

Powers, Alan, ed., *Robin Hood Gardens: Re-visions* (London: Twentieth Century Society, 2010).

Rawes, Peg and Beth Lord, *Equalities of Wellbeing* <http://www.equalitiesofwellbeing.co.uk/> (accessed 10 March 2016).

Rawes, Peg, Tim Mathews and Stephen Loo, eds, *Poetic Biopolitics: Practices of Relation in Architecture and the Arts* (London: IB Tauris, 2016).

Royal Institute of British Architects (RIBA), *The Case for Space: The Size of England's New Homes* (London: Royal Institute of British Architects, September 2011).

Robinson, Dickon, 'Fit for Purpose', in *Robin Hood Gardens: Re-visions*, ed. Alan Powers (London: Twentieth Century Society, 2010), pp. 120–30.

Shelter, *Little Boxes, Fewer Homes: Setting Housing Space Standards Will Get More Homes Built* (London: Shelter, April 2013).

Spinoza, Baruch, *Ethics, Treatise on the Emendation of the Intellect and Selected Letters*, trans. Samuel Shirley, ed. Seymour Feldman (Indianapolis: Hackett Publishing Company, 1992).

Survey of London, 'Public Housing in Poplar: The 1940s to the Early 1990s', in *Survey of London*, originally published by London County Council, London, 1994, pp. 37–54 <http://www.british-history.ac.uk/survey-london/vols43-4/pp37-54> (accessed 1 May 2015).

Survey of London, *Survey of London: Volumes 43 and 44, Poplar, Blackwall and Isle of Dogs*, ed. Hermione Hobhouse (London: London City Council, 1994) <http://www.british-history.ac.uk/survey-london/vols43-4/pp37-54> (accessed 1 May 2015).

Swenarton, Mark, *Homes Fit for Heroes: The Political and Architecture of Early State Housing in Britain* (London: Heinemann, 1981).

Wainwright, Oliver, 'Tories New Housing Design Guide Backs Tiny Unliveable Homes', *The Guardian*, 2 April 2015 <http://www.theguardian.com/artanddesign/architecture-design-blog/2015/apr/02/tories-new-housing-design-guide-backs-tiny-unliveable-homes> (accessed 5 April 2015).

Wainwright, Oliver, 'Revealed: How Developers Exploit Flawed Planning System to Minimize Affordable Housing', *The Guardian*, 25 June 2015 <http://www.theguardian.com/cities/2015/jun/25/london-developers-viability-planning-affordable-social-housing-regeneration-oliver-wainwright> (accessed 30 June 2015).

Wainwright, Oliver, 'Osborne's Planning Reforms Risk Slum House Building', *The Guardian*, 10 July 2015 <http://www.theguardian.com/artanddesign/architecture-design-blog/2015/jul/10/george-osborne-planning-reforms-risk-slums-house-building> (accessed 14 July 2015).

Wigglesworth, Sarah, 'Robin Hood Gardens', *The Architects' Journal*, 31 January 2011 <http://www.architectsjournal.co.uk/news/daily-news/-wigglesworth-unveils-robin-hood-gardens-rescue-rival/8610654.article> (accessed 14 July 2015).

Wilkinson, Richard and Kate Pickett, *The Spirit Level: Why Equality is Better for Everyone* (London: Allen Lane, 2009).

CHAPTER 3

Amorphous Continua

Chris L. Smith

The contemporary biomedical laboratory is the site of an intense interplay between the expressive force of architecture and the experimental contents of medicine: between the critical and the clinical. This chapter considers the near impossibility of the translation of the ideas and ideals of experimental biomedical science into architecture and the manner by which a number of contemporary laboratories are explicitly dealing with this impasse. The chapter will suggest that one key strategy engaged by the architects of biomedical laboratories has become the use of the 'empty' or 'floating signifier'. The idea arrived in the anthropology of Marcel Mauss and Claude Levi-Strauss to describe a word, event or gesture that is void of meaning and thus apt to receive any meaning: a signifier without a signified. An empty or floating signifier may refer to a concept, such as Mauss's account of 'mana', a stable and concrete word that is assigned to a phenomenon that is mobile and non-specific. Another example of an empty or floating signifier may be a non-linguistic sign, as in Roland Barthes's account of the Eiffel Tower, which, though stable as an architecture is without any particular meaning and thus in receipt of a multiplicity of meanings: national identity, city identity, scientific achievement, technological advancement, vision, state prowess, progress, romance, etc. This notion of the empty or floating signifier belongs also to the discourse of Gilles Deleuze's a-signifying semiotics. For Deleuze there is no sanctified relation between signifier and signified (that is, meaning). This is described as the *univocity* of the sign relation and this univocity plays itself out in the formulations of delirium, stutter and exhaustion in the *critique et clinique* project.

The empty signifier may be void of stable meaning but it is nevertheless a highly productive force in the architecture of the contemporary biomedical laboratory. As designers struggle to find referents to which their architecture may respond in the complex microscopy of biomedicine, a few have instead generated spaces that occupy the impasse itself. Charles Correa's Champalimaud Centre for the Unknown

(2007–10) in Lisbon, Portugal, is one such place (Fig. 8). Correa's biomedical laboratory, clinical health and education buildings, designed for the Champalimaud Foundation, attempt not to give an image for the brave new world of biomedicine but rather to open up the possibilities for thinking afresh.[1] The Centre for the Unknown partakes of 'the atmospherization or mundanization of contents' that Deleuze and his accomplice Félix Guattari speak of in *A Thousand Plateaus*.[2] The architecture here gestures to a horizon that is beyond us: non-anthropocentric and non-representational. It gestures well beyond the current biomedical endeavour. It gestures to the water beyond the Lisbon coast and to the cosmology of the sky. The content of medicine becomes vague as the expressive force of architecture gestures elsewhere. In this regard, architectural expression is itself disentangled from the ideas and ideals of biomedicine and yet responsive to the creative sense of possibility that the biomedical sciences hold. This chapter turns to the manner by which architecture may operate, as Deleuze and Guattari were to write of the empty or floating signifier, as an 'amorphous continuum that for the moment plays the role of the "signified"'.[3]

Figure 8 *The Centre for the Unknown*.
Source: author (2013).

The Centre for the Unknown

In his address at the 2010 inauguration of the Champalimaud Centre for the Unknown the architect, Correa, would suggest of the centre that

> it uses the highest levels of contemporary science and medicine to help people grappling with real problems: cancer, brain damage, going blind. And to house these cutting-edge activities, we tried to create a piece of architecture. Architecture as Sculpture. Architecture as Beauty. Beauty as therapy.[4]

In subsequent publications of the text of Correa's speech the phrase 'real problems' has remained underlined. It points, perhaps, to the disjunction that occurs in architectural thinking between content and expression – as if the latter were less real or important than the former. It also points to an anxiety that architecture merely houses or decorates the reality of the world. Correa's deferral to 'real problems' is an odd linguistic formulation given the abstract intentions and concrete outcomes of the Centre for the Unknown.

The centre is located on the coast of Lisbon, on the northern bank of the Tagus river, close to the point at which the Tagus joins the Atlantic Ocean. Basic and clinical research occurs alongside each other here. International researchers and specialists in translational medicine – in the areas of oncology and neuroscience – are brought together, in this centre, with clinical specialists. The 'translational' movement of experimental science through to clinical trial and application is one of the key aims of the centre and João Botelho, one of the centre's directors, suggests, 'the building is not neutral. It is part of this project.'[5] According to Botelho, 'we call it the Centre for the Unknown, because, likewise, our discoveries are from the realm of the unknown'.[6]

The centre covers more than 60,000 square meters and comprises three main structures. The first, and largest, is the Champalimaud Clinical Centre, a four-storey structure on the north side of the site. The building contains treatment and clinical facilities on lower levels and laboratories on upper levels. From the building lobby patients visiting the facility can get glimpses of the research laboratories above. The building can accommodate 440 scientists. Glass walls on all levels overlook a large indoor tropical garden that is accessible to staff and those receiving treatment (Fig. 9). The orthogonal interior spaces and garden are embraced by a sweeping four-storey wall. In plan, the wall appears as a large portion of an ellipse. This rounded smooth white stone wall is perforated with large ovoid (egg-shaped) openings.

A second structure, toward the south of the site, houses the auditorium, restaurant and the exhibition area on entrance level and a

Figure 9 Indoor tropical garden.
Source: author (2013).

conference centre along with Champalimaud Foundation administration offices above. This structure has a similar elliptical, or what may be described as a fragmented tear-drop, geometry. This structure is connected to the research labs by a twenty-one-metre long glass tube

Figure 10 The auditorium window with views to the river and the Belém Tower. *Source:* author (2013).

bridge at an upper level. The auditorium has a large ovoid window with views to the river and the Belém Tower in the distance (Fig. 10). This impressive Manueline sixteenth-century tower is associated with the Age of Discovery and the period of Portuguese global imperialism. The exhibition space on the ground level hosts events related to the Foundation's scientific endeavours and beyond. From the restaurant (*Darwin's Café*) a terrace extends toward the water.

A third structure is as much landscape as building (Fig. 11). It is more open than that that we tend to call 'structure' and more closed or controlled than that that we tend to call 'landscape'. It may be best referred to as a 'plaza'. This plaza glides between the two main buildings. It does so in such a way as to complicate the usual hierarchies of object and empty space, inside and outside. Occupying this space gives the sensation that it is indeed the plaza that generated the sweeping white walls that enclose the structures to the north and south. The plaza comprises an open-air amphitheatre and an open public landscape. The semi-circular stone amphitheatre hosts concerts and public performances with the river as a backdrop. Perhaps counter-intuitively,

Figure 11 This plaza glides between the two main buildings.
Source: author (2013).

the plaza slopes upwards toward two stone monoliths and the river. Correa writes:

> At the end of the ramp are two stone monoliths, straight from the quarry, as primordial as Stonehenge. When you reach the highest point, you begin to see a large body of water – which seemingly connects to the ocean beyond. In the centre of the water body, just below the surface of the water, is an oval shaped sculpture – made of stainless steel and slightly convex, so that it reflects the blue sky and the passing of the clouds above.[7]

The stainless steel spherical body in the water looks to be the back of a huge turtle, an emerging planet or a suspended air bubble expelled by a docile sea creature. Correa would suggest of this 'enigmatic object': 'it could be an island; it could be a Portuguese man-of-war'.[8]

The Floating Signifier

For Claude Lévi-Strauss the 'highly esoteric' ethnological and sociological work of Marcel Mauss provides an opportunity to reassess the manner by which 'meaning' operates.[9] Lévi-Strauss's introduction to the work of Mauss was published the year of Mauss's death in 1950, a time when anthropology was operating as a key field between the sciences and the humanities. Although Mauss had conceded to the idea that social phenomena were linguistically constructed, his work had led him to cultural products and events that were socially operable and yet fell short (or beyond) the logic of linguistics itself. It is the extra-linguistic quality and generous depth of Mauss's work that leads Lars Spuybroek to suggest that architects 'need a discipline that takes Mauss's force of things to heart'.[10] Mauss identified an 'embodied elemental force', a 'magical thinking' or 'mana' that was 'a representation which is singularly ambiguous and quite outside our adult European understanding'.[11] It is this idea that led Lévi-Strauss to the notion of the 'floating signifier':

> I believe that notions of the *mana* type, however diverse they may be, and viewed in terms of their most general function (which, as we have seen, has not vanished from our mentality and our form of society) represent nothing more or less than that *floating signifier* which is the disability of all finite thought (but also the surety of all art, all poetry, every mythic and aesthetic invention), even though scientific knowledge is capable, if not of staunching it, at least of controlling it partially. Moreover, magical thinking offers other, different methods of channelling and containment, with different results, and all these methods can very well coexist. In other words, accepting the inspiration of Mauss's precept that all social phenomena can be assimilated to language, I see in *mana, wakan, orenda*, and other notions of the same type, the conscious expression of a *semantic function*, whose role is to enable symbolic thinking to operate despite the contradiction inherent in it.[12]

The work would come as a challenge to the fundamentals of Ferdinand de Saussure's linguistics.[13] For Saussure, although the relationship between the signifier and signified was arbitrary, it was also fundamental. That is, one was never present without the other. The notion of a floating signifier meant that any 'primacy of the signifier', as Jacques Lacan later referred to it, was questionable.[14] In this sense, the submerged stainless steel spherical body at the end of the plaza of the Centre for the Unknown that may have been a turtle, a planet, an air bubble, an island or a man-of-war – may be reframed as: a turtle, a planet, an air bubble, an island *and* a man-of-war, etc. There is no need for any one sanctified meaning to dominate when this (partial) object operates as a floating signifier. The stainless steel spherical body signifies – but what is signified remains 'singularly ambiguous'. In a paper focused on information and asignification, Gary Genosko would suggest that 'triggering is the key action of particle-signs – signs that are partial, particle-like, and destratifying'.[15] The disconnection between signifiers and signifieds means that signs are operative in their own right: 'triggering' processes and interactions and associations beyond themselves. That is, a sign may be operative precisely because it is not connected to any fixed meaning.

Roland Barthes would write of the *mana* of architecture when he turns to the Eiffel Tower in his *Mythologies*. He would note a moment at which 'at the age of twelve, young [Gustave] Eiffel himself took the diligence from Dijon with his mother and discovered the "magic" of Paris'.[16] The Eiffel Tower itself would come to operate with the magic of the empty signifier. Barthes writes:

> This pure – virtually empty – sign is ineluctable, because *it means everything*. In order to negate the Eiffel Tower (though the temptation to do so is rare, for this symbol offends nothing in us), you must, like Maupassant, get up on it and, so to speak, identify yourself with it. Like man himself, who is the only one not to know his own glance, the Tower is the only blind point of the total optical system of which it is the center and Paris the circumference. But in this movement which seems to limit it, the Tower acquires a new power: an object when we look at it, it becomes a lookout in its turn when we visit it, and now constitutes as an object, simultaneously extended and collected beneath it, that Paris which just now was looking at it.[17]

Barthes would come to echo Lévi-Strauss's language in referring to a '"floating chain" of signifieds' and would note what Lévi-Strauss referred to as 'the inherent contradiction in it'.[18] It is this mobility of sign relations that would lead Barthes to introduce a sliding scale of iconic, motivated and arbitrary signs. The *iconic* tends toward one function, that is one singular link in meaning. The crucifix in Christianity or the crescent in Islam, for example. The *motivated* sign has a mobility to it, that is the sign may stand in place of multiple meanings. The safety pin may suggest the nappy of infants but in the nose of a punk takes on a different, and yet oddly connected meaning. The *arbitrary* sign is one that is near empty of fixed meanings and whereby the meaning of the sign is entirely given by the contingencies of its use. Pursing the lips as if to kiss may be such a sign . . . It can indicate love and affection or disdain depending entirely on the context. Barthes would suggest that 'every code is at once arbitrary and rational'.[19] The architect of the Centre for the Unknown is aware of the mobility of meaning. Correa would refer to something like Barthes's notion of the *iconic* as the 'primordial symbolism' that spaces, architectural spaces, open to the sky, can incite.[20] Correa would also note the arbitrary nature of some signs: 'the mud pot, used in an Egyptian village to draw water from the well, has completely different connotations when caught in a beam of halogen light at the Metropolitan Museum of Art'.[21]

Beyond Barthes' sliding scale, Lévi-Strauss's notion of the floating signifier suggests that there need not be any connection between a sign and meaning. Deleuze and Guattari would succinctly summarise Lévi-Strauss's finding: 'the world begins to signify before anyone knows *what* it signifies; the signified is given without being known'.[22] In *A Thousand Plateaus* signs fall into machinic systems of production. The philosopher and psychoanalyst identify multiple 'regimes of signs' through which meaning is constructed. Meaning comes to be a product of the forces of appropriation to which something ('a human, a biological or even a physical phenomenon') is subject.[23] Such regimes are operative only in a machinic or pragmatic sense. Their operation, appropriation and adoption are contingent upon the situations in which they arise and the pragmatism or what Brian Massumi calls the 'unruly pragmatism', of their deployment.[24] The mobility of the sign and the manner by which it 'floats' is very much a part of its pragmatic deployment. Deleuze and Guattari write:

> All signs are signs of signs. The question is not yet what a given sign signifies but to which other signs it refers, or which signs add themselves to it to form a network without beginning or end that projects its shadow onto an amorphous atmospheric continuum. It is this amorphous continuum that for the moment plays the role of the 'signified', but it continually glides beneath the signifier, for which it serves only as a medium or wall: the specific forms of all contents dissolve in it.[25]

In *Essays Critical and Clinical* (1993) Deleuze suggests that medicine (the clinical) may distil from art, and specifically literature (the critical), a symptomatology. Symptomatology is not concerned with the play between signifiers and signifieds, it is not concerned with origination, nor with 'giving a reason', nor with causation (as with etiology). Symptomatology is the study of signs – full stop. Symptomatology as such may be thought of as a logic of floating signifiers. Daniel W. Smith, the Deleuze scholar and one of the translators of the text, suggests that for Deleuze, 'while etiology and therapeutics are integral parts of medicine, symptomatology appeals to a kind of limit-point, premedical or submedical, that belongs as much to art as to medicine'.[26] Such a position resonates with Lévi-Strauss's suggestion that the floating signifier is 'the surety of all art, all poetry, every mythic and aesthetic invention'.[27] Deleuze turns specifically to literature (the critical) as a means of investigating this medical (clinical) notion. He suggests that writers do not account for, or necessarily represent, the world but rather construct or compose worlds from indeterminate symptoms. Writers compose the terrains, interactions, inversions: the sense of alternate worlds that constitute modes of existence. There is in symptomatology an infolding of contexts and selves and a constructing, a configuring, of both. A delirium. For Deleuze,

> the ultimate aim of literature is to set free, in the delirium, this creation of a health or this invention of a people, that is, a possibility of life. To write for this people who are missing . . . ('for' means less 'in the place of' than 'for the benefit of').[28]

I would argue that the architecture of the Centre for the Unknown is involved in a similar enterprise of health and the possibilities of life. Not in that it houses the contents of biomedical science but rather because of its expressive force. I will explore this proposition in three ways: in respect to the *inverted interiority*; the *dislocations* from reason and the profound *emptiness and exhaustion* that the Centre for the Unknown infolds.

Inverted Interiority

The entrance of the main building on the northern side of the Centre for the Unknown is clearly marked but visitors tend to be drawn toward the water, channelled along the plaza space. The plaza operates much like a central corridor to the site. It is as if the plaza generates the adjacent white walls and the spaces behind them, and not vice versa. The sense one has is that the heart or interior of this project is this open plaza. The clarity of this open central core about which multiple spaces unfold is a particular inversion that Correa played with in multiple unbuilt works in the 1970s, such as the farmhouse for Indira Gandhi (1972) and the Kapur Guesthouse (1976), and more recently the unbuilt Museum of Archaeology in Bhopal (1985). Such projects place 'the highest emphasis' on what Correa refers to as 'open-to-sky space'.[29] The patterning of courtyard spaces in relation to interiors,

Correa calls 'the Inside-Out Sock'.[30] That key dualism upon which architecture thrives – interior and exterior – is compromised here as it is compromised in the Mobius strip or the Klein bottle.[31] The open plaza of the Centre for the Unknown operates 'inside-out' in multiple respects. The plaza is largely unpopulated, empty and yet all the more desired because of this. The near-white Portuguese Lioz limestone cladding of the exterior of the Centre for the Unknown creates complex relations that empty or hollow out the site. On a summer day, the white walls and pavers underfoot produce a scintillating, near blinding, glare.

The glare operates much like Deleuze writes of light in German expressionistic cinema. In the text *Cinema 1: The Movement-Image* (1983) the philosopher describes a 'light which has become opaque, *lumen opacatum*', and suggests that 'from this point of view natural substances and artificial creations, candelabras and trees, turbines and sun are no longer different'.[32] Glare is the blinding and disorienting materiality of light. It has the ability to express and to obscure. Glare is a materialisation of light. The glare of the plaza space produces a *zone of indiscernibility* – where the clear and geometric distinctions of form are lost and multiple elements join. The glare that reflects off the rounded walls, the water and the plaza paving is not the glare of dialectical opposition: a glare between dark and light, or between matter and the immaterial, but is rather a glare with an incorporeal material presence of its own. Lévi-Strauss would refer to the 'apparently insoluble antinomies'[33] involved in *mana* and the glare generated by the Centre for the Unknown produces something similar. The material scintillates in the sun, making insoluble of walls, floors, water and sky. Glare operates as an expressive *quality* and what it obfuscates is *quantities*. The measureless and formlessness of glare plays itself against the most measured of quantities here: the geometries of the architecture and the precisions of biomedical research.

Dislocations

The site for the Centre for the Unknown is organised around a diagonal line that marks the course of the central plaza space. The line is an axis of a kind that points toward the river and the sea beyond. The logic here is not of the orthogonality of laboratory organisation, access, egress and the physical containment necessitated by biomedical research. The organisation of the floor plans makes the dislocations of content and expression in this building patently clear. The laboratories and clinical spaces are organised on a grid of columns and linear circulation. The internal planning is not unusual for a biomedical research laboratory and clinic. The emphasis is on efficiency and this striated form of organisation is one that may be associated with accountancy and economy. The internal gardens (the sunken garden, the infusion garden and the terrace garden) mediate between the refined efficiencies of the internal layout and the equally refined expressions of the exterior architecture (Fig. 9). If one were to remove the curved walls the building would almost certainly still stand and its function would be almost unaffected. However, it would be a radically different prospect. A radically different place. A radically different logic at play.

Correa would write of a '*Genius Loci*, the essential meaning of a site' and would suggest that it was 'Architecture's unique responsibility to express, to release, that meaning'.[34] Genius loci is the protective spirit of a place from Roman religion, but Correa is referring to the version of the idea that relates to a 'spirit of place' as the Norwegian architect and phenomenologist Christian Norberg-Schultz was to formulate it.[35] In stark contrast to the thesis of the floating signifier, for Schultz every place has a particular meaning that an architect must articulate or amplify. Schulz suggests that 'architectural history shows that man's primeval experience of everything as a "Thou", also determined his relation to buildings and artefacts. Like natural elements, they were imbued with life, they had *mana*, or magical power.'[36] The architect in such a formation is a communicator – a communicator with the forces of the earth and a soothsayer of 'essential form' and meaning.[37] A witch-doctor, priest, crypto-Platonist or astrologer. Correa refers to 'Architecture as a Model of the Cosmos – each expressing a transcendental reality, beyond the pragmatic requirements of the programme that caused them to be built'.[38] The architecture of the Centre for the Unknown is the generation of a particularised cosmos. An alien landscape. It's a negotiation of the site in its particularity but likewise generates something highly particularised or 'alien' to the place. Many of Correa's works, although highly mindful of context, remain clearly different to the structures that surround them. One cannot mistake the Centre of the Unknown for any other structure in Lisbon. This generation of difference is often written about in terms of Correa's penchant for the generation of 'microclimates'. However, there is something particularly singular to the centre that is far more expressive than functional, but no less operative.

In writing of the Eiffel Tower, Barthes suggests that 'architecture is always dream and function, expression of a utopia and instrument of a convenience'.[39] In such a delusion the architect is a communicator rather than creator, a discoverer rather than a constructor. We can forgive

Correa for indulging in the distillation of truths via metaphor when really what is at stake is a 'zero degree' architecture. Barthes would suggest that it was

> as if the function of art were to reveal the profound uselessness of objects, just so the Tower, almost immediately disengaged from the scientific considerations which had authorized its birth (it matters very little here that the Tower should be in fact useful), has arisen from a great human dream in which movable and infinite meanings are mingled: it has reconquered the basic uselessness which makes it live in men's imagination. At first, it was sought-so paradoxical is the notion of an empty monument – to make it into a 'temple of Science'; but this is only a metaphor; as a matter of fact, the Tower is nothing, it achieves a kind of zero degree of the monument; it participates in no rite, in no cult, not even in Art; you cannot visit the Tower as a museum: there is nothing to see inside the Tower.[40]

Emptiness and Exhaustion

It is this 'nothing to see' that makes a place such as the Centre for the Unknown so enticing, so productive. In engaging with the unknown the architecture does not generate a point of fixation but rather a point of departure. A gesture 'to infinity'.[41]

A space from which we float. The empty central space recalls the Salk Institute and engages what Correa refers to as the 'metaphysical aspects of the sky'.[42] In Correa's essay 'The Blessings of the Sky' the architect notes the singular and central value of the sky in 'human history', 'since the beginning of time'.[43] The sky becomes very much a key part of this architecture. Correa suggests that the sky operates as a metaphor 'for our relationship to something outside (and beyond) ourselves'.[44] If it is a metaphor it's a particularly obscure one. Obscure and empty. The plaza of the Centre for the Unknown generates a cacophony of symbols and references, but none are dominant. The overarching sense one has here is of emptiness. A pared-back, open, emptiness. The impotence of the architecture makes potent of the sensation.

In *A Thousand Plateaus* Deleuze and Guattari would suggest that 'the sign that refers to other signs is struck with a strange impotence and uncertainty, but mighty is the signifier that constitutes the chain'.[45] The odd alien minimalism of the plaza and the buildings that flank it generates the most intense of sensations. The space makes no attempt at a totalisable image for itself. That is, it doesn't rest in *this or that* metaphor or *this or that* symbol. Its material is less the discernible stone and curve than the indiscernible sky and glare. For Correa:

> We live in a world of manifest phenomena. Yet, since the beginning of time, man has intuitively sensed the existence of another world: a non-manifest world whose presence underlies – and makes endurable – the one he experiences every day. The principle vehicles through which we explore and communicate our notions of this non-manifest world are religion, philosophy and the arts. Like these, architecture too is generated by mythic beliefs, expressing the presence of a reality more profound than the manifest world in which we exist.[46]

The two fifteen-metre-tall monolithic stone columns that mark the step from land to river themselves bleed into the sky (Fig. 12). Correa would suggest that the columns 'announce the presence of the Infinite

Figure 12 Monolithic stone columns that bleed into the sky.
Source: author (2013).

Unknown that lies beyond'.⁴⁷ The ends of the columns have been painted blue, which lets them leave the earth entirely. There's a reflecting pond beyond the columns, beyond that the river and beyond that the sea. It's all finely graded – one element slides into another, producing an 'amorphous continuum that for the moment plays the role of the "signified"' as Deleuze and Guattari were to write of the empty or floating signifier.⁴⁸ Correa would write of the first time he saw this site:

> I knew that whatever else we did, the site must be structured along a powerful architectural diagonal axis, an open-to-the-sky space, going right from the entrance to the opposite corner, where you finally see the river beginning to merge with the ocean and the great unknown.⁴⁹

This amorphous continuum between the architecture and landscape, water and sky is highly productive.⁵⁰ The minimalist treatment of Correa's architecture is removed from the operations of the biomedical endeavour – and yet not entirely so. It's perhaps refraction rather than reflection. They breed fruit fly at the Centre for the Unknown, because of the ease by which they can be genetically manipulated, and they breed shoals of zebra fish, because of their uncanny ability to regenerate organs without scarring. The architecture is similarly malleable and generative – open to possibility and the contingencies of the unknown as it is open to the sky. One can imagine that Lévi-Strauss was speaking of the Centre for the Unknown when he writes of *mana*:

> Force and action; quality and state; substantive, adjective and verb all at once; abstract and concrete; omnipresent and localised. And, indeed, *mana* is all those things together; but is that not precisely because it is none of those things, but a simple form, or to be more accurate, a symbol in its pure state, therefore liable to take on any symbolic content whatever? In the system of symbols which makes up any cosmology, it would just be a *zero symbolic value*, that is, a sign marking the necessity of a supplementary symbolic content over and above that which the signified already contains, which can be any value at all, provided it is still part of the available reserve, and is not already, as the phonologists say, a term in a set.⁵¹

Notes

1. Correa worked with RMJM Hillier laboratory and clinical design architects and the Portuguese firm Consiste on the project. Correa was no novice when it comes to the design of research facilities. He had previously designed the Massachusetts Institute of Technology (MIT) Brain and Cognitive Sciences Complex in Boston, USA, the Jawaharlal Nehru (JN) Centre for Advanced Scientific Research at Bangalore, India, and

the Inter-University Centre for Astronomy and Astrophysics (IUCAA) at Pune, India. RMJM Hillier is an expert in the design of translational cancer facilities. Former projects include the Rutgers Cancer Institute of New Jersey, USA, The University Hospital Cancer Center at the University of Medicine and Dentistry of New Jersey (UMDNJ), the University of South Alabama Mitchell Cancer Institute, USA, the Louisiana Cancer Research Center, USA, and the University of Puerto Rico Comprehensive Cancer Center. The Champalimaud Foundation is a private organization established at the behest of the late Portuguese industrialist and entrepreneur, António de Sommer Champalimaud. The Foundation is composed of a Board of Directors, a General Council and a Scientific Committee. The Board of Directors is currently comprised of Leonor Beleza (former Portuguese Minister of Health and Vice-President of the Portuguese parliament), António Horta-Osório (Chief Executive (designate) of Lloyd's Banking Group) and João Botelho (former Head of Cabinet in two successive Portuguese governments). The General Council is composed of equally eminent figures including one former president of Portugal and one of Brazil. The Scientific Committee is composed of international renowned scientists including two Nobel Prize laureates, one of which, James Watson, chairs the committee.
2. Deleuze and Guattari, *Thousand Plateaus*, p. 112.
3. Ibid. p. 112.
4. Correa, 'Inauguration Speech'.
5. João Botelho in Sankalp Meshram's short film, 'Into the Unknown'.
6. João Silviera Botelho cited in Mays, 'Champalimaud Centre for the Unknown', p. 94.
7. Correa, 'Champalimaud Center for the Unknown'.
8. Correa in Meshram, 'Into the Unknown'.
9. Lévi-Strauss, *Introduction to the work of Marcel Mauss*, p. 1.
10. Spuybroeck, 'Charis and Radiance', p. 124.
11. Mauss, *General Theory of Magic*, p. 132. Refer to Lévi-Strauss, *Introduction to the work of Marcel Mauss*, p. 34.
12. Lévi Strauss, *Introduction to the work of Marcel Mauss*, p. 63; emphasis in the original.
13. Saussure, *Course in General Linguistics*.
14. Lacan, *Seminar III*, pp. 119–20.
15. Genosko, 'Information and Asignificaiton', p. 18.
16. Barthes, 'Eiffel Tower', p. 13.
17. Ibid. p. 4; emphasis in the original.
18. Barthes, 'Rhetoric of the Image', p. 39.
19. Ibid. p. 31.
20. Correa, 'Museums: An Alternate Typology', p. 328.
22. Deleuze and Guattari, *Thousand Plateaus,* p. 112; emphasis in the original.
23. Deleuze, *Nietzsche and Philosophy*, p. 3.

24. Massumi, 'L'Amérique de Deleuze', p. 56.
25. Deleuze and Guattari, *Thousand Plateaus*, p. 112.
26. Smith, 'Life of Pure Immanence', p. xvi.
27. Lévi-Strauss, *Introduction to the work of Marcel Mauss*, p. 63.
28. Deleuze, *Essays Critical and Clinical*, p. 4.
29. Correa, 'Blessings of the Sky', p. 20.
30. Ibid.
31. Cache, 'Plea for Euclid'.
32. Deleuze, *Cinema 1*, p. 51.
33. Lévi-Strauss, *Introduction to the work of Marcel Mauss*, p. 63.
34. Correa cited in Rubens, 'Champalimaud'.
35. Norberg-Schultz, *Genius Loci*.
36. Ibid. p. 50.
37. Ibid. p. 6.
38. Correa, 'Blessings of the Sky', p. 28.
39. Barthes, 'The Eiffel Tower', p. 6.
40. Ibid. p. 7.
41. Correa, 'Public, Private and Sacred', p. 9.
42. Correa, 'Blessings of the Sky', p. 27.
43. Ibid. p. 18.
44. Ibid. p. 27.
45. Deleuze and Guattari, *Thousand Plateaus*, p. 112.
46. Correa, 'Public, Private and Sacred', p. 1.
47. Correa cited in Rubens, 'Champalimaud'.
48. Deleuze and Guattari, *Thousand Plateaus*, p. 112.
49. Correa cited in Rubens, 'Champalimaud'.
50. Correa cited in Rubens, 'Champalimaud'. It is productive like the singularisation of an aesthetic that Guattari identifies in the Japanese city in which an asignifying semitiocs prompts 'a new poetry, a new art of living'. Guattari, *Machinic Eros*, p. 112.
51. Lévi-Strauss, *Introduction to the work of Marcel Mauss*, p. 64; emphasis in the original.

Bibliography

Barthes, Roland, 'Rhetoric of the Image', in *Image-Music-Text*, trans. Stephen Heath (London: Fontana, [1964] 1977), pp. 32–51.

Barthes, Roland, 'The Eiffel Tower', in *The Eiffel Tower and Other Mythologies*, trans. Richard Howard (Berkeley: University of California Press, [1979] 1997), pp. 3–17.

Cache, Bernard, 'A Plea for Euclid', *ANY: Architecture*, Vol. 24 (1999), pp. 54–9.

Correa, Charles, 'The Blessings of the Sky', in Charles Correa and Kenneth Frampton, *Charles Correa* (London: Thames & Hudson, 1996), pp. 17–28.

Correa, Charles, 'Museums: An Alternate Typology', *Daedalus* Vol. 128, Issue 3 (Summer 1999), pp. 327–32.
Correa, Charles, 'Inauguration Speech', 2010. Reprinted in Prasad Shetty, 'Interview' in *ARTIndia: The Art News Magazine of India*, Vol. 16, Issue 1, 2016 <http://www.artindiamag.com/quarter01_01_11/interviewPrasadShetty01_01_11.html> (accessed 1 June 2015).
Correa, Charles, 'Champalimaud Center for the Unknown', *e-architect*, 6 March 2014 <http://www.e-architect.co.uk/portugal/champalimaud-foundation> (accessed 2 May 2015).
Correa, Charles, 'The Public, the Private and the Sacred', in *A Place in the Shade: The New Landscape and Other Essays* (Delhi: Penguin, 2010), pp. 26–45.
Deleuze, Gilles, *Nietzsche and Philosophy*, trans. Hugh Tomlinson (London: Althone Press, [1962] 1983).
Deleuze, Gilles, *Cinema 1: The Movement-Image*, trans. Hugh Tomlinson and Barbara Habberjam (Minnesota: The Althone Press, [1983] 1986).
Deleuze, Gilles, *Essays Critical and Clinical*, trans. Daniel W. Smith and Michael A. Greco (Minneapolis: University of Minnesota Press, [1993] 1997).
Deleuze, Gilles and Félix Guattari, *A Thousand Plateaus*, trans. Brian Massumi (Minneapolis: University of Minnesota Press, [1980] 1987).
Genosko, Gary, 'Information and Asignification', *Footprint*, Vol. 8, Issue 1 (Spring 2014), pp. 13–28.
Guattari, Félix, *Machinic Eros: Writings on Japan*, ed. and trans. Gary Genosko and Jay Hetrick (Minneapolis: Univocal Publishing, 2015).
Lacan, Jacques, *Seminar III, The Psychoses*, trans. Russel Grigg (London: Routledge/Norton, [1981] 1993).
Lévi-Strauss, Claude, *Introduction to the Work of Marcel Mauss*, trans. Felicity Baker (London: Routledge and Kegan Paul, [1950] 1987).
Massumi, Brian, 'L'Amérique de Deleuze: un pragmatisme insoumis', *Magazine littéraire* Vol. 406 (February 2002), p. 56.
Mauss, Marcel, *A General Theory of Magic*, trans. Robert Brain (London and New York: Routledge, [1950] 2001).
Mays, Vernon, 'Champalimaud Centre for the Unknown', *Architect*, 7 April 2011 <http://www.architectmagazine.com/design/buildings/champalimaud-centre-for-the-unknown-features-a-glass-bridge_o> (accessed 18 June 2015).
trans.Meshram, Sankalp, 'Into the Unknown', 25 April 2015 <https://www.youtube.com/watch?v=mVdcNRNPQa4> (accessed 18 May 2015).
Norberg-Schultz, Christian, *Genius Loci: Towards a Phenomenology of Architecture* (New York: Rozzoli, 1991).
Rubens, Rick, 'Champalimaud', October 2010 <http://www.rickrubens.com/Champalimaud.htm> (accessed 18 June 2015).
Saussure, Ferdinand de, *Course in General Linguistics*, ed. Charles Bally and Albert Sechehaye, trans. Roy Harris (La Salle: Open Court, [1916] 1983).

Smith, Daniel W., 'Introduction: "A Life of Pure Immanence": Deleuze's "Critique et Clinique" Project', in *Essays Critical and Clinical*, Gilles Deleuze, trans. Daniel W. Smith and Michael A. Greco (Minneapolis: University of Minnesota Press, [1993] 1997), pp. xi–liii.

Spuybroeck, Lars, 'Charis and Radiance: The Ontological Dimensions of Beauty', in *Giving and Taking: Antidotes to a Culture of Greed*, ed. Joke Brouwer and Sjoerd van Tuinen (Rotterdam: V2 Publications, 2014), pp. 119–49.

PART II
Robotics

CHAPTER 4

Robots Don't Care: Why Bots Won't Reboot Architecture

Christian Girard

Robots don't care; they really don't care at all. Of course this goes without saying. But since the idea of *care* was central to the 3C Conference – Critical & Clinical Cartographies, a conference on Embodiment and Technology and Care and Design, this needs to be further explored. What does it mean when robotics is thought within the framework of '*the critical and the clinical*' – that is, an effort to replace dialectic by a promotion of differential thinking? How are robotics to be engaged with if we abandon fiction – be it *science*-fiction –and concentrate exclusively on technology? The aim is to stage a tension existing between a strong positivist attitude embracing contemporary technology and a materialist position insisting on an open range of different realisms, including metaphysical ones. In short, we are confronted with an overall tension where *critical and clinical cartographies* are both more needed and more difficult to produce than ever. An after-effect of this tension is the ever-shifting boundaries of disciplines, practices and knowledge.

A discipline such as architecture is a perfect candidate to be squeezed between antagonist positions: hasn't architecture been, from the start, trying to combine its artistic and scientific inclinations, with mixed success, depending on the century, the period or the architectural movement? And when robotics and automation enter the game, things tend to become quite hectic. So far, if robots don't care, we still care about architecture. At least because it is a given fact that architecture takes care of us. Architecture and/or design, urbanism, among other goals, must heal.

What could be the *point of view* of a robot on humans? Nothing sounds more true than the remark of an android-robot saying: 'Humans are inscrutable. Infinitely unpredictable. *This is what makes them dangerous.*'[1] When robots start to consider humans as dangerous, we enter a new era. Dangerous, humans certainly are, as history and everyday news confirm – dangerous to themselves and dangerous to nature – but the consciousness of this fact was left to humans only while animals sense this danger by instinct and experience. When artefacts have a sense of danger, and precisely of human danger, things will become different.

Again, robots don't care. However, paradoxically, they care – when humans use them to heal, or for surgery – but they also kill. They do care and it is even worse, they sometimes do care to kill. When they kill, it is due to an error or lack of prudence: factory workers are sometimes killed 'by' robots; this is extremely rare but it does happen. Robots are pretty good at killing, almost as good as humans since, so far they kill only by human command or mistake. They lack of course all the *humanity* that, per se, humans are able to put in killing one another. By 'humanity' we need to understand this mix of anger, hate, cruelty and fear and of complex psychological configurations only humans seem to hold. Automation and robotics have always been at the service of those perfectly *embedded* human urges to destroy. Are not some artefacts aptly called '*automatic machine* guns'?[2]

The *embodiment and technology* conference panel brought together entities that are at first considered as separate. The problem lies actually in this separateness. The November 2014 gathering at Delft University of Technology sorted out intricate questions and tried to merge things that are conceived as separate or opposed. Usually, bodies and machines are always opposed, just like flesh and technology. This is why we can talk about *embodiment and technology*. However, from the start one could think of *embodied technology* or *technologised embodiment*. Which, for architects at least, brings back the man-machine mythologies of the early twentieth century and even much earlier if we recall Leonardo and the Renaissance. Incidentally, are we now experiencing the second, third or even fourth *machine age*? Does it make any sense to even speak of 'ages'? How to reconcile such opposite concepts when disciplines are themselves atomised into different small perimeters? Andrej Radman, in his introduction to his dissertation *Gibsonism*, charters together on the conceptual map no less than 140 authors and draws a large cartography where he rightly insists on getting rid of the dichotomies, dualisms and oppositions.[3]

If breaking down every solidified dualism is a pursued goal, it is not usually attained. Out of this present condition it is expected that we formulate something about architecture today and, if possible, something sound about architecture in the near future. Where exactly do *embodi-*

ment, *care* and *design* meet, if indeed they meet, and if we decide they have to meet? Through which cartography? Why do we want them to meet in the first place? Such crude questions are quite common in epistemological endeavours, when an issue is the role given to specific concepts. Let us remember an occasion when a reunion of bright minds decided, through discussion, to change a specific cartography of concepts. Half a century ago, in 1962, a meeting discussed conflicting positions about the concept of 'information'. Norbert Wiener, Benoit Mandelbrot and Abraham Moles debated with philosophers such as Martial Gueroult, Jean Hyppolite, Lucien Goldmann and a host of scientists. What was the most important result of their exchanges? It turned out that almost to all participants, *'information theory'* in the sense given by Claude Shannon could no longer be a shared site of fruitful discussion. They at least agreed on one single idea, the necessity to do without a recent concept. Today, what do we have in common? The answer is computation, artificial intelligence, neurosciences, robotics, biomechatronics, biorobotics, biotechnologies, big data and so on. What question and what concept can we avoid or get rid of? Throwing away a notion, a concept, an ideology happens to be as crucial as creating new ones. A current problem, at least for architects, is that one of the concepts at risk is no less than that of architecture. Indeed, architecture seems to dissolve at a time when all disciplines are splitting into micro-disciplines, into small sites of research. This continuous segmentation brings about a greater porosity between fields of knowledge. Paradoxically, extreme specialisation on very narrow sites opens up a multiplicity of exchanges and a new porosity.

Today, catchwords and mottos of the 1980s and 1990s such as 'interdisciplinarity', 'transdisciplinarity' or 'pluridisciplinarity' are no longer necessary when everything is interrelated and intertwined to the extreme. Therefore, *embodiment, care* and *design* can meet by some kind of loose and fluid circulation of concepts and practices powered by big data and computational engines. In the meantime, architecture is at risk of dissolving itself less by lack of a disciplinary strategy than by a last desperate effort to maintain the discipline: holding the grip when it is too late with nothing left to hold on. For a long time, we have assumed that the autonomy of architecture was a truism, that it was a tautology that did not need to be sustained. At least history and theory would always be at hand to help us keep the flame burning. Now the game seems somewhat trickier. Can we give definitive credits to history? How can a theory of architecture be still sought after, when the idea itself of *theory* seems to be shaken everywhere, in every field of knowledge and practice? Architecture's autonomy has mostly relied on the dissimulation of its means and processes. The autonomy is minimal, minimalistic, reduced to a small fraction of the architecture produced and an even smaller percentage of the built environment. Architecture as 'Architecture' is rare and has always been. Architecture accounts for less than 5 per cent of the built environment worldwide. The architectural design project differs from other designed objects not only by its finality – a building rather than a chair or a shoe – but also by the cultural discourse its history carries. This stands – or stood – for the accepted storytelling; it remains to be seen how long it can last.

Everyone has a case study. Concerning the relation in architecture between hardboiled factors such as function, programme and construction, and soft ones such as fiction, all of them envisioned through the problematic of *care*, consider, for instance, the case of a monumental hospital facility built in Aachen (Germany, 1989), a machine to care and heal (Fig. 13) – an ongoing iconic megastructure, at least for me, which stands as a lesser-known offspring of Piano & Rogers's Beaubourg museum (1975) and, seemingly, of a specific earlier non-built project. For a long time, I have been relating this architecture with the famous Hans Hollein project of 1965: the Aircraft Carrier in the Landscape. The architects of the Aachen Hospital Klinikum have almost materialised Hollein's fiction.

On the one side you have a machine to care and heal (the hospital); on the other you have a machine to destroy and kill (the aircraft carrier). Both operate on the model of the city, holding hundreds of bodies. But the aircraft carrier, a moving city on the sea, is motionless in Hollein's photo-montage. By doing so, twenty-five years ago, the German architects proved in Aachen that architecture is working on its own fictions, even if unconsciously, at a level far more intense than the banal process of imagination and form recycling. We may ask ourselves what would take the place today of the dual system composed of the aircraft carrier of Hollein and of the Aachen Hospital? Perhaps a robot and a hand holding a drawing pen? They would symbolise machinery and body, or technology and flesh. This helps to understand why it is not unusual to find photographs of six-axes robots with a pencil at the end of their arms, when it's not even android bots effectively drawing on sheets of paper. Now hospitals everywhere try to look like office buildings or housing programmes, or both. And the machinery takes place inside the hospital, at the smallest scale possible, as close as possible to the interiority of the bodies. Medical mechanics and technologies tend to disappear. The machine embodies itself in the bodies. In the process architecture dismantles and disembodies itself in

WHY BOTS WON'T REBOOT ARCHITECTURE 127

1. H. Hollein, «Aircraft-Carrier in landscape», 1964. Collection du Museum of Modern Art, New York.

2. J. Kunz, P. Tröger, W. Weber (Weber, Brand & Partner), Hôpital Universitaire, Aix-la-Chapelle, 1984.

Figure 13 The parallel between Hans Hollein's famous montage and the built project in Aachen (Germany) was first drawn in the essay: Christian Girard, *Architecture et concepts nomades, traité d'indiscipline* (Brussels: Mardaga, 1986).
Source: Christian Girard.

nothing much, erasing its history, culture and modes of signifying. The time of architecture without content has come, and of content without architecture.[4]

Bots Don't Care Less

Robots will not reboot architecture, quite the contrary. Robots could kill off the need for architects as they have killed employment in dozens of trades. Robotics require new skills and kill existing skills. Robotics and automation at large are more than technology. They are far from new; what is new is their deployment across an ever-increasing spectrum – everywhere and all the time. Calculus is at the core of this condition. Calculus and coding are the technology. The rest is merely the extension of calculus and coding into materiality at every possible scale, and more often scales of the invisible, and invisible scales. You don't see the code when it is running. Therefore, embodiment requires and equals invisibility as much as indiscernibility. At the same time, no one ever sees inside her or his own body. Even though we perceive and feel, we do not perceive the totality of what is going on under the flesh. Technologies have always been articulated with the body, incorporated in a sense from pre-historical tools up to now.

Robots or architecture? 'Computation or revolution' is an even more perverted dilemma than the 'architecture or revolution' one offered by Le Corbusier.[5] If the choice between architecture and revolution was a non-choice, a false alternative, a corrupted montage of non-options, without a possibility to select one side or the other, since no common plane exists between the two poles, it is useless to bring together automation and architecture in the hope of just opposing them. Especially when the prospect of automated architecture is looming over the discipline.

Not So Abstract Machines

Because computation and automation encompass absolutely all *things*, whether concrete or abstract, there is no escape from the world of numbers. There is no way to leave the landscapes of computed reality that we are immersed in. We cannot go off-line, quit the grid or cut the power line. Talk of 'embodiment' or of 'incorporation': as humans we are to technologies based on computation, what we were to steam-based technologies, with a notable difference. Computation is not part of ourselves. The steam machine was not either. While not becoming, so far, a part of ourselves, computation is fabricating, constructing our selves, our collective and individual selves. When Kraftwerk sang 'We are the Robots' they had it right, provided that we forget their android robot look.[6] Anthropomorphic robots are only good for bygone science-fiction movies. The real change today is that science-fiction has been deprived of most of its raison d'être: computation and automation, artificial intelligence and robotics, together and tightly intertwined, jettison any need for science-fiction narratives. Fiction has become impoverished in terms of imagination compared to the power unleashed by computers. Consider the degree of sheer boredom one feels after half an hour of the most acclaimed science-fiction Hollywood movie released in the last decades from *Iron Man* to *Transformers*. Not only does nothing significant happen despite or because of the relentless action going on with sophisticated machines, techniques and inventions, all mixed and scrambled through spectacular visual effects, but nothing comes near to the everyday torrent of events that the media pour on us and the daily actions we have to take part in if we want to keep going in this world. Nobody ever admitted that *The Matrix* was a rather boring movie. In the meantime, quite ironically, science-fiction has become a field worth considering by philosophy and metaphysics.[7]

Let us not deny that technology has an impact on ways of thinking, and on philosophy itself. For philosophy to remain a major way of thinking, in parallel to the *'ways of making worlds'* in the sense given by Nelson Goodman, it can't avoid a direct confrontation with technology.[8] A romantic vitalism has seduced some digital architects, who embrace the ease with which algorithms produce what they take as faithful representations of nature. They see the formal exuberance as a positive asset working against a rationalised computed architecture. They forget how the tools they use are rigorously programmed and how they exist through an extremely rationalised process of calculation, be it with cellular automata, intelligent agents, evolutionary algorithms or emergent mechanisms. The digital simulation of life has indeed reached a high level of exactness and fine grain resolution of reality but this does in no way expel the very last remains of rationality. Quite the contrary: more than ever, rationality is still at work, if less visible. When it turns into a hyper-rationality, it leaves nothing untouched. The so-called Baroque relies on powerful mathematical tools. Leibniz was neither a dreamer nor an inventor of hallucinatory worlds. Excess is meant to be a sought-after value by architects pursuing an artistic agenda, or who believe that this is a meaningful agenda. But they keep forgetting how the Baroque itself could never happen without an excess of rationalisation, precisely of a geometrical and mathematical kind. This mistake drives an article titled 'The Visceral Materiality of the Digital and its Biological Poetics', which is a remarkable instance of plea for more of everything and a direct praise for what the authors call an 'architectural flesh':

> Freshly established research direction . . . taking advantage of advanced digital tools . . . it literally incorporates the . . . biological, corporal, visceral attributes when designing space but also when treating spatial conditions in general . . . we can see another approach, an organic one, closer to biology, attempting an organic approach of space. It generates a space which is able to respond to *more* psycho-somatic challenges, in a dialogue with instincts and *more* human, sensual readings of space and architecture . . . Surrealism always referred to a *more* 'internal' aspect of humans for creating art . . . a *more* general attempt for a reaction against the establishment of a so called 'digital rationalism' . . . It promotes a broader and *more* creative approach of digital design tools and the digital methodologies. The aim is the encouraging of a *more* poetic hue in architectural design, a hue that reveals *more* intuitive manipulation of space. The architecture . . . produced under this trend, resorts to a *more* narrative and plethoric documentation of 'another' space, potentially evident and inhabitable. This space is perceived *more* through intuition than through the proper decoding of a respectable, digitally coded design language.[9]

The 'less is more' lesson has been forgotten and replaced by this *more of more* rhetoric. These authors are pushing the divide to a point where it becomes a fake or a joke. We must recall here what Wolfgang Köhler's *Gestalt Psychology* showed in 1927 with the famous *maluma/takete* opposition.[10] Now that any curve can be controlled through computation – producing effortlessly what Albrecht Dürer was trying to draw with his serpentine drafting mechanical tools – such dualism loses all its strength. There is no more rationalisation in the *takete* line or shape than in the *maluma* one (Fig. 14). The *takete* order, the *takete* formal territory is just a specific kind of curved line, and is in no way less organic than the *maluma*: straight lines and figures abound in the natural world (plants, trees and crystals) and are present at many scales in the living world. A curve inflexion, a spline, has no pretence to organicity, ornament or baroque. To ask for '*more of this, more of that*' by asking more *maluma* is outmost naïve and superficial. And yet if indeed it is not totally absurd, it does not realise it, and is in a sense reinforcing the enemy by building such a decoy. It strives in the shadow of Modernism; it continues to think in the rhetoric of Modernism but upside down. It remains historically in that *History* of which Postmodernism was paradoxically a part. Instead of breaking with it, it pursues its dualism even while it seeks the end of all dualisms. Those who

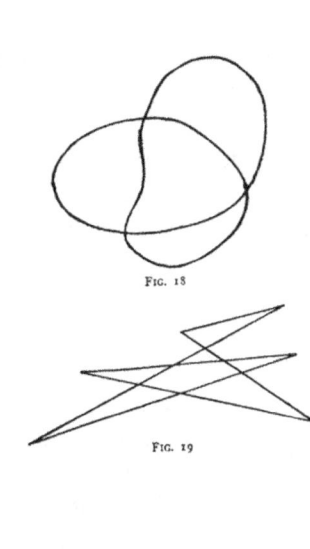

Figure 14 Maluma and Takete, Experiment by Wolfgang Köhler.
Source: Wolfgang Köhler, *Gestalt Psychology* (New York: Liveright, 1929), pp. 224–5.

keep insisting on the importance of the nonlinear weaken the nonlinear by transforming it into a slogan, a mere tagline in a pseudo-manifesto manner. So when we read a call for 'more body, more sense of touch, sound, more perception, more haptic, more flesh, more care' we can answer: bodies are still bodies, where is the problem? How can angular and straight lines be less or more 'organic' than supple, soft, curved, round, lines?

Hands-on or Hands-off Robots?

The thematic of the *hand* has gained momentum recently in synchronisation with the development of robotics and digital architecture. It is strongly associated with an ideology of craftsmanship. It offers a convenient entry back into a pseudo-phenomenological discourse where the fusion of body and mind is sought after, the hand replacing the main interface when design and architecture are the applied domains of practice. It is found in Richard Sennett's book *The Craftsman* translated in *Ce que sait la main La Culture de l'Artisanat* (What the hand knows),[11] where the chapters about architecture and computer-aided drawing (CAD) miss completely the point, whether considering John Portman's Peachtree Centre in Atlanta or Frank Gehry's Bilbao. Sennett wants hand and head to be back together again. The hand, and drawing by hand, are also promoted by Finnish architect and theoretician Juhani Pallasmaa.[12]

Embodiment is the argument. However, the argument can work in reverse mode. Computer-aided design–computer-aided manufacturing (CAD–CAM) has been and is still often considered as disembodied. This is a misconception, a prejudice. Just consider an architect working in front of their large computer screen working on a design. In what sense does he or she have no more 'body'? In an age of neurodesign experiments, when the design process hooks the brain and its electric brainwaves directly to the screen of the computer, the body–machine interface (or more precisely the body–computation interface), becomes even more intimate, to the point where some kind of body–computation copulation is on the way: a *compulation*, if you pardon the easy pun, is at work. The motto 'fuck the machine' is no more an aggressive anti-machine stance but a literal process.

Of more interest is the observation by Makato Sei Watanabe, a pioneer of algorithmic architecture, when he writes:

> Compared to sketches, programs benefit more from progress in technology. Eventually they will cover more areas than they do today and become far more useful. Sketches will probably not benefit from that kind of evolution . . . Over tens of thousands of years, the act of sketching has not changed.[13]

Watanabe is then at pains to create a real partnership between the digital and handicraft, between coding and drawing. He wishes to fill the gap between algorithmic design and hand drawing. He calls it 'induction design', but when his drawings join his 3D simulation on the same page, somehow the gap seems to widen. Two different orders of magnitude of precision cannot find a common site of discussion. This is why the *diagram*, which was so successful in architecture at the very end of the twentieth century, has regressed into advertising and communication.[14]

Towards Integral Simulation

To calculate a building, or for that matter, an architecture, means now to create a simulation of it. Calculating the airflow in a building, for instance, is computationally extremely expensive. Since we know that in a few years this will be extremely *inexpensive* or even free, things will be somewhat different. According to the epistemologist of simulation Franck Varenne a level has been reached where 'nature is no longer a good model of itself. Through computational simulation we can create a nature more natural than nature'.[15] When Gramazio and Kohler anticipate that 'the robot will enable a new architectural form of expression' they continue to think as standard architects and architectural historians who assume that 'expression' changes – and has to change – when technology changes.[16] If this remains globally true, such an argument is not needed, since it is much more accurate to focus on performative architecture than new forms and new expression: morphology at the service of performance has to be different to neutral morphology, but what we get cannot be reduced to what we see. It is first of all what we think and what the machine calculates that becomes exactly and literally what is built. And what builders and users see does not necessarily *express* the multi-level performances that the architecture will bring. One cannot see the outstanding structural performance of an insect able to carry its own weight multiplied by 100, and one cannot see, again, the intricate theatre of biological transformations taking place inside our own bodies and flesh. Medicine has access to this invisible scale and produces excellent cartographies of the human body, however only doctors can navigate those maps. Buckminster Fuller, among others, was well aware that architecture and urban design both happen in the invisible field where objective data are aggregated. Architecture can now be allowed to reach the invisible, deal with it and use it as a resource in the design and as a performative tool; the scale of the micro and even smaller is within reach. To keep visual representation as the first criterion of any architecture is a fallacy difficult to bear anymore.

Neither Abstract nor Concrete, Neither Craft nor Industry

The philosopher of technology Gilbert Simondon had some harsh lines against android machines or robots mimicking humans. He compares the role of a human organising machines with the role of an orchestra conductor. What do we understand by the openness of a machine? He observes that 'what resides in machines is human reality, human gesture fixed and crystallised in functioning structures . . . today's engine is a concrete engine, whereas yesterday's engine is an abstract engine'.[17] In 1958 Simondon linked abstraction to crafted and customised object: 'craft relates to the primitive phase of technical objects' evolution, which is to say the abstract phase: industry relates to the concrete phase. The customisation factor found in the work of the craftsman is unessential.'[18] Simondon despises customisation and cannot anticipate mass customisation through the digital, even as he endorses cybernetics and information theory both launched by his contemporaries Wiener and Shannon. For him, the design of the technological object cannot achieve an integral completeness in its relation to the environment: the complexity and the amount of data to be collected are way out of reach. So the *concreteness* of the technological object stops short of becoming a *completeness*. Of course, the context has changed and *big data*, total archive and real-time computing have radically changed the equation. Abstraction and concreteness are now closely articulated, as boundaries between artificial and natural, or immaterial and material, inorganic and organic are fading away. In a sense we have reached a point where it is no longer necessary to fight for the abolition of those oppositions.

The discourse of a Baroque and exuberant digital, and for an eccentric, voluptuous, sensual architecture is just another ideological attempt to continue separating and opposing what not only needs to be linked, but that is in fact already fusing together. The moderns needed the oppositional stance, the postmoderns also, as the journal called *Oppositions* reminds us.[19] It was followed by *Assemblage*.[20] Today, when material – and architectural – *assemblages* are fully automated, from the design to fabrication, we enter a sort of 'simulation as real' condition.

Very Literal

To overcome metaphor and to erase any metaphor, is a common claim found in every discourse. It is when a phrase such as 'this is not a metaphor' appears that I pay the most attention to any text: the denial of metaphor always hides a conceptual knot and a serious problem. For instance, Manuel DeLanda, in his *A Thousand Years of Nonlinear History*, after a first chapter explaining how human societies are to be compared with lava and rocks, asks: 'is it possible to go beyond metaphor and show that the genesis of both geological and social strata involves the same engineering strata?' He insists that 'all this [is] without metaphor'.[21]

Maybe for the first time, we now have the possibility to decide if we are in front of a metaphor or not: computational coding can tell if the simulated model of any phenomenon works in the same way as the phenomenon itself. The 'same way' means much more than an analogy or a metaphor; it is the identical process that goes on, or not – to the extent that soon it should become difficult to write or speak without saying something substantial and non-metaphorical. Or, to be brutal just for the sake of a thought experiment, computation will give the facts without the least metaphorical hint. We will speak and write about all other things, poetry, metaphysics or ontology, or fiction. However, architecture would go without all this fuzzy drapery of words and stick to its burden of acting on the physical world in the most precise, accurate and exact manner, through automated means from design to production. An automated and automatic architecture would be possible, that is, with no gap between thought and construction, which is even more exceptional than no gap between drawing and building. Integral simulation of reality via computation will soon erase the very last residues of metaphorical thought and replace it with a literal one. Literality, as opposed to metaphor, means that the computed is exactly what it computes, and what it computes is *reality*, with all the complexities, singularities, multilplicities, linear and nonlinear self-organizing and emergent qualities of reality. This is no small feat, in real time. Thus, technology, be it mechatronics, biotechnology, nanotechnology, bionanotechnology, robotics, nanobotics and so on, is no longer an artificial world separated from the world it seems to dominate but becomes embedded/incorporated in the natural-artificial world, on such a level that we no longer need concepts of *embodiment* and *incorporation*. In a sense we are close to having what I call the *WYSIWYG* of thought: the *What You See Is What You Get*: what you calculate is what it is, no less no more, with no loss (Fig. 15).

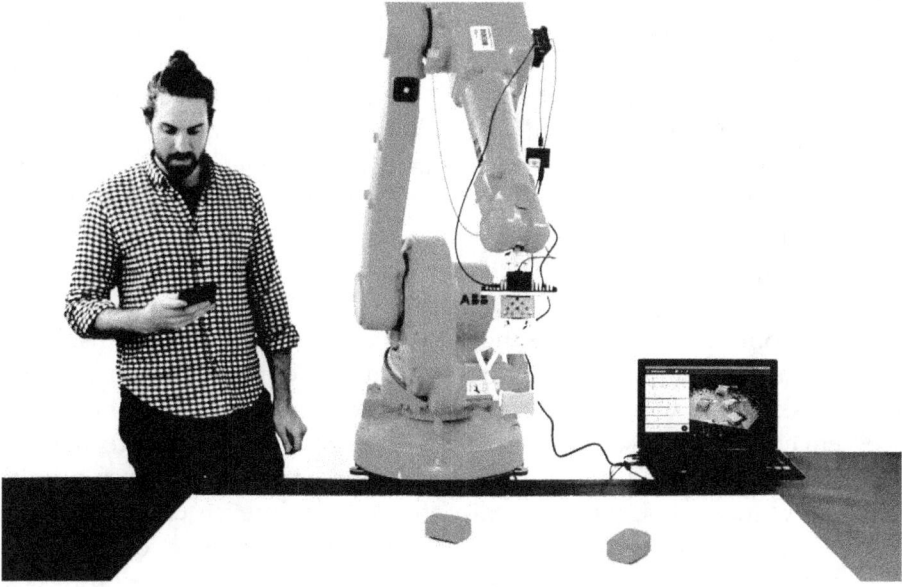

Figure 15 *DMR: A Semantic Robot Control Language*, Sebastian Andraos, Digital Knowledge Department, Final Thesis, ENSA Paris Malaquais, June 2015. Tutors: Ph.Morel/P.Cutellic.

Source: Sebastian Andraos, *DMR: A Semantic Robot Control Language*, ENSA Paris Malaquais <http://paris-malaquais.archi.fr/ecole/f/2015-pfe-dmr-a-semantic-robot-control-language-sebastian-andraos-dk/> (accessed 18 October 2016).

Architecture in a Time of Absolute Materialism

Are we, as architects, able to foresee a time when our work as 'something near enough' will finally become an absolutely material one?[22] Absoluteness, evolving from absolute data into absolute architectural execution and performance, could become the rule. In such a final evolutionary state, the residual autonomy that architects long to retain in architecture will shrivel and become vestigial at most. In other words, when designs attain an integral simulation of material reality, at any scale, from the visible to the invisible, then no difference will survive between the simulation and the simulated, between the virtual and the actual.

Thus, it would be a mis-judgement to consider the design 'immaterial' as opposed to the built project, considered as 'material': the design will encompass and grasp materiality with the same precision as the built project. It is something other than the so-called *virtual reality*; it is neither a *second life* experience with avatars, nor an augmented reality practice but true reality, no less and no more. Precursory ideas were indeed formulated two decades ago by architects promoting cyberspace architecture, with Marcos Novak as one of its main proponents.[23] Dismissed at the time, namely the 1990s, and submerged by practices aiming to create a continuum from file to factory, now standard, the cyberspace discourse has gained more credibility with the emergence of techniques enabling broad adoption. A comeback is underway as VR tools steadily mature for the mass market. However, a crucial difference in today's context is the momentum gathered by computational simulation.

By using heavy notions (such as *materiality*, which gives place to the even more complex one of *materialism*) within a philosophy or an ideology, one is at risk of losing one's bearings. Aiming at absoluteness when relativeness has been heralded as the condition of this world seems risky. One would rather leave it to religions, with all the chaos they engender, or to science when, for instance, it struggles to master *absolute* zero-degree temperature environments. To maintain a purely material stance asks, paradoxically, for highly intense conceptualisation, far from the usual 'concept' architects are so fond of using to describe whatever project they design. Exploring the sea of metaphysics may remain the task of philosophers, but architects still need a minimum of knowledge, culture and knowhow in such realms. They need it at least in order to understand where the thresholds are and in which specific contexts the practice of philosophy really helps to defend its autonomy. But it is exactly here where the very *reasonable* scheme starts to stall: the digitalisation of everything reshuffles and blurs the conventional limits on which the very idea of discipline is based.

The hypothesis of a *near absolute material architecture*[24] could prove to be useful since we cannot get away with a standard statement such as 'architecture is reaching a high level of control over matter and the material', which would indeed suffice to help bring a welcome overhaul of the discipline. This uneasy notion of 'absoluteness' opens up heuristic perspectives. The new rules of the game dictated by the power of simulation demand critical thinking that needs to consult theoretical practices such as philosophy and epistemology.

We predict that commissioners, developers, investors and politicians will increasingly reject approximation in buildings and will decline to buy sexy-looking but poor-performing projects. They will accept only thoroughly planned and built performative architecture where all the parameters are known, controlled, calculated and implemented. Let us trust Philippe Morel when he says:

> The time will come when architectural design and construction will only be taken seriously if entirely automated and checked by a computer. The prospect of this moment will be daunting to certain people, even though only a tiny percentage of buildings can actually be called architecture and global developments are governed not by this percentage but by the general majority.[25]

If architects want to participate both in the world of care and beyond, they must bring themselves near enough to an absolute material position where they take hold of automation, computation and an understanding of what is at stake with technologies. Robotics at large, as the most advanced embodiment of technology and technology of embodiment, will thus not reboot architecture unless architects start paying very close attention to what is going on. Without such scrutiny there is no way for them to find the exit. We often believe that robots run alone. And we believe that architecture needs brains. But automation needs brains, too. Thus automated architecture has to be a highly brainy trade. If architecture must focus on keeping an open space of undecidability and event in the design process, this should not be confused with some easy poetic licence and/or artistic bent, or whatever dandy-schizoid empty rhetoric by pseudo-avant-garde architects having misread Deleuze and Guattari. As an exquisite motto on a T-shirt warns us, 'Save The Drama For Your Mama', especially since we know that our *mother was a computer*, according to Katherine Hayles.[26]

Simulation, more exactly computational simulation, being one of the main places where the contemporary is at work, will give architecture a last chance by holding hands-free the potential of automation, from design to fabrication without losing any critical momentum. More than ever, theory in architecture is needed, if not architectural theory, provided that we understand 'theory' in a much more sophisticated way than what we have been accustomed to. If bots won't reboot architecture, architecture still has to take care of automation and robotics: real care.[27]

Notes

1. Wilson, *Robopocalypse*, p. 310; my emphasis. Cf. also *Robogenesis*. The most striking aspect of the novel is that the author does not invent anything. Wilson holds a PhD in Robotics from Carnegie Mellon University, Pittsburgh, USA, and makes a point in insisting that, in order to write his novel, he spent an extensive time in labs with artificial intelligence (AI) scientists. The novel just extrapolates a few years ahead of what scientists themselves work on presently. We tend to forget that scientists do make use of their imagination and project their research and finding in a time ahead. There is less fiction nowadays in science-fiction (SF) novels and movies, and more technological anticipation based on present achievements.
2. Recently, paleonthologists found new proofs of violence, war, torture and so on among Neanderthalian groups. The level of surprise among the contemporary scientists is as interesting as the discovered facts.
3. Radman, *Gibsonism*.
4. Geers et al., *Architecture without Content*, based on a series of studios run from 2011 to 2013 at the Graduate School of Architecture, Planning and Preservation (GSAPP) at Columbia University in New York City, USA, Accademia di Architettura Mendrisio, TU Graz (Graz University of Technology), EPFL(Ecole Polytechnique de Lausanne) and ENAC (Environnement Naturel, Architectural et Construit), a faculty of EPFL.
5. Morel, 'Computation or Revolution'.
6. Kraftwerk album *The Man-Machine, Die Mensch-Machine* (Kling Kmang-EMI Electrola-Capitol, 1978), track 1, 6:11, authors: Ralf Hütter, Florian Schneider and Karl Bartos.
7. An example is given by Meillassoux with his *Science Fiction and Extro-Science Fiction*.
8. Goodman, *Ways of Worldmaking*.
9. Tellios and Psaltis, 'Visceral Materiality of the Digital', pp. 104, 106.
10. Köhler, *Gestalt Psychology*.
11. Sennett, *The Craftsman*.
12. Pallasmaa, *Thinking Hand*.
13. Watanabe, 'Computer Programs versus the Human Brain'.

14. This fact has become common place up to the point where it can be found in a specialised field as plant design and planning: we see it explicitly expressed, for instance, in a marketing piece of the company Environmental intellect (Ei) 'P&ID = Piping and Instrumentation DATA (not Drawing) 24 July 2015 by Shane Kling. It's time we stop referring to P&IDs as piping and instrumentation drawings/diagrams and instead as piping and instrumentation DATA. We are on the cusp of a "paradigm shift" – where the "D" in P&ID will be a distant memory for those who reference diagrams or drawings. There will be numerous challenges to overcome in the "P&I DataShift", but I believe there has never been more justification to undertake such an initiative' <http://env-int.com/pid-piping-instrumentation-data-not-drawing> (accessed July 2015). Ironically, the last ones to capitalise, mainly for their communication, on diagrams, are architectural firms well trained for such practice when working for OMA/Koolhaas, such as BIG and their numerous less talented imitators.
15. Varenne, 'Nature of Computational Things'. Varenne's epistemology of simulation is a major source for whoever wants to tackle these questions. In his article 'What Does a Computer Simulation Prove?' he writes: 'when you read (Von Neumann 1951), you see that analog models are inferior to digital models because of the accuracy control limitations in the first ones. Following this argument, if you consider a prototype, or a real experiment in natural sciences, is it anything else than an analog model of itself? The test on the prototype is a real experiment. But is it something different and better than the handling of an analog model? So the possibilities to make sophisticated and accurate measures on this model – i.e. to make sophisticated real experiment – rapidly are decreasing, while your knowledge is increasing. These considerations are troublesome because it sounds as if nature was not a good model of itself and had to be replaced and simulated to be properly questioned and tested! It looks as if it was not possible any more to end a paper on simulation by reassuringly using the traditional word: "Simulation will never replace real experiments"' (p. 553).
16. Gramazio and Kohler, 'Thriving Digital Materiality'. 'With our thesis that the robot will enable a new architectural form of expression, the projects presented in this book reveal the widest possible spectrum of material articulation' (p. 18).
17. Simondon, *Du mode d'existence des objets techniques*, p. 12. 'Ce qui réside dans les machines, c'est de la réalité humaine, du geste fixé et cristallisé en structures qui fonctionnent.'
18. Simondon, *Du mode d'existence des objets techniques*, p. 24. 'L'artisanat correspond au stade primitif de l'évolution des objets techniques, c'est-à-dire au stade abstrait; l'industrie correspond au stade concret.'
19. *Oppositions* was an architectural journal produced by the Institute for Architecture and Urban Studies, Manhattan, New York, USA, from 1973 to 1984. Twenty-six issues were published during its eleven years of existence.

20. *Assemblage* was an architectural theory journal published by MIT Press from 1986 to 2000. Forty-one issues were published in total.
21. DeLanda, *A Thousand Years of Nonlinear History*, pp. 59, 62.
22. The idea of 'something near enough' comes from the title of Jaegwon Kim's essay *Physicalism, or Something Near Enough*.
23. Novak, 'Liquid Architectures in Cyberspace'.
24. This obviously bears no relation with Aureli's *Possibility of an Absolute Architecture*.
25. Morel, 'Computation or Revolution', p. 84. Morel continues: 'So architectural robotics should mainly target this group, not only because it can, but because it has no choice.' Morel is again astute when he observes: 'The romantic associations between architecture and technology deflect industrial economy problems onto an aesthetic theory of widespread creativity. The theory holds that robots and machines are the Neo-Ruskinian gadgets of those types of architect who are incapable of accepting their own obsolescence, as convinced of their own infallibility as booksellers as they are of the need for a greater human touch than the algorithms of Google and Amazon. Of course, reality proves every minute that the opposite is true, and a mere glance at the situation architecture finds itself in shows the extent of this misjudgement.'
26. Hayles, *My Mother Was a Computer*.
27. Cf. Girard, 'L'architecture, une dissimulation'.

Bibliography

Andraos, Sebastian, *DMR: A Semantic Robot Control Language*, ENSA Paris Malaquais <http://paris-malaquais.archi.fr/ecole/f/2015-pfe-dmr-a-semantic-robot-control-language-sebastian-andraos-dk/> (accessed 18 October 2016).

Aureli, Pier Vittorio, *The Possibility of an Absolute Architecture* (Cambridge, MA: MIT Press, 2011).

DeLanda, Manuel, *A Thousand Years of Nonlinear History* (New York: Zone Books/MIT Press, 1997).

Geers, Kersten, Joris Kritis, Jelena Pancevac, Giovanni Piovene, Dries Rodet and Andrea Zanderigo, eds, *Architecture without Content*, five volumes (London: Bedford Press AA Publications, 2015).

Girard, Christian. *Architecture et concepts nomades, traité d'indiscipline* (Brussels: Mardaga, 1986).

Girard, Christian, 'L'architecture, une dissimulation. La fin de l'architecture fictionnelle à l'ère de la simulation intégrale' (Architecture, a Dissimulation: The End of Fictional Architecture in an Era of Integral Simulation), in *Modéliser & Simuler. Epistémologies et pratiques de la modélisation et de la simulation*, Tome 2, sous la direction de F.Varenne, S.Dutreuil, Ph. Huneman, M. Silberstein (Paris: Editions Matériologiques, 2014) pp. 246–92.

Goodman, Nelson, *Ways of Worldmaking* (Indianapolis: Hackett Publishing Company, 1978).

Gramazio, Fabio and Matthias Kohler, 'Thriving Digital Materiality', in *The Robotic Touch*, ed. Fabio Gramazio, Matthias Kohler and Jan Willmann (Zurich: Park Books, 2014).

Hayles, Katherine, *My Mother Was a Computer: Digital Subjects and Literary Texts* (Chicago: University of Chicago Press, 2005).

Kim, Jaegwon, *Physicalism, or Something Near Enough* (Princeton: Princeton University Press, 2005).

Kling, Shane, 'Electronic P&ID Markups that Save Time and Money Today (Using Free Software)', 1 July 2015, posted on the blog *Environmental Intellect* <http://www.env-int.com/blog-stream/electronic-pid-markups-that-save-time-and-money-today-using-free-software> (accessed 2 May 2016).

Köhler, Wolfgang, *Gestalt Psychology* (New York: Liveright, 1929).

Meillassoux, Quentin, *Métaphysique et fiction des mondes hors-sciences* (Paris: Editions Aux Forges de Vulcain, 2012); trans. Alyosha Edlebi, *Science Fiction and Extro-Science Fiction* (Minneapolis: Univocal Publishing, 2015).

Morel, Philippe, 'Computation or Revolution', in *Made by Robots Challenging Architecture at a Larger Scale*, A.D. May/June 2014, Profile 129, guest edited by Fabio Gramazio and Matthias Kohler (London: Wiley, 2014), pp. 76–87.

Novak, Marcos, 'Liquid Architectures in Cyberspace', in *Cyberspace: First Steps*, ed. Michael Benedikt (London: MIT Press, 1991), pp. 225–54.

Pallasmaa, Juhani, *The Thinking Hand: Existential and Embodied Wisdom in Architecture* (London: John Wiley & Sons, 2009).

Radman, Andrej, *Gibsonism: Ecologies of Architecture* (Delft: TU Delft, 2012).

Sennett, Richard, *The Craftsman* (New Haven: Yale University Press, 2009).

Simondon, Gilbert, *Du mode d'existence des objets techniques* (Paris: Aubier, [1958] 1989), trans. Ninian Mellamphy, *On the Mode of Existence of Technical Objects* (Minneapolis: Univocal Publishing, 2016).

Tellios, Anastasios and Stylanios Psaltis, 'The Visceral Materiality of the Digital and its Biological Poetics', in *Surface: Digital Materiality and the New Relation between Depth and Surface*, ed. Nikolas Patsavos and Yannis Zavoleas, EAAE European Association for Architectural Education Transactions on Architectural Education no. 48, EAAE Subnetwork on Architectural Theory Meeting in Chania, September 2010 (Athens: Futura Publications/Technical University of Crete, 2013), pp. 104–11.

Varenne, Franck, 'What Does a Computer Simulation Prove?' in *Simulation in Industry, Proceedings of the 13th European Simulation Symposium*, ed. Norbert Giambiasi and Claudia Frydman, Marseille, France, 18 October–20 October 2001 (Ghent: SCS Europe Bvba, 2001), pp. 549–54.

Varenne, Franck, 'The Nature of Computational Things. Models and simulations in Design and Architecture', in *Naturaliser l'architecture Archilab 2013*, ed. Frédéric Migayrou and Marie-Ange Brayer (Orléans: Editions HYX, 2013), pp. 96–105.

Watanabe, Makoto Sei, 'Computer Programs versus the Human Brain', *L'Arca International* Vol. 271 (July 2011), pp. 2–33.

Wilson, Daniel, *Robopocalypse* (London: Simon & Shuster, 2011).

Wilson, Daniel, *Robogenesis* (London: Simon & Shuster, 2014).

CHAPTER 5

The Convivial ART of *Vortical* Thinking

Keith Evan Green

Ask a young child to draw a robot, and (almost without exception) he or she responds by delineating with pencil or crayon a humanoid robot – symmetrical, boxy and awkward, part us, part machine. Indeed, this representation of robotics is the one that most adults maintain: robotics defined by anthropomorphism, most manifested as humanoid robots. It follows that a robot intended to care for you is presumed to be, again, a humanoid-robotic servant, or maybe a friendly, animal-like robotic companion, such as a furry white seal or a plush teddy bear. This presumption comes partly from media – the many humanoid robots featured in movies, television shows and comic strips – and partly from that ancestor of the humanoid robot, the robot before electricity. Indeed, one third of Lisa Nocks's history of robotics, *The Robot: The Life Story of a Technology*,[1] is devoted to such mechanical contrivances resembling human beings and animals – the precursors of electronic offspring that, for Nocks, is almost exclusively that artificial breed recognisable as the humanoid robot.

Without face or fur, with no or few words, behaving in a manner that is only vaguely familiar to us, an altogether different kind of caring robotics is being realised within my Architectural Robotics Lab (Cornell University, New York, USA) and in situ within the Roger C. Peace Rehabilitation Hospital (Greenville Hospital System University Medical Center, South Carolina, USA). Very much a continuing project undertaken by this partnership, *home+* (Fig. 16) is a suite of networked, distributed robotic furnishings integrated into existing domestic environments (for ageing in place) and healthcare facilities (for clinical care). *home+* is our reply to the question, How can our everyday environments be outfitted with assistive robotics promoting independent living?

A response to this question, coming from my trans-disciplinary design research team representing Architecture, Human Factors Psychology,

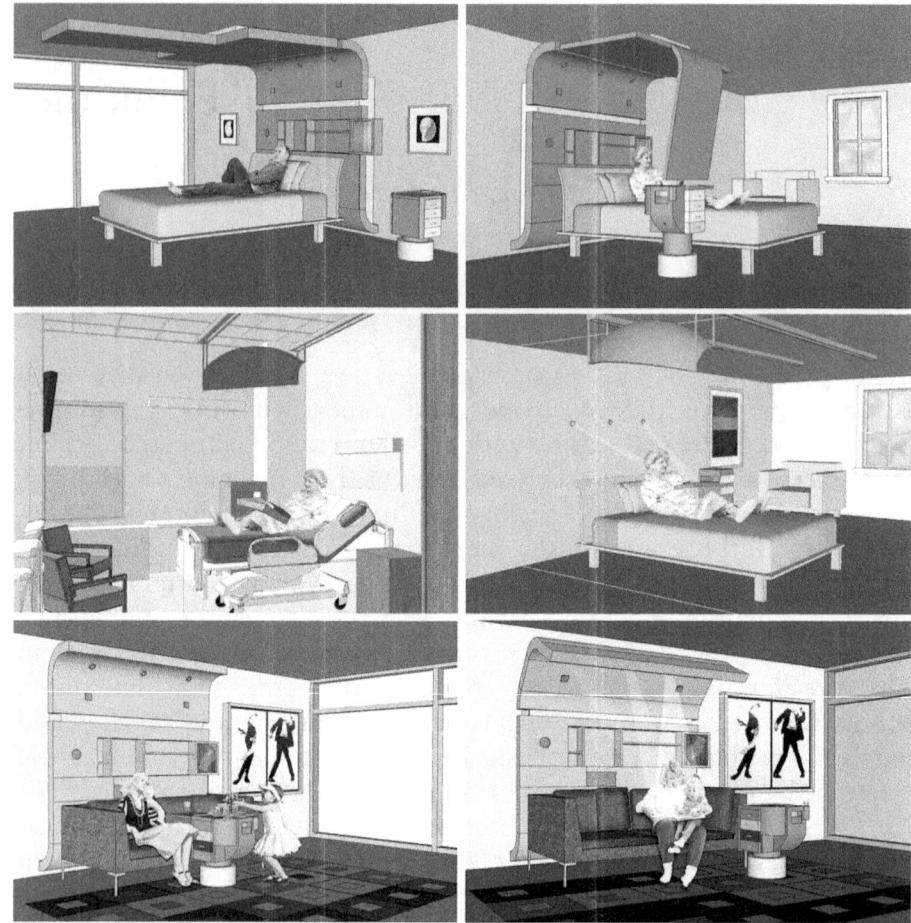

Figure 16 *home+*, in concept, at home and hospital.
Source: author.

Robotics, and Medicine, has to date been manifested as several digital simulations as well as to-scale and full-scale prototypes of the home+ suite of furnishings.[2] In conceptualising *home+* initially, the team quickly prototyped the *home+* concept as an integrated vision, in all its complexity: the design of various furnishings interacting with one another and their users in a specific built environment, a studio apartment that we had designed purposefully for this research. Unlike the typical user engineering approach, we elected to realise this early prototype before we had acquired sufficient data to responsibly guide it (in a measured, industrial design and engineering sense). The virtue of our quick and

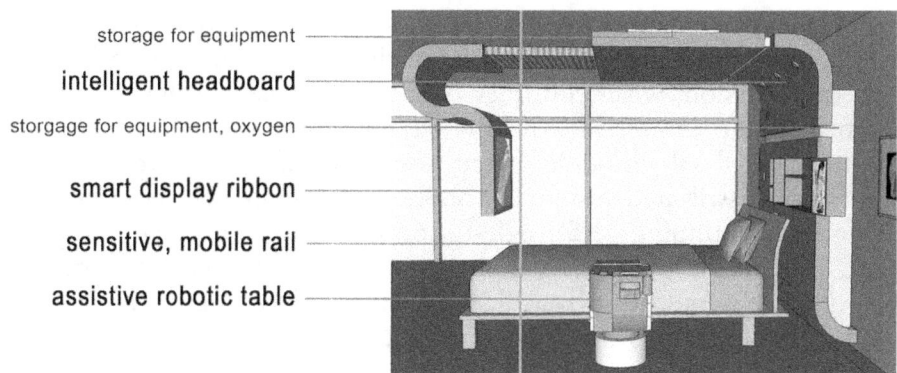

Figure 17 *home+* components.
Source: author.

early prototyping activity permitted us to get past the inevitable mishaps and unfruitful pathways to a *home+* design.

1. An *Assistive Robotic Table* gently folds, extends and reconfigures to support therapy, work and leisure activities.
2. An *Intelligent Headboard* adapts to the profile of the patient's back, morphing its supportive surface periodically to alter the patient's position as a vehicle to ensure against bed sores. The *Intelligent Headboard* also offers storage, accommodates an oxygen tank as required and provides intelligent lighting for reading, work activities and ambient illumination.
3. A *Sensitive, Mobile Rail* is not unlike the vertical tubular-metal post found in metro cars – except that this one moves with you, providing more or less assistance, like the arm of a friend, as the user moves between one location within the dwelling to another location, tracing the most common paths travelled in the course of a day. As envisioned, the *Sensitive, Mobile Rail* recognises how much assistance you may need, and 'backs off' from supporting the user when it senses the user is managing fine with less of its help.
4. *Intelligent Storage* manages, stores and delivers personal effects (including medical supplies), and communicates to caretakers when eyeglasses and other belongings are not moved over a period of time.
5. A *Personal Assistant* (not shown in these figures) retrieves objects stored around the room and away from the bed. The robot uses a vision-based recognition system via wireless communication to ensure that the robot retrieves the correct item.

These five components of the *home+* suite of furnishings recognise and communicate with one another in their interactions with humans and, we envision, with still other *home+* components. Figures 16 and 17 provide an impression of how the system may look and behave (and not its fixed, pre-determined design). Indeed, towards realising the larger, distributed system of robotic furnishings, the research team persists in continuing iterative design and evaluation of the *home+* concept, generating alternative, conceptual visualisations of *home+* and prototyping numerous physical, functioning prototypes of its individual components, identified here and otherwise.

The Assistive Robotic Table (ART)

As the research team wished to focus its effort not only on the big vision of the *home+* concept but on a more developed and refined component of it, we dedicated considerable efforts on the *Assistive Robotic Table* (ART), given its relative complexity within *home+*. Our aim was to develop this key component as a fully functional prototype developed with the participation of clinical staff.

ART (Fig. 18) is a hybrid of the typical nightstand found in homes, and the over-the-bed table found in hospital rooms. We envision ART integrated into the domestic routines of its users, even as users transition from home to clinic and, it is hoped, home again. Our research team hypothesises that users employing ART as part of their domestic

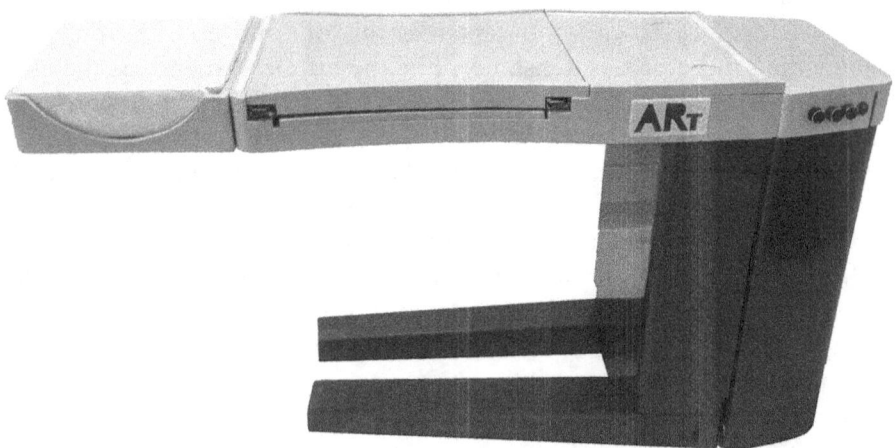

Figure 18 The assistive robotic table (ART) – our final, fully functioning prototype.
Source: author.

landscape will live independently, longer. Moreover, we expect ART to free familial caregivers from performing certain arduous tasks for ART's target populations, allowing caregivers to devote more energies to meaningful, human interaction with ART's users. In medical facilities, ART aims to augment the rehabilitation environment by improving patient well-being, rehabilitation and staff productivity (in this vexing moment of limited resources).

Along with the larger *home+* environment to which it belongs, ART benefits from the convergence of advanced architectural design, computing and robotics largely absent from prior efforts in assistive technologies. In particular, this *enabling technology* is not distributed everywhere in the physical environment but *where it is needed*. It is not intended to be invisible (as per ubiquitous computing) but *visible, attractive and integral* to the home and the patient room *by design* and it is not meant to serve as a means for surveillance but rather as environmental support that recognises and dignifies what people can do for themselves.

These attributes of ART and the larger *home+* environment are of a kind identified by Donald Norman as 'the next UI breakthrough', defined by 'physicality', and accomplished with 'microprocessors, motors, actuators, and a rich assortment of sensors, transducers, and communication devices'.[3] In broad theoretical terms, Nicholas Negroponte anticipated ART and the larger *home+* environment in the 1970s in his vision of a 'domestic ecosystem' that regulates aspects of 'environmental comfort and medical care'.[4] More recent inspirations for ART include Malcolm McCullough's plea for 'architecturally situated interactions', which 'permit the elderly to "age in place" in their own homes'.[5]

Physically, ART is a significant development of the over-the-bed table commonplace in hospital patient rooms. What distinguishes ART from the conventional over-the-bed table is its novel integration of physical design and functioning, coupled with an interactive human–object interface (Fig. 19). Integral to ART is a novel, plug-in, *continuum-robotic* therapy surface that helps patients perform upper-extremity therapy exercises of the wrist and hand, with or without the presence of the clinician (refer to Figs 21 and 24). Considered in depth shortly in this chapter, ART's *therapy surface* was recognised as a key requirement of ART, following our early conceptualisation of the larger *home+* vision, and as a significant outcome of our ethnographic investigations in the clinical setting (or in the 'wild', as human–computer interaction (HCI) defines territories situated outside the lab). These ethnographic investigations suggested the promise of the *therapy surface* for rehabilitating

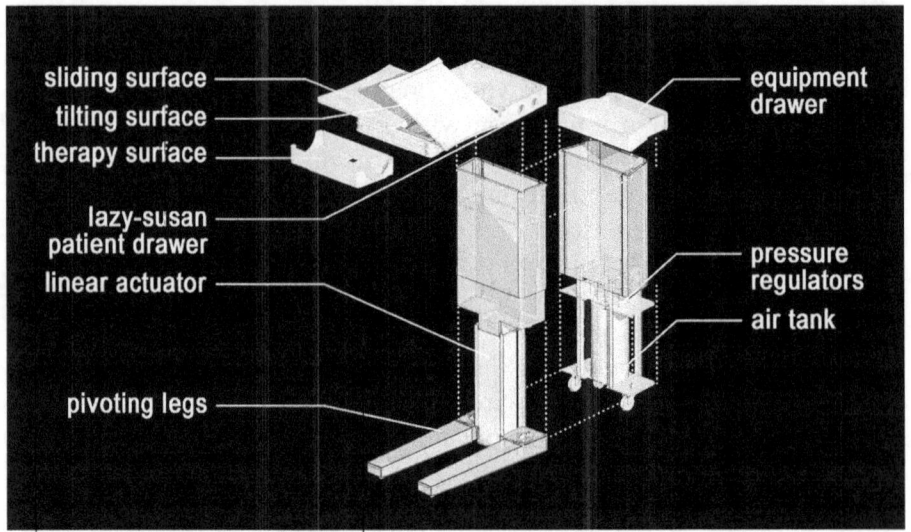

Figure 19 ART's key components.
Source: author.

in particular the upper extremities (upper limbs) of post-stroke patients. Beforehand, it is useful to become acquainted with ART through a scenario, a hypothetical narrative telling the happy story of Amy meeting ART for the first time. Amy had a stroke and has right hemiplegia (paralysis of the right extremities) and aphasia (speech and language problems). After treatment in the hospital, Amy returned home, fitted with *home+*, the same suite of robotic-embedded furnishings that supported her in her patient room. *home+* has uploaded Amy's preferences as acquired from her hospital stay, and modifies these preferences and those of her caregivers over time to best support Amy's recovery. Amy relies on the *Assistive Robotic Table* of *home+* every day (Fig. 20): its *therapy surface* helps Amy rehabilitate her arm; ART tilts and changes height to best accommodate Amy's activities; ART's non-verbal lighting cues remind Amy to take her medications; and ART learns and adapts to the gestures, a form of nonverbal communication that Amy performs with her right arm as she regains more capacity in moving it. ART logs Amy's reading time as a wellness metric, and ART initiates storage of Amy's reading glasses when she has finished reading. These functions and others help Amy improve more quickly. ART's components recognise and partly remember, communicate with, and cooperate with her, her caregivers and the other components of *home+*, empowering Amy to remain in her home for as long as possible, even as her physical capabilities alter over time; and, in more grave circumstances, affording

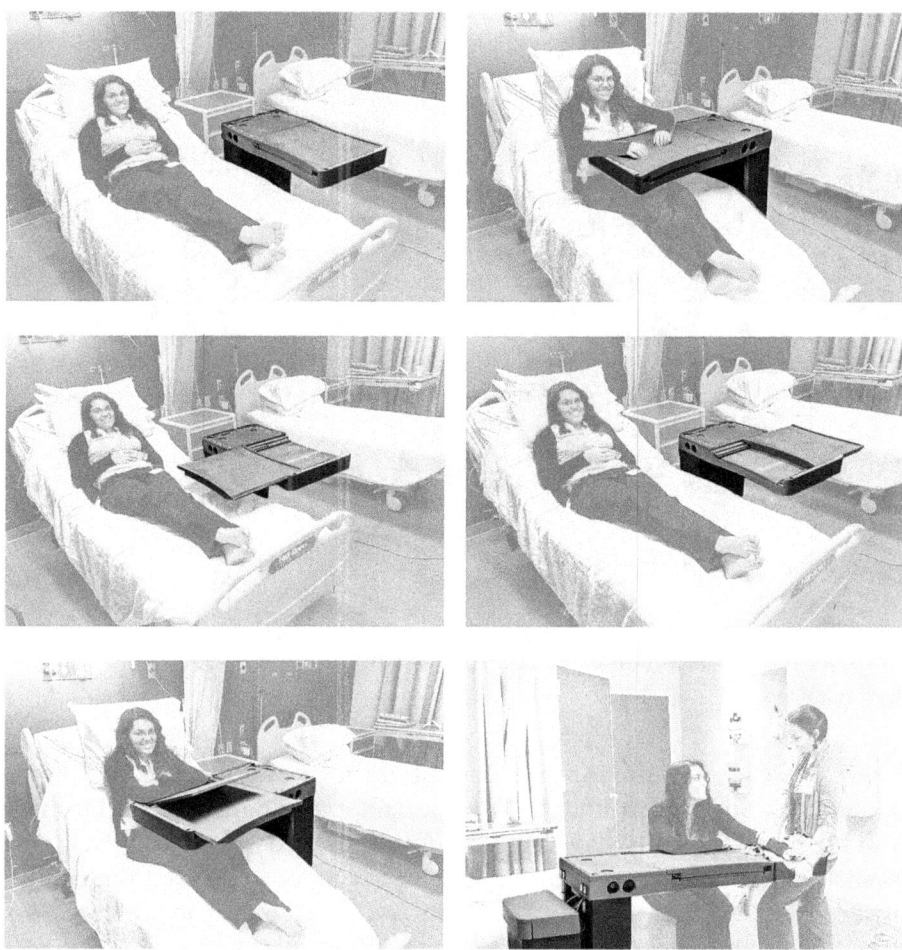

Figure 20 In the hospital with ART and 'Amy': six instances.
Source: author.

Amy some semblance of feeling 'at home' as she moves to an assisted facility also equipped with *home+*.

While this storytelling acquaints us with ART, we also know from life experience that good stories don't always translate well to the real world. Accordingly, in developing the full-scale, full-functioning ART prototype, my research team needed to subject ART to the clinical environment where healthcare assistance is delivered and received (see Fig. 21). In so doing, we conducted five iterative phases of research.[6] We involved healthcare physicians and occupational, speech and physical therapists. We conducted these research activities within the aforementioned studio apartment of our own design, our purpose-built *home+* lab at the

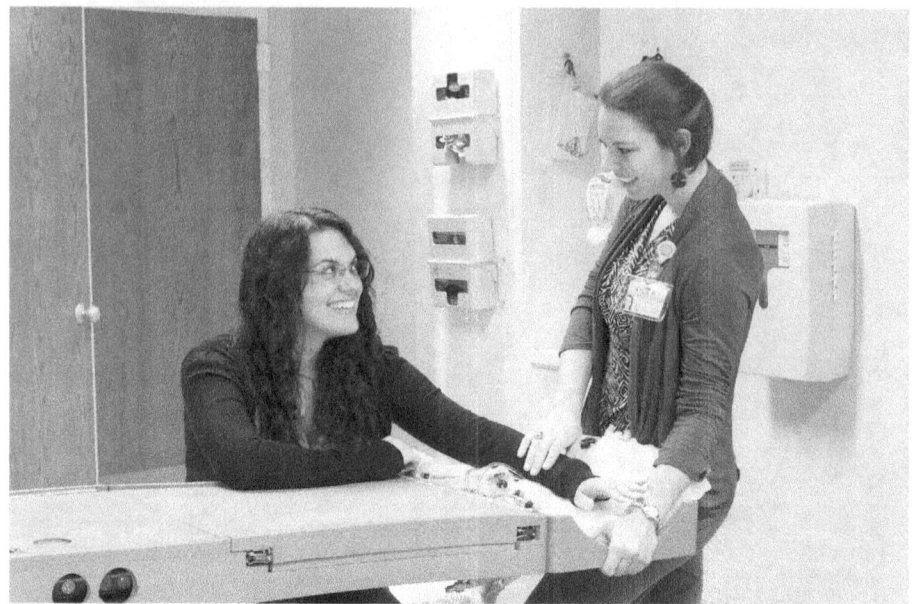

Figure 21 ART in the hospital, undergoing evaluation.
Source: author.

Roger C. Peace Rehabilitation Hospital of the Greenville Health System (Greenville, South Carolina, USA). As patient populations of rehabilitation hospitals typically have a high number of post-stroke patients, and as such patients partake in therapies employing over-the-bed tables, the post-stroke population was a particularly apt target for our research on the design of a forward-looking, over-the-bed table.

Convivial or Enslaving

There is no doubt that assistive technologies like ART and the larger *home+* suite of robotic furnishings to which it belongs is urgently needed. The global population continues to age, healthcare costs have been increasing, and there is a smaller segment of society to both care for and pay for the well-being of older and clinical populations. It is estimated that by 2025, in the United States as in many nations around the globe, there will be a shortage of physicians and nurses, and by 2040 there will be more than 79.7 million adults in the United States alone who will be 65 or older – a 92 per cent increase from 2011. *home+*, the response from my research team, strives to reduce the burden on healthcare staff (to deliver their services) and a decreasing tax base (to pay for this service), while meeting patient needs and ensuring the well-being of an increasingly older population.

Given these alarming trends, few would doubt the need for such assistive technology; however, the character of it – its behaviours and appearance – is a topic of continued debate. The lively debate over the character of caring robotics was anticipated in a remarkable book, *Tools for Conviviality* by Ivan Illich, in which the author recognises two trajectories for new technologies: one aiming 'to extend human capability', and the other, used 'to contract, eliminate, or replace human function'.[7] Illich classifies the former as 'convivial tools' that foster 'self-realisation'.[8] Convivial tools, writes Illich, 'enable the layman to shape his immediate environment'.[9] 'Manipulatory tools', instead, reduce 'the range of choice and motivation' for the user, destroying his or her capacity to creatively, 'playfully' engage society.[10] For Illich, the convivial tool is exemplified by the public library, an information machine providing resources openly and freely to the public.

In *Tools for Conviviality*, Illich refers to the historical hypothesis that machines were meant 'to replace slaves'. For the discussion of this chapter, note that the word robot comes from the Czech *robota* meaning 'forced labor'.[11] What both astonishes and distresses Illich is that machines, intended to mostly 'replace slaves', ultimately 'enslave' those they intend to free.[12] Illich locates the blame for this unexpected turn in the designers of such machines; that, too often with machines, it is left to their 'designers to determine the meaning and expectation of others' in human–machine interaction without the participation in the design process of a representative group of likely users.[13] Illich's plea is instead to

> give people tools that guarantee their right to work with high, independent efficiency, thus simultaneously eliminating the need for either slaves or masters and enhancing each person's range of freedom. People need new tools to work with rather than tools that 'work' for them.[14]

Indeed, *Tools for Conviviality* was published in 1973, and perhaps recent history suggests that those who design and build machines strive, as my design research team has, to become more enlightened (more convivial) in their approach to design. Co-design, iterative design, human-centred design and participatory design are common terms employed by a not insignificant part of the academic community and industry entrusted to design today's machines. This more inclusive process of designing the most complex and sensitive of artefacts, caring robots, is fundamentally what this chapter hopes to capture.

Robotics that Give

Clearly, home+ and its key component, ART, represent a breed of robotics quite apart from humanoid robots. Whereas the humanoid robot is meant to resemble us and is (given its appearance) expected to behave like us, home+ is a suite of furnishings distributed through an everyday physical space – robotics made integral with furnishings in familiar domestic rooms that, at least typologically, are familiar to us. As well, many humanoid robots (and most industrial robots) rely on rigid links to move, whereas key parts of ART and home+ depend on softer means to more fluid movements, as afforded technically speaking by pneumatics (compressed air, the pressure of which is digitally regulated) and less frequently tendons (steel cables pulled by digitally controlled motors). ART's *therapy surface*, in particular, requires a delicate but exacting 'touch' to impart confidence to patients and their caregivers that it can assist them.

To meet these demands, rigid-link robotics is not a likely candidate. The shortcomings of rigid-link robotics (commonplace in factories and often used for humanoid robots) are comprised of a series of heavy links, few in number, actuated by electric motors. The movements of rigid-link robots are correspondingly stiff but precise, which contrasts significantly with the less distracting, nuanced, and even graceful movements exhibited by ART's *therapy surface*, actuated by pneumatics. A rigid-link, electro-mechanical system is contrastingly heavy and – worse for unstructured environments – unyielding upon contact with another physical mass. In simplest terms, these rigid-link robots don't 'give' when they collide with people and their physical property.

Consider, then, two of the most unstructured environments inhabited by two of the most vulnerable human populations, both of which define the *home+* context: dwellings inhabited by older people, and healthcare facilities occupied by patients. In these two demanding contexts, the *home+* suite of furnishings must not only move in a manner that is safe, comprehensible and pleasing to their users, but also be able to 'give' – to be compliant upon contact with people and physical things. In such an intimate setting as *home+*, robotics will inevitably collide with people and their physical property; consequently, they must be made compliant to avoid costly and potentially grave consequences.

Compliant, fluid, graceful kinematics is characteristic of the robotics classification called *continuum robotics* referenced earlier as exemplified by ART's *therapy surface*. An emerging subfield, *continuum robotics*, describes the kinds of robotics with smooth, compliant backbones that render their movement fluid, natural and more life-like. Overall, the smooth movement and softness of *continuum* robots lend themselves to intimate and elicited interactions with users, while stiff and hard, rigid-linked robots are better suited to the demands for strength, repetition and accuracy of industrial applications. As a flexible, continuous 2-D *programmable surface*, ART's *therapy surface* represents a new class of *continuum robotics* capable of contributing to the formation of physical space within the built environment scale.

Recognise that the therapy surface of ART bends in more than one direction (it morphs across its surface). Consequently, the *therapy surface* can assume a multitude of shapes that aren't stiff like commonplace industrial robots, or limited by bending in section alone. Rather, ART's *therapy surface* smoothly morphs in two dimensions (like the sea). Given its continuous and highly manoeuvrable backbone, a 2D continuum robot like ART's *therapy surface* is classified as *invertebrate*. The control of such a robot – the use of sensors, actuators and algorithms to configure it – is a difficult matter. Unlike the control of rigid-link robots, the control of continuum robots often involves controlling for not only bending but also the extension/contraction of the physical mass, as compared to simply controlling the hinging of an axial joint in the typical rigid-link robot, comprised of a limited number of joints.

Let's consider the bending in two different continuum robots in concept: one of these is *one-dimensional* (1D – it bends in section only), and the other is *two-dimensional* (2D). If you take two short segments of a continuum robot and connect them, head-to-head, you then have a continuum robot in the form of an extended line (much like an elephant's trunk). The bending that occurs within each segment of such a linear robot is defined as *one-dimensional*. (A well-known example of this kind of robotic appendage is the *OctArm* developed by Ian Walker, my close research collaborator.) To make the kinematics of a one-dimensional (trunk-like) continuum robot relatively less cumbersome to define, it *theoretically* reconfigures as the 'essentially invertebrate' snake just described, having a much larger number of smaller links that bend at distinct and well-defined points in such a way that each of its segments bends at constant curvature. If, instead of connecting segments head-to-head, you connected them in such a way that they form a surface (for example, three segments to make a triangular surface, four segments to make a quadrilateral surface), and then co-join several such surfaces, you have the continuum robotic surface exemplified by the compliant *therapy surface* of ART. The bending that occurs within each surface of this continuum robotic surface is defined as *two-dimensional*. As such, the kinematics for such a continuum robotic surface is highly complex, as a continuum robot has, theoretically, an infinite number of joints across a surface, rather than along a line (as in *OctArm*) and, consequently, an infinite number of ways in two dimensions to assume its goal configuration. To make the kinematics of a two-dimensional continuum robot surface (like ART's) relatively less cumbersome to define, it *theoretically* reconfigures as a surface of many snakes joined to one another, at their sides, so that stretches of this surface has constant curvature. Consequently, the problem of determining the shape of this robotic surface, its kinematics, is simplified compared to a determination made on the basis of the curvatures of a considerable number of very short segments of the surface. But even when characterising the ART surface (and continuum robotics generally) as *essentially invertebrate* robots exhibiting constant curvature, the number and complexity of terms involved in their kinematics can be formidable. (The kinematics for such a surface is reported in a technical paper authored by our research team.)[15]

A more tangible way of understanding continuum robotic surfaces, and so, ART's *therapy surface*, is through the qualitative study of functional, physical prototypes. To realise ART's *therapy surface*, my research team built a working prototype from a square surface of conventional foam (36 cm on a side), two McKibben actuators, some zip ties to fasten the actuators to the foam surfaces and a Kinect from Microsoft's computer gaming system. McKibben actuators are artificial muscles (that is, bladders) that expand and contract as they are filled or depleted of compressed air. According to the recipe advanced jointly by my lab and by the co-joined lab of my close collaborator, Ian Walker, we built our own McKibben actuators using high-temperature, silicone rubber tubing, expandable mesh sleeves, nylon reducing couplings and nylon, single-barbed, tube-fitting plugs (refer to Fig. 23). Digital pressure-regulators control the precise pressure of air delivered to the actuators. Flexible tubing connects the system of actuators, regulators, the air compressor (a common air tank will do) and the control computer (a common laptop will do). We attached two actuators to the square surface of sheet foam in four different orientations: running the actuators in parallel, forming a perpendicular ('Swiss') cross with them and making 45 and 90 degree angles of them. We marked twenty-five points on the square surface that we tracked with the Kinect sensor, each point being either 8 cm or 10 cm from the previous point. For each muscle arrangement, two sets of data were collected. The first set of data measured the depth (the distance of the point on the surface from the sensor) of the *un-actuated surface*. The second set measured the depth of the points on the *actuated* surface by using a computer programme that allowed the user to select feature points in two different image frames. The selected points in the first image frame captured the depth for the *un-actuated* surface at each point, and the selected points in the second image frame captured the depth for the *actuated* surface at each point. With respect to the surface, the Kinect sensor was positioned directly in front of it, at the same height of it and parallel to its surface. To calculate the distance that each point on the surface moved in the x-direction (that is, the height that the surface moved), the actuated data was subtracted from the un-actuated data. Once the x-distance data had been calculated for each point, the x-distances for the same points were calculated using the appropriate (that is, muscle-arrangement dependent) kinematic model; then, the mean square error (MSE) between the two distances was calculated. Additionally, the MSE between the same kinematic data and the physical data for an un-actuated surface, which would have x-distances of zero, was calculated for comparison. This MSE would represent the worst-case scenario and should, therefore, be larger than the MSE for the actuated surface. If all this is a bit difficult to follow, Figure 22 conveniently visualises the outcome that the theoretical kinematics for the different muscle arrangements reasonably approximated the data we collected from testing the physical surface.

The analytic exploration just described guided the more complex design of ART's *therapy surface*. Rather than two actuators bending a square sheet of foam, the *therapy surface* (Fig. 23) was actuated by multiple actuators and required exacting behaviours that match five therapeutic exercises for the upper limbs performed by post-stroke patients, and guided by physical therapists working without any medical devices during physical therapy sessions.

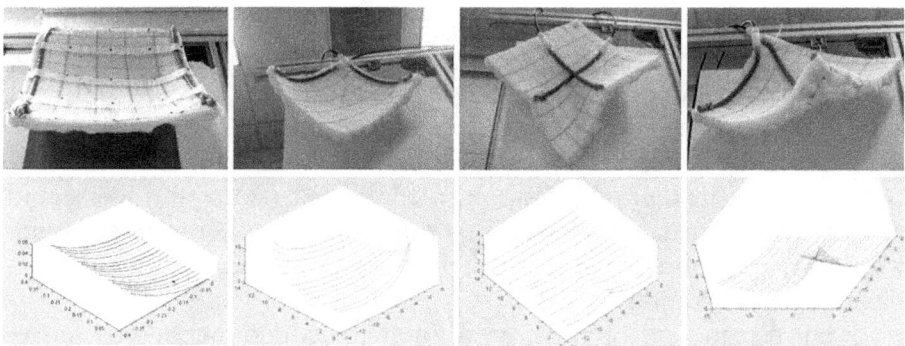

Figure 22 Evaluating the therapy surface's kinematic models: qualitative versus simulated results.

Source: author.

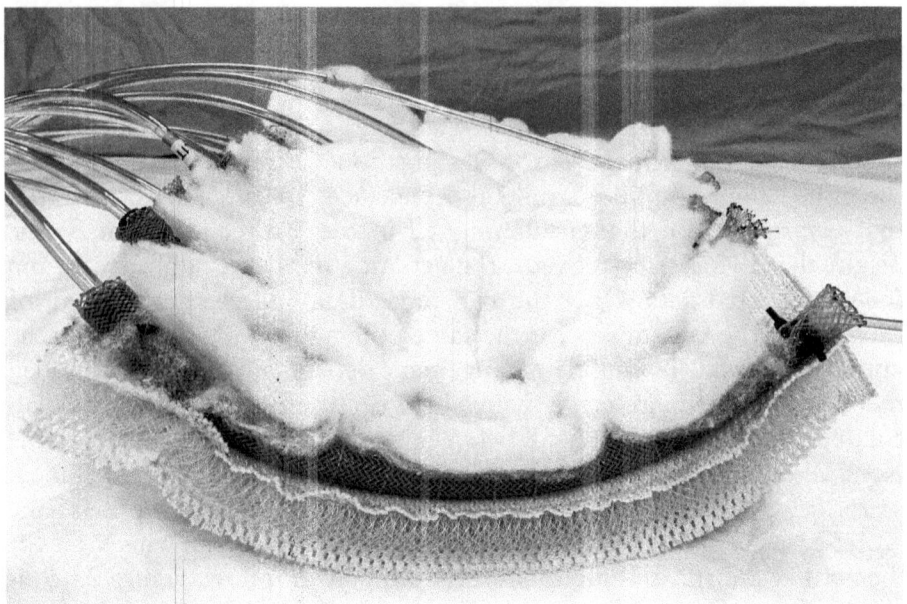

Figure 23 ART's therapy surface – working prototype.
Source: author.

With these starting points in mind, the research team iteratively designed and evaluated the therapy surface in tandem with the larger five-phase ART investigation (referenced earlier), translating the early studies (with the foam square and two actuators) into a rehabilitative robotic surface. Notably, in phase one, specifically through observations of therapy sessions, the research team drew three inferences that would significantly impact the therapy surface's design requirements. First, we inferred that our continuum surface could perform many therapeutic exercises, including wrist flexion and extension, forearm pronation and supination, flexor synergy, shoulder flexion, shoulder rotation and cross-body movements as well as arm cupping for support. Second, we inferred that these same exercises were sufficiently commonplace in therapy sessions so that our therapy surface would prove useful. Third, we inferred that our surface could provide enough variability in the movements of these exercises to accommodate various injury levels and types.

In our design development, we ultimately divided the therapy movements into those that can be achieved by our continuum robotic surface (wrist extension and flexion, forearm cupping, forearm *pronation* and

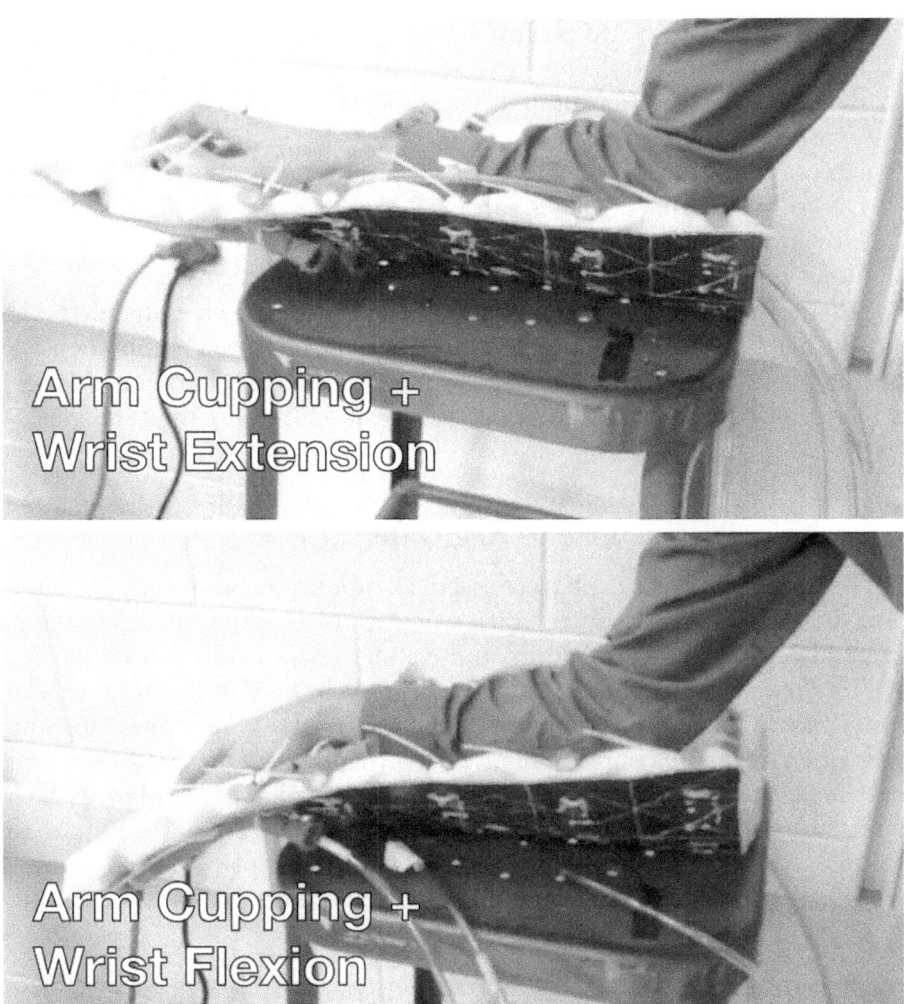

Figure 24 ART's therapy surface performing a therapy.
Source: author.

supination) and those that can be achieved simply by the movement of ART on which the continuum surface was mounted: flexor synergy, shoulder flexion, shoulder rotation and cross-body movements. To assess how well the various behaviours of the robots matched the expectations of the therapists for each of these exercises, we prepared and presented a video of the various therapy movements performed by the prototyped surface to a group of therapists, and asked them the following:

Is this how you would perform wrist flexion?
Is this how you would perform wrist extension?
Does this device offer enough variability for various patient types?
Would you use this device?
Do you think a patient would use this device?
Do you think this device is safe?
How does this device improve therapy sessions?
What information would you want this device to gather?
What other therapy movements do you think the device can perform?

The five iterative phases of design research resulted in a comprehensive prototype of a continuum robotic surface for stroke rehabilitation therapy that garnered positive and promising responses from the clinicians.

The *Vortical*: Thinking about Architecture and Engineering Differently

ART's *therapy surface*, characterised as smooth, continuous, flowing and dynamic, suggests a very different way of thinking about surfaces in architecture like walls, floors and ceilings. As already considered for ART's *therapy surface*, soft, smooth, fluid reconfigurations are essential to its success as it supports and enhances everyday human activities in, very often, intimate confines. Figure 25 provides a summary of four 'soft robotic' efforts, conceived by me and my close collaborator, Ian Walker, which are counter to a rigid-link, robotics approach. Similarly, Figure 26 captures in graphic, even macabre terms our concept of *home+*, where we imagine a humanoid robot being dismembered as limbs, head and

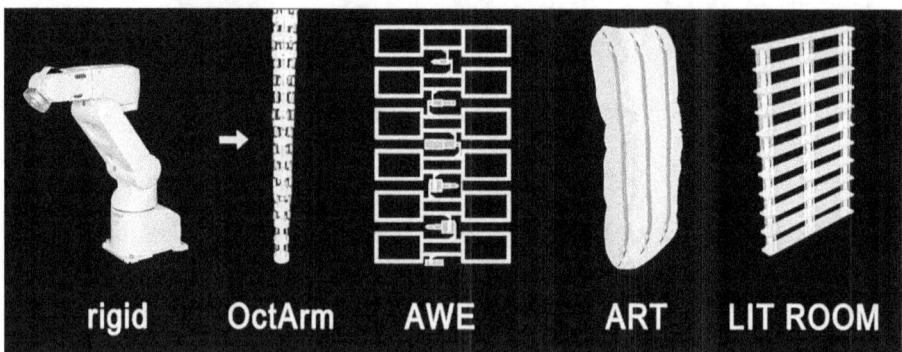

Figure 25 A summary of four 'soft robotic' efforts from the research labs of the author and I. Walker, which are counter to a rigid-link robotics approach (far left) (diagram by the author).

Source: author.

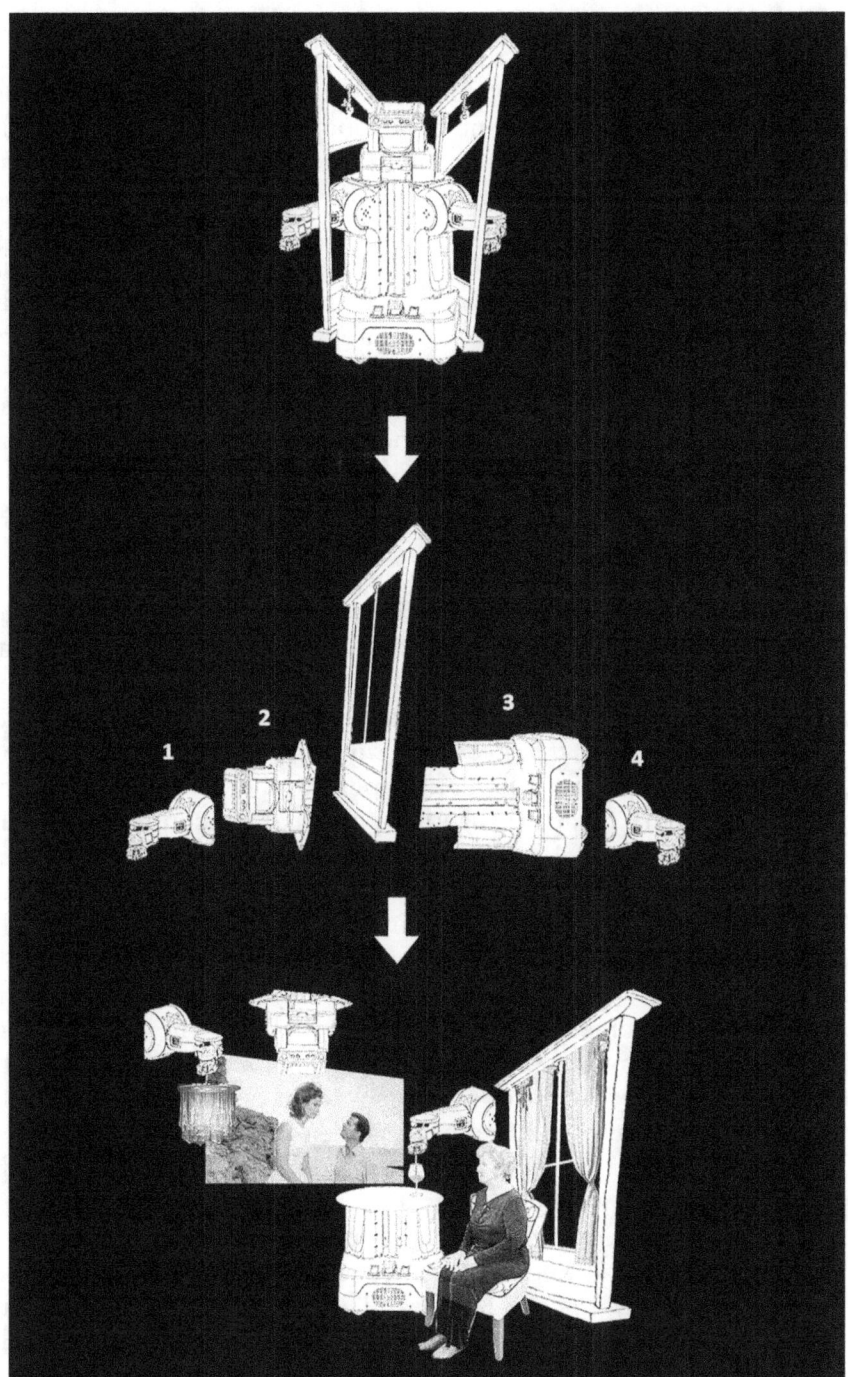

Figure 26 From a humanoid robot to a *living room* (diagram by the author).
Source: author.

torso that we re-package and fine-tune as a *living room* full of lively, distributed furnishings, each one tuned to specific tasks and playing its part, each one knowing something about the others and something about you.

By now it should be evident that robotics embedded in (that is, integral with) the built environment (from the scale of furniture to that of the metropolis) opens up a very different way of conceptualising architecture. The notion that the physical mass of an architectural work is set in motion by robotics, reconfiguring its spatial quality, disrupts our fundamental understanding of Architecture as *firmitas*, in its being firm, constant, solid, static and stable. When robotics is integral with the physical fabric of architecture (at wide-ranging scales) is classified as continuum rather than rigid link, its *firmitas* (that is, firmness, stasis, constancy) is arguably more upended, as the architectural work is not only moving but, in a way, quivering as if it were alive, and shape-shifting into something that seems foreign to what it was at an earlier point in time.

Before ending this consideration of ART's *therapy surface*, it is therefore useful to ponder the questions suggested by it, and the larger ART artefact, and our vision of the *home+* to which they belong, and finally to the World Wide Web and the Internet of Things as speculated in the concluding remarks of this chapter: *What kind of place is this?* and *How does it reflect our current state* (or, you may say, *disposition*)? In a curious way, French philosopher Gilles Deleuze and psychoanalyst Félix Guattari together ponder the second question literally and abstractly, in *A Thousand Plateaus*. In the twelfth 'plateau', exploring the relationship between the State and its military, Deleuze and Guattari posit two models of thinking and acting that, strangely, provide a compelling perception that applies well to ART and *home+* – the kinds of places they make, and how they reflect, interact with, and somewhat become us.[16] These two models are the *authoritative model* exemplified by the State, and the *vortical model* exemplified by the military. Architectural Robotics, as offered here, is very much the latter, and Architecture and Engineering, very much the former.

The *authoritative model* is characterised by: the straight line, the parallel (laminar) flow, solid things and a spatial environment that is closed, striated and sedentary. In its propensity to edify and monumentalise, Architecture and Engineering, argue Deleuze and Guattari, have 'an apparent affinity for the State' and its authoritative model. The *authoritative model* is manifested in the built environment as 'walls, enclosures and roads between enclosures', 'bridges', 'monuments' and, generally, a 'metric plane' that is *'lineal, territorial, and numerical'*.[17]

In contrast, the *vortical model* is characterised by: the curvilinear line, the spiral (vortical) flow, the flow of matter and a spatial environment characterised as open, smooth, flowing and fluid.[18] Free flowing and measureless, the *vortical model* is manifested in the natural environment in the desert and the sea.[19] Outside of its manifestations in nature, the *vortical model* is exemplified by the intermezzo of music, the space of Zen, and in the built environment (in a rare exception) as the impossibly slender, sky-reaching Gothic cathedral.[20]

More generally, Deleuze and Guattari characterise the vortical space as 'vectors of deterritorialisation in perpetual motion'.[21] Additionally, Deleuze and Guattari, referencing phenomenologist Edmund Husserl, describe the vortical space as a 'fuzzy aggregate', 'a kind of intermediary', 'vague' and 'yet rigorous'.[22] Something of the same is central to the thinking of Manuel Castells in his concept of the 'space of flows'.[23] For Castells, as for Deleuze and Guattari, the fluid, vortical 'space' is a way of thinking and making that finds validation in our propensity for 'neurological play', in the lure of controlled 'ambiguity' that 'engages and challenges the brain to allow multiple meanings'.[24]

If this philosophical equation between form and thinking seems itself vague and ambiguous, we need only walk the linear boulevards that cut through Paris, Rome and Washington, D.C. to connect urban form to figureheads (Napoleon III, Mussolini, George Washington). Today, we have not the grand boulevard but rather the kingly quarters of the private domain, exemplified by Apple's new Cupertino headquarters. Accommodating some 12,000 Apple employees, this five-billion-dollar building designed by Foster + Partners (with Arup as engineers) takes the form of a perfect glass circle defining a perfectly circular courtyard. No matter how elegant its glass skin, this closed, *authoritative* form – an apple without the bite – chillingly evokes Dave Eggers' aptly titled *The Circle* that chronicles a tech employee caught in the web of a colossal, oligarchical Information Technology (IT) company.

The other side of this equation, *the vortical*, surfaces in the architectural writings and works of Robert Venturi (often times, in collaboration with his wife, Denise Scott Brown) in his theory of a *'both-and'* architecture, rather than an *'either-or'* architecture: an architectural work characterised as *both* complex *and* contradictory, *both* high-culture *and* low-culture, *both* contemporary *and* drawn from history.[25] Much like Venturi but from the perspective of robotics and not architecture, Rodney Brooks envisions a future in the *both-and* – both organic and mechanical.[26] Suggesting that the convergence of human intelligence and machine intelligence will come only by way of a new conceptual framework, outside that of digital computing, Brooks calls for a bio-technical body, and sees its emergence in our becoming increasingly more robotic (by way of, for instance, advanced prostheses) while robots are becoming more biological (in their capacity to sense, adapt, learn, move and otherwise actuate).[27] Presumably for Brooks, we will meet one day in the middle: human-machine, machine-human.

But unlike Venturi's dialectical, physically static, and aestheticised *both-and* architecture, and taking a more tempered view of Brooks' prophetic, bio-technical body, *home+* and its ART suggests a very different way of thinking about architecture and robotics today, whereby architecture is more than an aesthetic search or a stylistic path, and robotics, more than a technological quest. Something at the threshold between architecture and robotics and their long-standing concerns is captured in Deleuze and Guattari's 'Concrete Rules and Abstract Machines', in which they define a novel 'technological plane', as

> not simply made of formed substances, aluminum, plastic, electric wire, etc., nor of organising forms, program, prototypes, etc., but of a totality (ensemble) of unformed matters which present no more than degrees of intensity . . . and diagrammatic functions which only present differential equations.[28]

With ART as its key component, *home+* represents a start to realising the dream of a *living room*, a distributed suite of robotic furnishing that 'come to life' in the ways De Chirico dreamt – furnishing that know about us, and know about one another. Furnished with this information about itself and its surroundings, the *home+ living room* reconfigures, smoothly and softly, to support us in our everyday lives. But more than this, the convergence of architecture and robotics, as manifested as *home+*, promises new vocabularies of design, and new, complex realms of understanding ourselves in this dynamic, expanding ecosystem of physical bits, digital bytes and biology.

Speculations on Scaling Up

Had Ivan Illich written *Tools for Conviviality* today, he would surely have recognised today's *starchitecture* shaped by a short list of globe-trotting architect-stars as epitomising his classification of the manipulatory tool, an architecture designed by individual geniuses with little or no participation by its likely inhabitants, funded by the wealthiest, mostly private patrons, and servings as iconic status symbols – works of art of the largest physical scale that anchor a portfolio beheld by those who can pay for it. Likewise, Illich today would surely have recognised the computing *apps* and operating systems shaped by Apple, Microsoft, Facebook, and other oligarchical IT entities as epitomising the manipulatory tool: tracking what we do, whom we know and what we prefer – a control system ripe for abuse. Can we envision another future where architecture and computing converge, a more convivial and caring future outside starchitecture and these operating systems and applications, outside the World Wide Web, outside the Smart City or *e-topia*? May we say – instead, in a bit of a silly way – that the future where architecture and computing converge resides on *cyberPLAYce*, at once local and global, physical and digital?

While it may not be possible to disconnect from the Internet for long, and arguably it may not be desirable to do so, imagine a future that relies less on the *net* and more on a cyber-physical *mesh network*: a digital-physical communications network far looser than the Internet and its associated apps and operating systems – less formalised, less oligarchical, more decentralised, more community-responsive, more ad hoc, more localised and more fine-tuned to individual needs and opportunities. Like a wireless mesh network, this more expansive, playful plan for the convergence of architecture and computing, *cyberPLAYce*, is populated by cyber-physical nodes of activity that may or may not form a vague connectivity with one another.

What does cyberPLAYce look like? How does it behave? How do we design it? Indeed for any relevant designer of architecture, computing systems and their hybrids working today, 'the primary design question is how to architect a complex system to be extendable to multiple arbitrary scales in time and space'.[29] On cyberPLAYce, the design response may look and behave something like this: distributed cyber-physical nodes decoupling from and re-engaging the Internet to relative degrees, as determined by local situations defined partly by human–machine interaction and partly by individual and community control. How such nodes receive, transmit and interpret signals from one another is defined as 'interoperability on a systematic level'.[30] Such interoperability across cyberPLAYce has the potential to provide more apt, more assistive, more augmenting computing resources to individuals and their local communities. What the oligarchical entities receive in return for relinquishing a bit of pervasive control to these more localised mesh networks is focused, localised tutoring of their Artificial Intelligence (AI) platforms, as well as the knowledge required to deliver more focused content to individual cyber-physical 'neighbourhoods'.

To make CyberPLAYce more tangible, let's refer to a modest interactive and adaptive built environment just underway in my research lab that would especially benefit from such a network. LIBRARY-CUBED is a physically reconfigurable, technology-rich library unit, ten foot on a side, that can be installed in branch libraries or other public buildings in underserved communities. Accordingly, a single LIBRARY-CUBED, a single 'architectural robotic' node, can exhibit a variety of connectivity conditions: it may function off-line; or it may join the Internet for increments of time or continuously; or it may form a mesh network, for increments of time or continuously, with other LIBRARY-CUBEDs or, maybe, LIBRARY-CUBEDS and some services at the library, staffed by librarians; or it may form a mesh network (as above), which in turn connects to other mesh networks for increments of time or continuously. In this computing-future, the information transmitted across individual nodes doesn't flow evenly throughout the mesh network, but 'hops about' to neighbouring nodes as affinities arise from interactions between itself, people and the things around and inside it: that amalgamation of physical bits, digital bytes and biology, at a given instant. As well, recognise that while such an underserved community may never see the likes of a library designed by a *starchitect*, it may nevertheless have 'high-design' delivered to it in the form of this compact, architectural-robotic 'package'. Such a package, installed in an existing civic building in underserved neighbourhoods, is capable of reconfiguring in ways that offer many of the architectural experiences that may fill the much larger volume of the main library, designed by the starchitect, found in the wealthiest cities. This cyber-physical system is of a kind that can create a small home – *your home, your neighbourhood* – for architecture and computing resources.

While the overwhelming tendency in architecture and computing is to globalise, *cyberPLAYce has the potential to domesticate*. On a typical day on cyberPLAYce, cyber-physical artefacts of all scales and capabilities collectively create a commodious, convivial home for you, situating your meaning within a broader networked world that yet provides you some locus of control. This future where architecture and computing converges, promises a robust, dynamic, strategically decentralised, ecological, bottom-up platform for the cooperative interaction across computing, the built environment and people. This would be Illich's convivial society, 'designed', in his words, 'to allow all its members the most autonomous action be means of tools least controlled by others'.[31]

Notes

1. Nocks, *The Robot*.
2. This aspect of the research is elaborated in Threatt et al., 'Vision of the Patient Room'.
3. Norman, 'Next UI Breakthrough'.
4. Negroponte, *Soft Architecture Machines*, pp. 127–8.
5. McCullough, *Digital Ground*, p. xx.
6. These five phases of research are further elaborated in Threatt et al., 'Assistive Robotic Table'.
7. Illich, *Tools for Conviviality*, xii.
8. Ibid. p. 24.
9. Ibid. p. 34.
10. Ibid. p. 29.
11. The term 'robot' was coined in K. Čapek's 1920 play, *R. U. R.* ('Rossum's Universal Robots').
12. Illich, *Tools for Conviviality*, p. 10.
13. Ibid. p. 21.
14. Ibid. p. 11.
15. Merino et al., 'Forward Kinematic Model'. In this section, I'm also indebted to Jessica Merino for her MS thesis work in our lab. Merrino, *Continuum Robotic Surface*.
16. Deleuze and Guattari, *Nomadology*. (*Nomadology* is the English translation of a chapter from *A Thousand Plateaus: Capitalism and Schizophrenia*.)
17. Ibid. pp. 88, 51, 22, 45, 30 and 63 (the last, with italics by Deleuze and Guattari).
18. Ibid. p. 18.
19. Ibid. p. 48.
20. Ibid. pp. 50, 45, 22.
21. Ibid. p. 62.
22. Ibid. p. 96.
23. Castells introduced the concept of 'space of flows' in his *The Informational City*. With respect to the consideration of this book, see also Castells, *Rise of the Network Society*.
24. Mallgrave, *Architect's Brain*, p. 149.
25. Elaborating this theory is a canonic book for the discipline of architecture, Venturi, *Complexity and Contradiction in Architecture*. Ignasi de Solá-Morales and his concept of a 'Weak Architecture' is arguably a better reflection of vortical thinking than Venturi's dialectical both/and (see Solá-Morales, 'Weak Architecture').
26. Brooks, 'I, Rodney Brooks, Am a Robot', p. 73. In this instance again, there is a marvelous analogue in architectural thinking found in Rykwert, 'Organic and Mechanical'. Here Joseph Rykwert traces the co-mingling of the terms 'organic' and 'mechanical' in architectural thought, beginning

with Vitruvius's treatise (where 'the Latin oganicus did not mean anything very different from machinicus'), to Gottfried Semper, Owen Jones and John Ruskin (all of whom cultivated 'ideas about a new way of imitating nature, or relating the organism to the built form') (pp. 13 and 17). In the conclusion of his paper, Rykwert laments that while architects have long been preoccupied with nature as inspiration for decoration and form, there is not yet a 'theory of architecture based on a direct appeal to . . . the nature that biology and chemistry study' (p. 18). Perhaps some aspects of this chapter represent small steps in the direction that Rykwert had anticipated.
27. Brooks, 'I, Rodney Brooks, Am a Robot', pp. 73–4.
28. Deleuze and Guattari, 'Concrete Rules and Abstract Machines'.
29. Horváth, 'What the Design Theory of Social-Cyber-Physical Systems Must Describe, Explain and Predict?', p. 116.
30. Kominars, 'Interoperability Case Study'.
31. Illich, *Tools for Conviviality*, p. 20.

Bibliography

Brooks, Rodney, 'I, Rodney Brooks, Am a Robot', *IEEE Spectrum*, Volume 45, Issue 6 (June 2008), pp. 71–5.

Castells, Manuel, *The Informational City: Information Technology, Economic Restructuring, and the Urban-Regional Process* (Cambridge, MA: Blackwell, 1989).

Castells, Manuel, *The Rise of the Network Society, The Information Age: Economy, Society, and Culture* (Malden, MA: Wiley-Blackwell, 2010).

Deleuze, Gilles and Félix Guattari, 'Concrete Rules and Abstract Machines', *SubStance*, Vol. 13, Issue 3/4 44–45 (1984), pp. 7–19.

Deleuze, Gilles and Félix Guattari, *Nomadology: The War Machine* (New York: Semiotext(e), 1986).

Horváth, Imre, 'What the Design Theory of Social-Cyber-Physical Systems Must Describe, Explain and Predict?', in *An Anthology of Theories and Models of Design: Philosophy, Approaches and Empirical Explorations*, ed. Amaresh Chakrabarti and Lucienne T. M. Blessing (London: Springer, 2014), pp. 99–120.

Illich, Ivan, *Tools for Conviviality* (London: Marion Boyars, [1973] 2009).

Kominars, Paul, 'Interoperability Case Study: Internet of Things (IoT)', in The Berkman Center for Internet & Society Research Publication Series (The Berkman Center for Internet & Society), 31 March 2012 <https://cyber.harvard.edu/node/97248> (accessed 12 April 2016).

Mallgrave, Harry Francis, *The Architect's Brain: Neuroscience, Creativity, and Architecture* (Chichester: Wiley-Blackwell, 2010).

McCullough, Malcolm, *Digital Ground: Architecture, Pervasive Computing, and Environmental Knowing* (Cambridge, MA: MIT Press, 2004).

Merino, Jessica, Anthony L. Threatt, Ian D. Walker, and Keith E. Green, 'Forward Kinematic Model for Continuum Robotic Surfaces', in *Proceedings of IROS 2012: the 2012 IEEE/RSJ International Conference on Intelligent Robots and Systems*, Vilamoura, Algarve, Portugal, October 2012, pp. 3,453–60.

Merrino, Jessica, *Continuum Robotic Surface: Forward Kinematics Analysis and Implementation* (Clemson, SC: Clemson Univerisity, 2013).

Negroponte, Nicholas, *Soft Architecture Machines* (Cambridge, MA: The MIT Press, 1975).

Nocks, Lisa, *The Robot: The Life Story of a Technology* (Westport, CT: Greenwood Press, 2007).

Norman, Donald A., 'The Next UI Breakthrough, Part 2: Physicality', ACM interactions, July and August 2007, pp. 46–7.

Rykwert, Joseph, 'Organic and Mechanical', *Res: Anthropology and Aesthetics*, Vol. 22 (Autumn 1992), pp. 11–18.

Solá-Morales, Ignasi de, 'Weak Architecture', in *Architecture Theory since 1968*, ed. C. Michal Hays (Cambridge, MA: MIT Press, 1998), pp. 614–23.

Threatt, Anthony L., Jessica Merino, Keith E. Green, Ian D. Walker, Johnell O. Brooks et al. 'A Vision of the Patient Room as an Architectural Robotic Ecosystem', in *Proceedings of IROS 2012: The 2012 IEEE/RSJ International Conference on Intelligent Robots and Systems*, Vilamoura, Algarve, Portugal, October 2012, pp. 3,322–3.

Threatt, Anthony L. et al., 'An Assistive Robotic Table for Older and Post-Stroke Adults: Results from Participatory Design and Evaluation Activities with Clinical Staff', in *Proceedings of CHI '14, the ACM Conference on Human Factors in Computing Systems* (New York: ACM Press, 2014), pp. 673–82.

Venturi, Robert, *Complexity and Contradiction in Architecture* (New York: Museum of Modern Art, 1966).

CHAPTER 6

Emotive Embodiments

Kas Oosterhuis

We Robots

First, I wish to reposition the human condition as we usually feel it and perceive it. I do feel the need to mentally reposition ourselves in the evolution of human society so far. My opening statement is that from the moment we started using tools as the extension of our bodies, we entered a robotic condition. From the moment we started to use a hammer to nail hard material into somewhat softer material we not only extended ourselves with tools, but we also started to interlace ourselves to systemic relationships between things. Using tools is where the Internet of Things and People really started. Let me give a few examples to illustrate this opening statement. The point that I wish to make is that we can no longer discuss human society without describing the intrinsic relationships between people and things/systems; we must take into account all things, tools, systems and environments. I abandon the anthropocentric view and look at things from levels up into the macro-world, and simultaneously from levels down into the microscopic universe.

Let us imagine a cabinet-maker in the workshop, which already represents a huge step in the evolution of the society we take part in. I have to put emphasis on 'taking part in' since already in this situation we can no longer maintain the position that humans have full control of this environment, that the cabinet-maker has full control over the tools that are used and the space they are working in. The cabinet-maker plays a role in the systemic relationships between things and the self. There would be no nail if there would not be something like a hammer, and there would be no hammer if not a person to hold the hammer. Things like hammers and people live in a symbiotic relationship, and in the context of today I would even say that things and people live in a systemic and indeed robotic relationship. The person is an integral part of the system, but is not controlling the system. The cabinet-maker co-evolves with the tools, the materials that are prepared for further operations and the space that one is working in. In this workshop condition the person is needed to fire the action, to pick up the hammer and to force down the nail into the wood. Seen from the perspective of the workshop as a complex environment the person is used to activate the tools and to put together the product. This we can visualise by speeding up time in a time-lapse movie, and we actually see the cabinet come into being without even the trace of the cabinet-maker. He will appear as a ghostly appearance as to feed the workshop, basically functioning as the energy and the information that is used to compose the piece of furniture in that specific environment. The workshop is a metabolic cell that generates cabinets using the workforce of people as carriers of energy and information. Information is processed via the bodies and brains of the cabinet-makers to the tools and materials. The human body functions as an information hub in a much larger system of propagating information. Information feeds upon the human body to be processed, demonstrating the verifiable non-anthropocentric view on society. The higher purpose of the existence of human beings is neither their own lives, nor the further development of their own species; I believe people exist only to function as a carrier to process information and to propagate processed information. In a similar fashion the car body is a carrier to distribute information, which in its turn is carried by people. The purpose of the Internet is not to be convenient to people, but to distribute information on a global scale. Information thus has become highly successful in evolutionary terms, and is soon expected to massively radiate out of our solar system and into nanoscales as to propagate further, perhaps until galaxies and quarks will find out that they are part of the same logic. In that larger system of things and people the mundane workshop is a cell of a larger body that is called the city, which in its turn is a hub in society at large.

How robotic is that? How robotic is such deep mutual relationship between people and things? We do not need a robotic arm or a robotic vacuum cleaner to mark the emergence of robotic environments. Robotic environments started to form right away after the dawn of systemic intelligence, extending versatile human bodies with mobile tools and instruments. The fact that these tools are not fixed to the human body does not make them less of a direct extension. It is exactly this mobility that is the fascinating aspect of 'recent' (millions of years) evolution. Things and systems are deeply linked to bodies without a physical connection. It is exactly for that reason that I am not impressed by performances with explicit robotic extensions that are fixed to one's body. To me that represents a pitiful misunderstanding of what evolution of information is actually doing with us, and what humans are doing with evolution. Similarly, I am not impressed or even annoyed by constructing robots with the intention to copy human behaviour, as to walk, grip, sniff and snooze.

Now imagine the driver car highway system. We can look at this system from different perspectives, from the perspective of the highway, the car and the driver. Let us first look at it from the viewpoint of the highway. It is fair to say that the highway processes the cars, that the highway actively digests a variety of cars, trucks, vans, motorbikes, basically any vectorial body that is fit to join the industrious traffic swarm scanning the global highway network. This process of digesting cars never stops; there is not a single moment that the highways and other roads stop processing. Sometimes they may process more cars, and then fluctuating to less car movements, but the highway is always in action. Even when the road is blocked for maintenance it must be seen as a form of action. And sometimes the 'hungry' highway system may be temporarily blocked by traffic jams, yet always offers a bypass for their mobile 'prey'. New highways are being constructed all the time, some old ones are fading away and new and more intense networks are made as to facilitate the evolution of information. Indeed, highways are information processing machines, delivering data carriers from A to B and back again. Such a network of highways is very much alive. It never stops pumping cars through its veins, and is bound to increase its capacity in the coming decades, eventually to make place for a new intelligent transport system exchanging packages of information in a more efficient way. What drives the system? Who is the driver? Who is the driven? Are we humans really the drivers of this system? Or are we – as is the position I will take in this paper – basically a form of data for the transport system, packages of information that are carried by our bodies, which in their turn are hosted by the cars, which in turn are processed by the highways. Driving a car, one may have the illusion that one is in charge, that one is free to choose where to go. But in fact the journey is very much predetermined, governed by precise rules of conduct. Even more so when automated vehicles will become the norm. Free will is relative, which both applies to the human driver, to the car rolling on the highway, as to the highway processing the cars. I consider the road system to be a real-time behaving input – processing – output machine. Each perspective is just as valid as any other point of view.

The car has no meaning without the driver. The highway would not exist without the existence of cars. The human body would not be considered a driver without the idea of a car. Drivers, cars and highways form one complex interwoven system, fully co-evolved in their doings. Drivers, cars and highways, all of these can be considered robotic, as they execute certain mutually related tasks following a strict yet slowly evolving set of rules. Robots that work. 'We' robots.

The Body

Architecture is considered in this context as the interface between the flow of people and the building components, the buildings themselves and the built environments at large, including infrastructure like highways, airports and harbours. Humans are embodied in their bodies, but what about buildings and infrastructure? Are they contained as bodies, too? The logical consequence of the operational metabolism of buildings is that buildings must indeed be considered to possess a body. Buildings feed on material they are absorbing, basically everything that enters the semipermeable membranes of the house, as there are people, water, gas, electricity, data, air. Buildings process that information and convert it to something new or different. People rearrange things in the house, wear them out, throw them away, bring in new products. Water is injected in the house, rubs hands, bodies, forks and knives and waters the plants and leaves the house in a slightly contaminated condition. Gas feeds the oven and the gas boiler, is burnt and thrown out of the chimney. Electricity feeds the lights, TV set and other devices. Data feed your computer. Air is let in as to dilute the contaminated air. All these processes are in essence based on information exchange between the smallest nanoparticles. People leaving the house are uploaded with new information as distributed by television, radio or the Internet, and they disseminate this information to other people, via notes, messages, spoken word and gestures. Information has been processed by the building, with people as the enzymes facilitating that process. Without people operating the body of the house that house would be a dead body. But viewing the bi-directional process from the other end is more telling. Homes are a necessary attribute of society. Homes have their own metabolism. Arrays of homes congregate into buildings, along streets, which again have a metabolism of their own, not functioning apart from people but intrinsically interwoven with the operations people execute on the buildings. So we are talking about a co-existence of homes and family, of buildings and communities, of city and society. We as people have become deeply linked to the built environment, in such a way that the one cannot be seen separately from the other. They have built up a bi-directional relationship, ever evolving so as to feed the propagation of information.

In a similar way of looking at the co-existence of people and things the infrastructure of highways and secondary roads are processing vehicles in a non-stop process. Vehicles are like packages of information, carrying and delivering information to remote destinations, information in the form of people, who in turn are carriers and processors of information, or in the form of books, goods, stories and designs. The infrastructure is a huge extensive globally distributed information processing machine, interacting with other such machines operating at different scales, like the aviation network, the containership network, the bicycle network, the invisible communication networks and radio and television networks. The highway network is a machine that executes complex series of actions, and is programmed by its design to host the swarm of vehicles in a complex manner. Complexity is based on simple rules; the outcome of the process looks complex when viewed from outside the system. All such complex networks must be considered to constitute a body, whether physically contained in a body or ephemerally contained in software and rules of conduct.

Now that buildings, infrastructures and their programmes have been identified as bodies, the question that must be raised here is whether these building bodies and those infrastructural bodies are robotic by nature. Nature as we know it is considered here as a computation, based on information exchange between the acting players, whether big or small. How do these building bodies actually operate? Can they be stopped at all? Is there an on/off button? Are they being operated or do they operate? Can they die? Are they born? How are they produced? The most likely answer: by the people – workshop/factory complex adaptive system. How did they come into existence at all? The most essential definition of a robot is that it is a constructed body that 'works'. Here we are: houses work, buildings do work, cities work, infrastructure works. Robots do not need to resemble people to bear the name robot. I will adopt here the broader definition of what a robot means to us and to our society. The narrow definition is according to the *Oxford Dictionary*: 'A machine capable of carrying out a complex series of actions automatically, especially one programmable by a computer'.[1] Since I consider the global highway network a 'machine' that feeds on vehicles, which in their turn are programmed by people and gas, and since as seen from a non-anthropocentric viewpoint they cannot be stopped to do so, hence automatically carrying out their actions, and while they are programmed to execute their tasks of facilitating the flow of vehicles, the highway network must be considered a robotic body. We live with them, live in them and cannot live without them. Our lives and the lives of the robotic environments are deeply interlaced and form a co-creative steadily evolving being.

A Day in the Life

This paragraph intends to elaborate upon the intrinsic relationship between people and machines that surround them and are operated by them. Whether we like it or not, you are bound to be intimately linked to a machine or system of some sort for most of the day. You are caught in machines, which in turn are caught in systems, and it feels so natural that you don't feel that you are taken hostage. Let's experience a day in the life, and simply register how and how long you are communicating with some version of a machine. Or, in other words, to what extent you are an analogue robot dancing with a mechanical slash digital robot. The tour of a normal day featuring extensive robotic interplay. A typical day starts with the alarm clock. Just this seemingly innocent fact links a person to a global system of time construction, time manufacturing, time processing and time management. Setting and obeying the alarm clock means accepting the intrinsic interlacing with a global system, meaning accepting one's position as a player in the same game, on level playing field with the robotic alarm clock. Setting the alarm clock, organising your agenda, living up to its input, linking you to other people's time management, imprisons you voluntarily for hours in a day to globally arranged robotic time fabrication systems, both in terms of hardware and software. Then at the dawn of the day I switch on my espresso machine. The entire process of operating the machines to fabricate the cappuccino and to consume it chains me voluntarily for at least fifteen minutes, to be repeated several times a day. Then I may take a shower, brush my teeth using the electric machine and again I am linked to a very complex system of infrastructural water-treatment installations, to a range of sanitary products, indirectly to some factories worldwide to produce them, to resellers that sell them to me, to advisors and dentists that advise me on my demand, again linking me in a robotic fashion to such sanitary environments. Should I go on? When analysing a full day one is almost 100 per cent intimately related to machines, to systems, to procedures. It is only the variety of cross-linking and entanglements with machines, systems and procedures – in short with robotic environments – that makes one feel self-determined. But in the end every citizen is ultimately dancing with machines and procedures full time. Who is the robot? The position I take here is that both the analogue biological player and the mechanical and digital machine are bodies that are interacting with each other. Neither of them can live without the other. There is no life as we know it without being so intimately linked to mechanical and digital bodies, while the machined bodies cannot exist and evolve without them being linked to analogue players. Living with products links you to the product industry. Watching TV links you to the entertainment industry. Using your cell phone links you to the information and communications technology (ICT) industry. Sitting on a sofa links you to the domestic industry. Paying your bills links you to the financial industry. Driving your car links you to the automotive industry. Living in your house links you to the building industry. Walking on the pavement links you to the infrastructure industry. Filling your gas tank links you to the energy industry. Playing tennis links you to the leisure industry. There is virtually not a single moment in your life that you are not linked to people and things. There is a strong mutual interdependency between people on the one hand and things and systems on the other, which makes me believe that nature and products must be seen as one integrated ecosystem, where old and new nature are not seen as different ecosystems that are fighting each other. There will be no winners, no losers, only the further propagation – and as of now – unpredictable transcription of information. Taking it to the extreme, nature as we know it, enhanced with product life, may eventually evolve towards information in another guise. We are temporary carriers and processing units of information, just like quarks, atoms and molecules. We are feeding an explosive increase of information content of the crust of Earth, eventually leading to popping out of the globe into deep interior and exterior space.

The Body Plan

Bodies have a plan. They evolve according to a plan. They live a life-cycle plan, and they love it when a plan comes together, and eventually they fade out, are destroyed, or simply disappear. From here on I will focus on building bodies, and leave infrastructural and organisational bodies out of further considerations. How do building bodies come into existence? How are they constructed? How do they operate? How do they survive or die? In earlier writings I have proposed to give any design proposal legal birthright as from the very first conceptual idea.² Meaning that no one owns the design except the design – and later the built building – itself. The building design grows up to transform into an explicit, often physical identity, as a player in the fabric of the city, naturally with the right to 'speak'. Un-built designs keep their ephemeral identity, and live their lives in the minds of people, in texts, in images. Hyperbodies grow according to a body plan, which is not a blueprint, but rather a genetic code, instructing simple rules of cell division and cell specification. Hyperbodies are informed bodies, similar to hypertext as an informed evolution from plain text, and hypersurfaces as digitally augmented surfaces of building bodies. As always symmetry plays an important role in the making of bodies according to their body plans. Although not noticed by architectural critiques, de facto all of my designs whether small or big – Waterpavilion, our many housing projects, iWEB, A2 Cockpit, Bálna Budapest, LIWA tower – are symmetrical in their basic lay-out (Figs 27–32), only allowing for a local asymmetry when it comes to responding to specific local conditions. Sometimes rotational symmetry is used as in the Waterpavilion, where at one side the feature line of the fin goes up towards the nose, at the other side going up towards the connection with the alchemist style NOX sector, indicating where the openings are, the main side entrance and the emergency exit, both detailed as cut-outs on the sides of the body and skin as to maintain the fluidity of the body (Fig. 27). We see similar detailing in car bodies,

Figure 27 Waterpavilion, Neeltje Jans 1997, design Kas Oosterhuis – ONL.
Source: ONL.

Figure 28 Web of North-Holland, Floriade Haarlemmermeer 2002, design Kas Oosterhuis – ONL.
Source: author.

Figure 29 A2 Cockpit in Sound Barrier, A2 Highway Leidsche Rijn, Utrecht 2005, design Kas Oosterhuis – ONL.
Source: author.

Figure 30 Whale, mixed-used cultural complex, Budapest 2012, design Kas Oosterhuis – ONL.
Source: author.

Figure 31 Liwa tower, Abu Dhabi 2013, design Kas Oosterhuis
Source: Gijs Joosen.

Figure 32 A2 Sound Barrier, Utrecht 2005, design Kas Oosterhuis – ONL.
Source: Henk Meijers.

where the door – always embedded as a side entrance as to go with the flow – is cut out from the body skin and the body structure. Understanding body plans and their inherent detailing is crucial to understand the nature of robotic environments. I will explain why robotic buildings simply need body plans. As pointed out before, humans and their tools/vehicles have co-evolved, whereas the evolution of humans mainly took place by enhanced connectivity to things and other people in their brains, by evolving language by labelling things and tools, since they already had the mobile body to free themselves from a fixed coordinate on Earth. The co-evolution means a mutual dependency of humans and their tools, vehicles, building bodies and societal systems. People spend most of their day connected to clothes, machines and tools, and de facto always being connected to buildings and transport vehicles. Reviewing a day in the life, counting how much time you have spent brushing teeth, making coffee, listening to the radio, sending e-mail, driving a car or bike, using keys to lock and unlock spaces/buildings, checking your cell phone for Facebook or Twitter, cooking food, getting dressed, sitting on chairs, writing code, working with robots, drinking beer and performing differential calculations, you will feel how deeply one is embedded in the built environment. You are basically always interacting with some sort of system or machine. The systemic links between tools, factories, distribution channels, selling points, buyers, users and communities are so strong that using such a tool means being an active part of that system. That global system of humans–tools/vehicles/buildings interaction is a continuous operation, never stopping, the systemic components always in the process of being made and always in the process of being used, and therefore that system must be considered to be processing information, temporarily 'on air' until it is replaced by a further evolved system.

Specification of Components and Spaces

The building components of the building body are the constituting cells of their bodies. They have evolved over a long period of time, and are getting smarter over time. Having left the period of structural brickworks far behind – although some die-hard old school architects still regret this – we are now entering a phase in the building industry where the constituting building components are becoming smart informed components, both in the design process and in the fabrication process and in the process of operation. Building components have started to talk to one another, and some building components will be equipped with actuators so as to act in real time on specific requests from externally and internally changing conditions. 'Smart' in the design process means that the designs are programmed, scripted, generated, bred and co-created on the basis of open collaborative design systems. I have developed applications for climbing walls and acoustic ceilings (Figs 33 and 34). These applications offer an open platform for designers and users alike to design their own climbing wall and their own acoustic ceiling, as long as they follow the simple rules as set by the designers of the parametric software. For the climbing wall and the acoustic ceiling, the solution space for the designers is extended to include nonstandard complex geometries, therewith offering a maximum of design freedom for the co-creators. The nonstandard design paradigm is an inclusive approach, allowing for traditional rectangular designs as well. Each of the constituting components in such an open-design system is principally unique in its shape, dimensions and properties; such a smart design component is 'only that, only there, and only then'. The body as a result of the design process consists of thousands of such unique linked components. For example, the A2 Sound Barrier consists of about 40,000 unique steel components and in the region of 10,000 unique plates of glass (Fig. 32). Together they form the body of the structure that communicates freedom of expression and synthetic integrity at the same time.

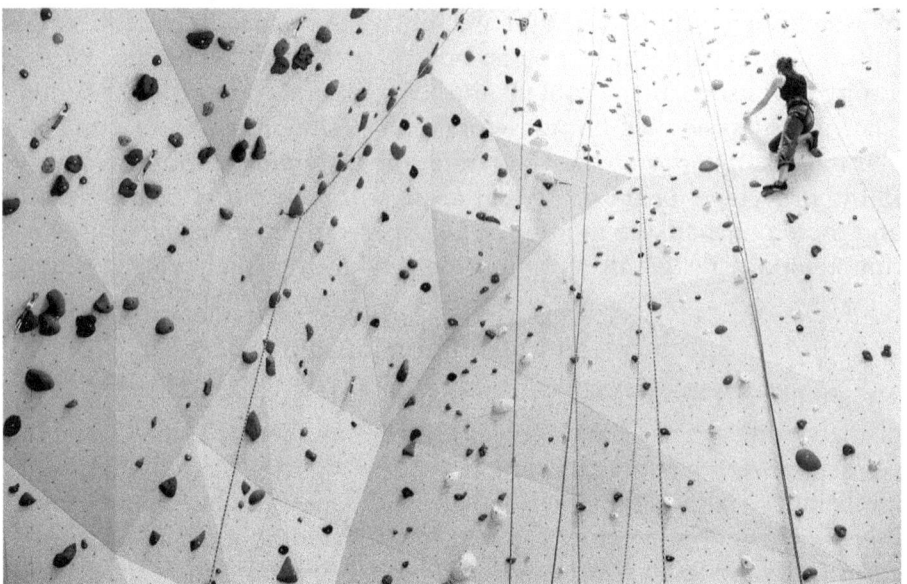

Figure 33 Climbing Wall, Amsterdam 2012, design Kas Oosterhuis – ONL.
Source: author.

Figure 34 Acoustic Ceiling, Ecophon Helsingborg 2015, design Kas Oosterhuis – ONL.
Source: author.

Digital Design to Production

Bringing the discussion back to daily work, it is necessary to discuss the principles of the digital 'design to production' and 'design to operation' processes. Digital design to production means nothing less than a revolution in the building chain of built environments. Architecture will never be the same. Digital design to production is by definition based on code. Designing with code unravels new horizons both in the world of design and in the world of production. Taking advantage of the architecture of the Computer Numerical Machine (CNC) machine – these machines read code, not drawings or 3D models – an immediate relationship between design and machine can be established. Writing code forms the basis for parametric design methods as well, meaning that a systemic design to production procedure paves the way for an unprecedented nonstandard, extreme customisation of the design and production of the individual building components, the only limitation being the imagination of the designer and the physical constraints of the machines. Freedom of expression here does not imply extra costs; exactly the opposite is true in my designs. The design to production methods I have designed and executed have proven to be time efficient and cost-effective. The A2 Car showroom is built for a mere 750 €/m^2, while the A2 sound barrier in which the Cockpit building is embedded, has been built for a mere 450 €/m^2. All components are different and yet competitive with the standard sound barrier as promoted by the Ministry of Infrastructure. The cost advantages are both in the design process, which literally saves 20 per cent in each phase of the design process, from preliminary design to execution, and in the production process, while no information has to be redrawn and nothing gets lost in translation, that is two times more effective than a traditional linear chain of design, engineering, tendering, contracting and manufacturing process. Today's CNC machines are capable of dealing with difference much more than is typically used by the designers. The machine really does not care whether it produces series of the same or series of unique components; it all depends on the code it is reading. The ultimate freedom of architectural expression can be disclosed when the designer produces code instead of using traditional 3D modelling, which is basically an optimisation of a traditional linear design process, design techniques that need to be translated into code. Many possibilities are not taken into account or will get lost in translation, when not thinking as a programmer-designer. Clearly architectural programming must form the core of teaching in courses and design studios at architectural schools from day one when the high-school students enter the faculties. This is what Hyperbody has introduced since the establishment of the TU Delft Architecture Department chair in 2000. Students will need to get acquainted with thinking like a programmer, to write code, and to instruct the machines in the most direct way possible. Designing and producing will naturally become an iterative creative process, where the design is almost instantly informed by the potential and constraints of the production (and operation) process.

Emotive Embodiments

The future of architecture is just this, precisely here and exactly now. The unique building components that together constitute the building body keep their unique identity over time, meaning that that they can be addressed in the design, production and operation process individually to perform. Each building component is potentially an actuator, either processing data in an otherwise solid cell/component, or actuating the cell/component by changing its shape, dimension and/or performance. Actuating cells of the building body are relatively dumb, processing only a few simple algorithms, yet in sync with their immediate neighbours. The Hyperbody research group at TU Delft and ONL have designed and produced a wealth of interactive prototypes, all of them considered today as robotic embodiments, realised decades before we referred to them as robotic installations. Now we consider these interactive installations robotic environments, to fight the idea that the robots would look like humans and copy human behaviour. Recently, ONL's Nonstandard Architecture (NSA) Muscle project was already archived in the Centre for Contemporary Architecture in Montreal, as part of a show curated by Greg Lynn.[3] ONL and Hyperbody have developed CNC-driven design to production methods to realise in practice the complex geometry that gives shape to their design concepts. As demonstrated above, I consider existing buildings, also the most primitive adobe houses, as a simple form of a robotic environment, since they are processing materials, people and data without ever stopping to do so. Now when building components are no longer made of mud, brick or steel, and become tiny robotic environments in themselves operating in a swarm, I suggest labelling them emotive robotic embodiments in order to differentiate them from the static shells of the environment. These emotive robotic environments are so complex in their behaviour that they are experienced by humans as living things, as emotive embodiments.

Society of Robots

The word 'robot' typically is used to indicate that something 'works', something that acts, functions in real time, something that is processing information. Now when we consider built environments as environments that 'work' while processing data continuously and are composed of components that 'work' while processing data in real time, we can rightfully state that we ourselves live as soft bots in a robotic world. A robotic world that is handmade robotic since millennia, machined robotic since centuries and additionally programmed robotic in the twenty-first century. Whether we like it or not, we live in a society of robots and robotic systems and people are their co-evolving teammates, living as embodied people in the Internet of Embodied Things.

Notes

1. 'Robot', *Oxford Dictionary* online, <https://en.oxforddictionaries.com/definition/robot> (accessed September 2016).
2. Oosterhuis, Kas, *Hyperbodies: Towards an Emotive Architecture* (Basel: Birkhäuser, 2003).
3. *Archeology of the Digital, Media and Machines*, exhibition in CCA Montreal, 2015.

PART III
Medicine

CHAPTER 7

Ecologies of Corporeal Space

Katharina D. Martin

In the last two hundred years, medical tools and devices have been a determining factor in establishing the body as a site for the production of new images and new fields of meaning. One crucial aspect is the gaze of the doctor, which is observing, intruding and productive at the same time. This gaze, professionalised with the help of technical instruments, is part of the reciprocal relationship between corporeal space and diverse and changing systems of knowledge. The contingent historical development of symptomatology and medical treatment is correlated to changing technical conditions. This means: the study of symptomatic signs shows its changing modalities in correlation with the historical circumstances created by spaces and tools. As such, it is important to consider technical development and medical practices, as well as knowledge production, in terms of their coeval progression.

In the following chapter I will first present several epistemological aspects of (clinical) medicine and its practice. Next, I introduce an ecological notion of corporeal space, understood as a multi-layered 'milieu'. This is followed by an investigation of the various technical implications within the medical context from the angle of the interconnectivity between the different milieus or environments formed by matter and signs. The (human) body is active but stable; it is an ecological state, or, more specifically, a stasis based on continuous variation. My argument aims to demonstrate that the different medical techniques and instruments function as a membrane, thus as a permeable but selective barrier between various corporeal spaces and their different overlapping milieus. These different fields or milieus are constituted by the combination of signs, which do not yet form a system of knowledge, but rather an arrangement of significant but a-signifying signals.

Surface and Depth

Regarding the historical changes within medical science and in particular the development of the clinic, one has to acknowledge the significance of the epistemological analyses by Michel Foucault in his *The Birth of the Clinic*, first published in French in 1963. For Foucault this publication was not merely about the century of the clinic's 'birth', but also, as stated in its subtitle, about an *'archaeology of medical perception'*.[1] Key aspects of his analysis are space, language, death and the act of seeing.[2] His study presents the development and methods of medical observation in a period of crucial changes. It recalls the history of classical medicine and the rise of the clinic, and presents in detail the techniques of mapping symptoms and anatomical spaces. The chief goal of nosology as part of theoretical medicine in the eighteenth century was a comprehensive classification of all diseases. This motivated medical actors to let a disease unfold in the most free and natural way, after which they could describe the changes of the symptoms in detail and classify the disease. The natural space for being ill was the everyday environment of people's life: the home.[3] In the course of the eighteenth century, however, it became a common practice to take an ill person who lacked a supportive family out of his home and into a publicly financed hospital.[4] The clinic, serving as both a hospital and a place for education, evolved into the preferred and most neutral site for the observation and treatment of people suffering from a disease. This marked a shift towards an understanding of medicine as an objective science and practice.[5]

In classical medicine it was common to observe the patient thoroughly or, put more precisely, medical actors described and categorised medical symptoms. The 'tableau' of classical medicine basically involved a straightforward set of classifications and structured data on families, genera and species. The medical gaze was deployed to observe pathological abnormities, after which they would be arranged, structured and put into a specific 'order' with the help of language.[6]

A tableau, including extensive descriptions and structured information, became part of the interplay between the spoken and perceived. The similarity between a phenomenon and the symptoms classified resulted in a further step, in the essential moment of recognition of the disease in the tableau. Once the particular manifestations of some illness were linked to specific coordinates on the tableau, it took over the space of the body. The information on the outer body and its symptoms made its way into the flat language tableau, after which the disease became apparent in corporeal space. These acts of classifying involved perceptions led to a productive mode of thinking in terms of particular codes. By connecting the medical gaze and language, a new medical system of knowledge was produced.[7]

In the middle of the eighteenth century, the dissections of dead bodies on a regular basis resulted in a growing amount of anatomical knowledge. Anatomical-pathology developed quickly as a body of knowledge and clinical diagnostics underwent great changes. If in classical medicine the dead body, regarded as the opposite of a healthy body, was believed to be of no further use to medicine, in anatomical-pathology the corpse served as a great source of knowledge for all further diagnostics. Earlier, in nosology, a disease was merely a bundle of characteristics on the surface of the body, but now the body could be horizontally and vertically penetrated to uncover the layered depth of its bulk.[8] The simple gaze of the doctor expanded into a comprehensive anatomical-clinical apparatus, based on the senses of sight, touch and hearing, which allowed one to map the living body.[9] As a result of this exploration of the inner space of the dead body, the living body turned into corporeal space as well. As Foucault states:

> For us, the human body defines, by natural right, the space of origin and of distribution of disease: a space whose lines, volumes, surfaces, and routes are laid down, in accordance with a now familiar geometry, by the anatomical atlas.[10]

A disease was no longer defined as a virtual ideal scheme and theoretical tableau placed on the body. Rather, a disease was now embedded in the flesh and locatable in corporeal space.

Surgical interventions on the living body were not so much motivated by a desire for knowledge, but by the need to act within the corporeal space and treat malfunctions. In the early eighteenth century, most surgical procedures were amputations, which had to be done very quickly to prevent the patient from dying from excessive pain or loss of blood. With the discovery of anaesthetics in about 1845 surgery in the modern sense became possible.[11] A patient under narcosis could be operated without pain, thus time was less of an issue and more complicated surgery became possible. The living body was silenced and the patient became merely a physical object. The living and fleshy organism, suspended in unconsciousness, could now be opened and entered without interferences. In surgery, the patient is cut open wide, wide enough to see, to access, and touch the organs. In combination, anaesthetics and scalpels make it possible for the hands of the surgeon to enter the corporeal space and operate within it. Drawbacks of this invasive procedure include the damage resulting from the incision, the risk of infections and the time needed for recovery.

When surgery grew into a regular medical practice, anatomy became an irreplaceable part of medical education. To develop necessary skills and good judgement, a practicing physician needed not only to be taught established biological facts but needed also to undergo anatomical training. The recognition of an illness was not just based on cognitive concepts but on hands-on experiences generated by exploring the human body. The recognition of an illness was not just based on cognitive concepts but on hands-on experiences generated by exploring the human body.[12] The combination of cognitive knowledge and experience helps to give form to an illness, a process that Rachel Prentice calls object formation.[13]

When Prentice writes about her experiences as observer during surgery, she describes the process of recognising 'arthritis', which before was merely a linguistic object to her. When seeing the pinkish bone of an opened up knuckle, it was not clear to her what she had seen. Even when the surgeon pointed out that this discolouration was a sign of degraded cartilage, the pinkish bone just took its form as 'arthritis', after she had seen another patient's open knuckle, showing a healthy white bone. For Prentice in this moment the illness became a material object.[14]

Going into the depth of the body enables object formation, which in turn will expand the language-based cognitive concept of an illness. At this juncture lies one important reason for the anatomical training for future surgeons. In anatomical training a profound understanding of the body can be established. The corpse is mapped in three dimensions and the physicians operate with coordinates constituted by different planes and depths.[15]

During her research Prentice engages not only with anatomy within the laboratory. The means of anatomical education are diverse and in the US National Library of Medicine she discovered two fully digitalised cadavers, a woman and a man, created from an image database extracted from two persons.[16]

This Visible Human Project (VHP) is based on a highly complex technique, which makes use of computed tomography, magnetic resonance imaging (MRI) and a photographed cross-section of the frozen corpses.[17] As a result, a very detailed and fully digitalised anatomical map was created. After given public access to these three-dimensional and interactive body replicas, the two-dimensional representations became outdated. Prentice stresses the fact that the cognitive model of education, characterised by rationalising logic, is moving into medicine to promote quantification, precision and 'efficiency'. Bodies are being translated into 'informatic "body objects," digital and mathematical constructs that can be redistributed, technologized, and capitalized'.[18]

Although she agrees that vision is a significant component of medical work, Prentice claims that its significance can be overstated. She underpins her assertion using the example of an anatomist who states that 'he gauges his progress in opening the spine more by sound than by sight'.[19]

In *Bodies in Formation* Prentice distillates the specific dimension that degraded within medical education: the hands-on-experience, the moment of facing a dead person, cutting into its body as an experience that includes multisensory richness. In the history of medicine, we can account for an important shift by Andreas Vesalius, who put practice before text. In contrast, recent developments show that 'tactile experience is on the verge of being replaced by the visual'.[20]

Although surgery has become a widely established and successful medical discipline today, it is not yet possible to operate and navigate within the living human body without damaging any tissue. Ongoing research concerned with medical techniques has been driven by the aspiration to find enhanced methods that would guarantee secure navigation within the anatomical area, avoiding all unnecessary penetration. One example is the endoscopic technique that already emerged in a primitive form in the nineteenth century.[21]

Using tools similar to binoculars or telescopes, which were based on the reflection on light from the outside, they entered the natural openings of the body. One way to reach the stomach, for instance, was by passing a straight, static tube through the oesophagus.[22] The technical progress of optics and light were decisive for the further stages of endoscopic tools. Indeed, the invention of the Edison lamp made it possible to bring artificial light into the body that in turn brought a whole new perspective for the medical gaze. The subsequent anatomical and biomedical knowledge production relying on this new invasive visibility is a direct consequence of historically situated, technical capabilities.

It is clear that regardless of whether we are dealing with the outside or the inside of the body, the question of visibility was, and still is, a crucial aspect in diagnostics and treatment within medical practice. The ground-breaking and fascinating discovery of the X-ray technique in 1895 made it possible for the first time to show details of the inside of a living body and also record images of it. For decades the photographs were exploited in popular culture, and since one of the first bone images showed Bertha Röntgen's hand, the X-ray of the female hand became a fetish object.[23]

It has always been common to apply technical apparatuses in unconventional ways, linked to the world of magic and occultism. The X-ray technique, for instance, has also been used in attempts to capture ghostly apparitions.[24] In this sense it is also possible to interpret media in a wider context than the one proposed by the cybernetic diagram.[25] The model of communication that assumes an information source sending a message to a transmitter, which is subsequently conveyed in an unaltered fashion to the receiver and its destination, is perhaps all too familiar. But, a medium is an agent between different areas of meaning; it submits, produces and/or shows new combinations of signs, which are to be understood as new information. The discussions around concepts of social constructivism and scientific realism entwine around exactly this question of agency and mediation. It was Karan Barad who introduced the phrase 'agential realism', seeking to tie together these epistemological and ontological issues.[26] She bases her arguments on Niels Bohr's interpretation of the discontinuity of quantum mechanics[27] and states that 'measurement is a potent moment in the construction of scientific knowledge – it is an instant where matter and meaning meet in a very literal sense'.[28]

Regardless of the numerous metaphorical interpretations and beliefs that came along with the sudden fact that it was possible to see the most intimate 'inner self', the X-ray technique was soon mainly understood as an objective, diagnostic tool. When in the late 1920s the X-ray technique made its way from the artistic studios of photographers into clinics, technologically produced medical images began to serve as an important tool for diagnosing illnesses at an early stage. If the exploration of the dead body made it possible to establish a map of the living corporeal space, the X-ray produced an additional, new space, separated from the patient's body: patients could now be diagnosed without being present. Increasingly, the personal experience of the patient was rated as subjective and unreliable.[29] As it became possible to show indications of a medical disorder within the corporeal space of the X-ray image, one no longer needed the physical space of an actual patient to be present. When Prentice identifies object formation she clearly differentiates it from the act of objectifying the patient. Illness as an object has different social implications than an ill person as an objectified patient. Within the biomedical realm there are many voices[30] who argue that the physicians are not able to express adequate motions, that they are distant and show an objectifying behaviour towards the patient. The professional relationship between practitioners and patients often seems characterised by a dehumanising reduction of the patient to a cluster of symptoms.[31] This in turn will create problems because it gives too little agency to the patient to understand and deal with their own pathology. However, so Prentice argues, there is the need for an objectification of the patient's body parts, which does not include objectifying the patient or weakening their authority over their own body. The advantages do also apply to the patients: to objectify the pathological body part as cause for illness makes surgery easier to bear by creating necessary distance to the painful procedures.[32]

With the advent of the abovementioned medical technologies 'the doctor's subjective sensorial impressions' has increasingly been replaced by '*supposedly* objective visual evidence'.[33] As van Dijk states: 'X-ray pictures, like other mechanical reproductions, always yield mediated perspectives, as their meanings are always shaped by the knowledge and feelings of their interpreters'.[34] With the help of a machine and based on the patient's body a new corporeal space had been produced. These new kinds of images correlate directly with the production of knowledge since they 'mold as well as reflect visual reality'.[35] The medical system of knowledge is pervaded by power, and to this day it is difficult to gain access to this system. Whoever has had the experience of having his body X-rayed will remember the 'moment of truth', when the doctor puts up the X-ray photo, looks at it and formulates a diagnosis. Even if the physician explains the image, or even if there is a clear fracture of a bone that is easy to spot, it is still obscure to see this encrypted image of one's own internal organic structures. One has to agree that these pictures are less transparent than they may seem to be. They need to be carefully interpreted by trained specialists before becoming a reliable source for medical diagnostics. Similar to the tableau with its tables, X-ray is 'a representative technology creating an *illusion* of unmediated, objective reality'.[36] At this point in medical history, the question of mediation began to present itself much more clearly. As indicated, to *mediate* means not merely to transmit, but to convert through diverse channels, and between different milieus. The media are always reshaping signs and reformulate information, even if one does not recognise the impact of the media's productive force. In this respect it is hardly surprising that the doctor who looks and listens has to give way to the specialist qualified to handle the technology and read their images. Despite the fact that a technical image of the body such as an X-ray is strongly mediated, it is commonly seen to be superior to any subjective form of perception.

Parallel to the rise of this kind of technologically mediated diagnostics, the endoscopic procedure swiftly developed as well. By having more sophisticated optical equipment, a safe and reliable internal light source and a flexible cable that could be passed between organs, the technique would prove to be more than just an extension of the eye of the doctor: it became also an extension of the hand of the surgeon. To have to make an incision to be able to enter the body with an instrument was a highly intrusive step. Anaesthesia was needed as well, which is why the procedure did not become common before the twentieth century. At that same time the former seeing device became a real instrument that functioned as an extension of the hand of the surgeon. Nowadays the technique is in many ways applied as a routine procedure. The entrance hole for the surgical device is small, and the instrument is even able to show video images. As of the 1980s the recorded images of the internal body proliferated not only within the medical world; they also made their way into study curriculums of architecture and design schools, into art exhibitions, private homes and on the Internet.[37] During the actual surgical procedure the camera feed is shown on a monitor. To work with these displaced images requires great skill and a lot of concentration on the part of the surgeon, since this additional source of visual information has to be observed and interpreted to then being transferred and practically applied onto the patient's body.[38]

The general map of the physical space of the human body is no secret anymore, and with the evolved technical media, highly detailed information can be gathered. Each measuring machine produces its own characteristic image and particular encryption. This in turn has prompted a need for specialists with diverse technical and visual skills. Digitally augmented reality is a recently developed technique of interactive training exercises and image-guided surgery. Due to a high image quality and interactivity, training software can be of great benefit to students. Besides that, it can be an advantage to be led by a 'virtual' body during surgery. Throughout endoscopic operations, the gaze of the surgeon is directed away from the body because the camera feed is shown on a monitor. This cognitive performance demands great discipline and extensive training. The projections of the augmented reality on the other hand make it possible to merge this additional information and help to reduce the workload or even shorten the time of narcosis. Already before the surgical procedure, data about the patient's physical condition has been processed and can be used. Computer tomography images or X-ray images are visually prepared and then projected directly on to the relevant body part. Tracking devices attached to the skin are connected to a wireless mouse, and they allow one to switch between different visualisations of the internal corporeal space.[39] This visualisation technique was only recently improved by a method to produce an overlaying image with transparency and spatial presentation. The *Edge Overlay* visualisation aims to provide depth cues when viewing sealed objects.[40] When showing occluded objects, it was very effective to preserve the context of occluding structures by rendering just the edges. A perception of depth is achieved through including a 'window' without determined frames: in a certain area around the central image, the 'tissue' becomes more transparent and therefore produces a spatial appearance. Thus, very little of the occluded object is obscured by the thin edges, but there are enough visual cues to give a compelling sense of depth. In other words: the clinical information is processed and prepared in order to be turned into a new projected image, making it possible to perceive a three-dimensional space. This field of the optical dimension is a field of mixed realties, and functions as membrane between the internal corporeal space and its cognitive and digital correlation. Based on the physical body of the patient, a seemingly decrypted internal image appeals to the gaze 'immediately' and instantly.

Homeostasis and Allostasis

To analyse the specificity of the entanglement between medical practice, its media and corporeal space, I will introduce the notion of 'milieu' as methodological instrument. Claude Bernard, the founder of modern physiology, scorned classic nosology and wanted to establish an 'experimental medicine'. His research of animal physiology was based on vivisection, surgery on living organisms conducted for experimental purposes. Bernard's work has been recognised primarily for his concept of the constancy of the 'internal environment'. This *milieu intérieur* is geared to stabilising and maintaining the uniformity of the organism's conditions, so that it can pursue a free and autonomous life.[41] For instance, the *milieu intérieur* ensures a steady body temperature, and helps the body to adjust to the oscillating climate changes of the external environment, the *milieu cosmique*.[42] This process, also called homeostasis, describes the sufficient regulation of the physiological adaptations necessary for internal stasis. The control of temperature, pH, glucose, protein, oxygen, sodium and calcium are important examples of these regulatory responses to the systemic physiological requirements. As argued in recent research in the field of neuroscience, however, Bernard misjudged the environmental context and overrated the separation of the internal milieu from the external world.[43] Suggesting that the concept of homeostasis was defined too narrowly, it proved impossible to explain all the observed adaptations by the organism. Today's research emphasises that a viable stasis cannot be accounted for by physiological adjustments only, as behavioural ones are equally relevant. The concept of 'rheostasis', for instance, includes a wider range of biological systems, taking into account variations tied to context, season and surroundings. Considering reactive and predictive homeostasis, it does include physiological and behavioural regulations, giving rise to a notion of 'physiology of change'.[44] An alternative neuro-scientific concept, 'allostasis', was introduced to acknowledge the change of state as a prerequisite for viability. Allostasis comprises both the behavioural and physiological processes that maintain internal parameters for the essential requirements for life. The concept acknowledges the impact of an external (social) space, and it is considered a plausible hypothesis for connecting events that may seem to be unrelated at first glance.[45]

If we want to follow up on the notion of a connected internal and external milieu, we should turn to Jacob von Uexküll and his concept of *Umwelt* (environment). In his theoretical biology (*Theoretische Biologie*)[46] and theory of meaning (*Bedeutungslehre*),[47] Uexküll emphasises the importance of a subject-oriented epistemology, which he based on his biological research.[48] A significant aspect is the reciprocal relationship between an autonomous organism and its geographical environment, on which each unique milieu is based. He interrogates the specific living arrangements in and around an oak tree, and establishes his idea of many different environments that cross-cut one another. A subject has its own milieu (*Umwelt*) and is simultaneously an object in several foreign milieus.[49] Especially well known is Uexküll's example of the tick, which was enthusiastically taken up and propagated by Gilles Deleuze and Félix Guattari.[50] The tick reacts on the perceptive signs produced by the hunted animal. The prey's smell, its skin temperature, as well as the resistance of its hair, are the few relevant signs that compose the tick's *Umwelt*. A unique composition of perceptive (noticed) and effective (produced) signs constitute a particular *Umwelt*, and the reciprocal connection between different organisms form a functional cycle and overlapping milieus.[51] Based on many different examples, Uexküll demonstrates that the functional cycle is as meaningful relation, organised by perceptive and effective signs.[52] The main point is that the organism does not respond merely to causal impulses, but to perceptual signs or meaningful signals.[53] Each subject's *Umwelt* includes carriers of meaning,[54] and together they maintain a reciprocal relationship and a shared field of meaning. Surrounded by these vital counterpoints the subjects 'internal front' does entail imprinted images of them.[55] Uexküll elaborates this aspect when he describes the spider's web. The fly, never seen before by the spider, is present as a primal image. This *Urbild* is the form on which the spider is able to build the perfect web for the hunt.[56] The web is a well-made mould of the fly, so to speak, and it would not exist in this way without the fly's concrete characteristics. Every subject is considered a carrier of meaning, who produces 'formative melodies'[57] and corresponding properties.[58] A subject is formed into a 'recipient of meaning'[59] and foreign motifs are taken on. These different formative melodies mutually influence one another. Deleuze and Guattari admire Uexküll's theory of transcoding based on music: nature as music, which includes possible passage and bridging based on rhythm and melodies, and that always produces a surplus value. This rhythmic plane does not exist in its pure form; in reality it always shows itself mixed.[60] For Deleuze and Guattari a subject's *milieu* has to be established by organising a limited space and keeping the 'forces of chaos' outside. Within the chaos the milieu is a stabilising centre; it is a home and a field of certainties.[61] In this context it may be helpful to note that milieu or environment should not be confused with a fixed geographical space that can be easily determined. Although the 'milieu' is part of the space, it is defined by an assembled, multi-layered realm of matter, signs and meaning. The cycles of meaning (constituted by perceptive and effective signs) each have a certain mode or style and their own particular semiosis.[62] But there are additional elements that help to grasp the moment of ecological change with respect to processes of exchange. In the analyses of Francis Bacon's paintings, Gilles Deleuze identifies three core elements: structure, figure and contour.

> This contour, as a 'place,' is in fact the place of an exchange in two directions: between the material structure and the Figure, and between Figure and the field. The contour is like a membrane through which this double exchange flows.[63]

Deleuze does not talk about a fixed system with respect to form, with boundaries that stabilise a self-preserving and organised system against a hostile and fluctuating environment. His concept contradicts system theories, which rely on the differences between a complex environment and an operated superior order.[64] The contour is a permeable boundary enabling processes of exchange between the different layers. The three aspects identified by Deleuze – form, structure and contour – are ecological components of a milieu. As Deleuze and Guattari write: 'the living thing has an exterior milieu of materials, an interior milieu of composing elements and composed substances, an intermediary milieu of membranes and limits, and an annexed milieu of energy sources and actions-perceptions'.[65] To conceive not only the human body as a milieu, a corporeal space that is changing its reciprocal relations, but also give heed to its counterparts as formative motifs and carrier of meaning, implies a shift in the angle of this investigation. We can now look at the inflections concerning corporeal space, medical practice and the involved media.

Symptomatology

I will attempt to apply the ecological concept of milieu and revaluate the examples of medical techniques discussed above. In medical practice we deal with differently coded fields and one can recognise that the ecologies of corporeal space concern the gaze and the language of the doctor, the intruding scalpel and hands of the surgeon, as well as the mediated images of the body. In each case one finds a certain kind of porosity between diverse fields of meaning. Next to the act of cutting the skin and touching the organs, there are many other intersections between surface, internal space and environment. Tableaus, X-ray photographs, brain scan images or even the printed curve of an electrocardiography are highly induced new arrangements of signs correlating between the physical body and the particular abilities of the measuring instrument. In each case one finds the corporeal space expanded into different milieus.

First, there was the directed gaze of the doctor, near to the body, mapping the outside space and its symptomatic signs, followed by the crucial step of converting the body's code, customising its signs and translating it. By translating symptoms into language these are introduced into a different field of meaning, thus into a new milieu. During this process, certain recognisable collections of symptoms are identified and named. In so doing, diseases are configured, and a new knowledge system, including a particular concept of illness, is produced. It is indeed fascinating that until today one finds a reciprocal dimension within this kind of diagnostic processes. Identification of a set of symptoms of a patient gives rise to a diagnosis. From there on the patient has a disease, which becomes evident in his corporeal space. The body, besides having its fleshy milieu with its own symptomatic signs, is now also part of a differently coded field of signs: the system of illnesses. Just as the fly is the counterpart for moulding the spider's web, the patient's body is 'framed' by the system of medical knowledge. In most cases the subjective symptoms of the patient and the correlated technically induced images are counterparts. The X-ray photo, for instance, offers a two-dimensional image of the three-dimensional body. The photos are made with the help of non-visible radiation. The radiation actually has to cross the body to shape the image. In that sense, the body functions as a counterpart par excellence. But it is the machine and its particular technical functioning that is responsible for the characteristic shape of the image. There is a reciprocal relationship at work between the patient's body and the X-ray machine and the X-ray image shows its own productivity due to the conditions of the technical device. As a result, one has a correlating technically induced body image, which is indeed geographically separated from the physical space of the patient's body, but nevertheless showing clear indications for shared milieus.

The actual cutting of the skin to open the body may be the most obvious act of crossing between milieus. We know that a certain stasis is necessary to keep the organism alive. This securing state, however, seems to be based on constant changes and adjustments. A living organism has to deal with diverse environmental situations, caused by changing seasons with many weather and climate variations. Furthermore, there is a strong influence by changing social and physical interactions with other organisms. Each situation is balanced by a combination of different means dependent on each unique subject's abilities. During and after an open surgery, the body reacts strongly but very often it is also able to cope with the situation and restore a 'healthy' stasis. Still, it is an understandable desire to want to enter the internal milieu of the organism without opening the skin (contour or membrane). In fictions one can find fantastic stories of travelling through the body. The internal milieu is turned into 'space', which can be entered for an adventurous trip. These kinds of narratives are usually about a crew and a ship shrunken to a microscopic size, which enters the body through a natural opening. The travellers will follow the blood circulation system through the organs and may later use a tear to exit the body again.[66] The internal organic space is presented as the unexplored and infinite outer space. This genre of fictions reflects on the scientific challenge of visiting the internal space of the body unnoticed. The endoscopic technique in fact comes quite close to this ideal. If an incision is needed, it tends to be very small, while the instrument is sterile and flexible. To place the view of the doctor inside the body appears almost like an unnoticed visit. The secret and technical gaze inside is nevertheless a productive one, reflected in the parallel video feed on a monitor. The surgeon's gaze is not directed towards the patient's body; he or she is guiding the endoscopic instrument with the help of the image displayed on the monitor. In that sense the surgeon operates within the corporeal space of the video feed. Based on the camera's mediation, the patient's body manifests itself in an additional geographical place, and becomes part of a different system of signs.

Today's endoscopic surgery replaces images not only geographically; it also translates them into digital form. Thus, the patient's corporeal space does not only exist in its fleshy form, but also as technical physicality in a video feed or a digital illustration. In the case of these new graphical images, all visual information is based on the digital system as a particular way of computation. The digital form is without further expression or flexible relationships and always formatted. It should be stressed that it is possible to directly address a particular pixel without having to traverse the precursor. The pixels, due to their continuous addressability, are more text than image and the computer graphics are therefore quite easy to manipulate.[67] In fact, it is not just possible to manipulate images, but also to produce entirely new images. Of course, these new images are an integral part of reality, and shall not be dismissed as somehow 'virtual' and therefore a less relevant part of the world. But there are particular complications between the digital and analogue domains that reveal themselves in the area of the optical and the sensory realm.[68] In the Visual Human Project the corpses have been literally scanned slice by slice, to be turned into a high-quality simulation. The virtual anatomy instruction is mainly visual, and van Dijk coins the term 'eyes-on-experience' since no further senses are involved. However, she emphasises that the digital cadavers and their virtual dissection offer a 'new body of knowledge'.[69] Within the VHP a seamless cross-over between three-dimensional images of organs, tissue and bones is possible. The perspective can be changed by mouse movements, and images can be rotated and manipulated. The interactive and virtual bodies are not merely represented but rather simulated and the VHP claims to constitute 'unmediated inscriptions'[70] of cadavers. The medical education with help of the VHP may not equate with a hand-on-experience, but it does appear perfectly suited for contemporary medical practice, in which physicians depend every day on representational technologies.[71] Furthermore, van Dijk concludes that the digital cadavers 'constitute a distinct continuation of age-old anatomical practices'.[72] One should give attention to the image as part of a mixed reality. Since the optical field as part of the sensory realm is a place where transcoding proceeds, one should acknowledge the image as an element of mixed realities. Body simulations and augmented reality behave like a membrane between digital code and human perception. Even if the computer graphics are entirely based on digital code, by generating an image that can be seen and spatially perceived, the digital information provides analogue stimuli. The digital realm based on code produces actual corporeal spaces while at the same time maintaining their milieu constituted by digital structures.

The world of medical practice, with its intruding instruments and diagnostic visualisation machines, is marked by an array of intersections between different milieus. There are many examples of the organic body pervading its corporeal space into the technical milieu. One should be wary to claim completeness when investigating these ongoing ecological processes. To grasp the multiplicity and the infinite character we do not need to pursue exhaustive historical research, but merely point to observable intersections. I would like to refer once more to Uexküll, who acknowledges that the incredibly large number of different milieus must lead to confusion, and that focusing on one set of arrangements is the methodologically correct response.[73] This chapter is a speculative but nevertheless practical and realistic approach towards an ecological understanding of the body. The ecology of corporeal space does not end with its own organic area and symptomatic signs. Next to the signs of the flesh, there is an array of symptoms that is based on medical instruments and media, systems of language and the rhythm of the digital code. The body as an ecological form is able to expand its range from a collection of subjective signs into the encrypted field of medical knowledge and digital formations.

Notes

This chapter is a revised and extended version of the paper 'Ecologies of Corporeal Space', in Radman and Kousoulas, eds, *3C: International Conference Proceedings*, pp. 39–50.

1. Foucault, *Birth of the Clinic*.
2. Ibid. p. ix.
3. Ibid. pp. 39, 40.
4. Ibid. p. 40.
5. Ibid. pp. 109ff.
6. Ibid. pp. 60ff.
7. Ibid. p. xviii

8. Ibid. pp. 135ff.
9. Ibid. p. 164.
10. Ibid. p. 3.
11. Brandt, *Illustrierte Geschichte der Anästhesie*.
12. Prentice, *Bodies in Formation*, p. 5.
13. Ibid. pp. 16ff.
14. Ibid. pp. 16f.
15. Ibid. p. 22.
16. Ibid. p. 21.
17. Van Dijck, *Transparent Body*, p. 119.
18. Prentice, *Bodies in Formation*, p. 20.
19. Ibid. p. 14.
20. Van Dijck, *Transparent Body*, p. 123.
21. Ibid. p. 66.
22. Ibid. p. 66f.
23. Ibid. pp. 84, 89.
24. Ibid. p. 94; Thacker, 'Vermittlung Und Antivermittlung'.
25. Ibid. p. 306. See the communication model by Claude Shannon and Warren Weavers.
26. Barad, 'Meeting the Universe Halfway', p. 167.
27. Ibid.
28. Ibid. p. 166.
29. Van Dijck, *Transparent Body*, pp. 84, 87, 89.
30. Prentice, *Bodies in Formation*, pp. 17ff.
31. Ibid. p. 17.
32. Ibid.
33. Van Dijck, *Transparent Body*, p. 86 (my italics).
34. Ibid. p. 99.
35. Ibid. p. 98.
36. Ibid. p. 98 (my italics).
37. Ibid. p. 71.
38. Tönnis, *Augmented Reality*, pp. 132–40.
39. Ibid.
40. Ibid. p. 133.
41. Bernard, *Leçons sur les phénomènes de la vie*, pp. 112ff.
42. Ibid. p. 117.
43. Schulkin, *Rethinking Homeostasis*, p. 2, n. 1.
44. Ibid. p. 16.
45. Ibid. p. 21.
46. Uexküll, *Theoretische Biologie*.
47. Uexküll, *Foray into the Worlds of Animals and Humans*, pp. 136ff.
48. Rüting, 'History and Significance of Jakob von Uexküll', p. 49; see also Buchanan, *Onto-Ethologies*.
49. Uexküll, *Foray into the Worlds of Animals and Humans*, pp. 128ff.

50. Ibid. pp. 44ff.; Deleuze and Guattari, *Thousand Plateaus*, p. 51.
51. Uexküll, *Foray into the Worlds of Animals and Humans*, p. 145.
52. Ibid. p. 150.
53. Ibid. p. 164; Bains, *Primacy of Semiosis*.
54. Uexküll, *Foray into the Worlds of Animals and Humans*, pp. 139ff.
55. Ibid. pp. 159, 164, 189, 202.
56. Ibid. pp. 158ff.
57. Ibid. p. 198.
58. Ibid. p. 202.
59. Ibid. p. 198.
60. Deleuze and Guattari, *Thousand Plateaus*, p. 314.
61. Ibid. p. 311.
62. Paul Bains speaks in the case of a whole functional cycle of 'a semiosis'. Bains, *Primacy of Semiosis*, p. 63.
63. Deleuze, *Francis Bacon*, p. 12.
64. Balke, 'Auf dem Rundgang Bilder des Lebens'.
65. Deleuze and Guattari, *Thousand Plateaus*, p. 313.
66. Van Dijck, *Transparent Body*, pp. 65f.
67. Kittler, 'Computergraphik'.
68. Colebrook, *Blake, Deleuzian Aesthetics, and the Digital*, pp. vii–xxxvi.
69. Van Dijck, *Transparent Body*, pp. 119, 125.
70. Ibid. p. 125.
71. Ibid. p. 124.
72. Ibid. p. 120.
73. Uexküll, *A Foray into the Worlds of Animals and Humans*, p. 133.

Bibliography

Bains, Paul, *The Primacy of Semiosis: An Ontology of Relations* (Toronto: University of Toronto Press, 2014).

Balke, Friedrich, 'Auf dem Rundgang Bilder des Lebens bei Franics Bacon, Gilles Deleuze und Martin Heidegger', in *Struktur, Figur, Kontur: Abstraktion in Kunst und Lebenswissenschaften*, ed. Claudia Blümle and Armin Schäfer (Zürich; Berlin: Diaphanes, 2007), pp. 317–38.

Barad, Karen, 'Meeting the Universe Halfway: Realism and Social Constructivism without Contradiction', in *Feminism, Science, and the Philosophy of Science*, ed. Lynn Hankinson Nelson and Jack Nelson (Dordrecht: Kluwer Academic Publishers, 1996), pp. 161–94.

Bernard, Claude, *Leçons sur les phénomènes de la vie communs aux animaux et aux végétaux* (Paris: J.-B. Baillière, 1878).

Brandt, Ludwig, *Illustrierte Geschichte der Anästhesie* (Stuttgart: Wissenschaftliche Verlagsgesellschaft, 1997).

Buchanan, Brett, *Onto-Ethologies: The Animal Environments of Uexküll, Heidegger, Merleau-Ponty, and Deleuze* (Albany: SUNY Press, 2008).

Colebrook, Claire, *Blake, Deleuzian Aesthetics, and the Digital* (London: Bloomsbury Academic, 2012).
Deleuze, Gilles, *Francis Bacon: The Logic of Sensation* (London: Continuum, 2004).
Deleuze, Gilles and Félix Guattari, *A Thousand Plateaus: Capitalism and Schizophrenia II* (Minneapolis: University of Minnesota Press, 1987).
Dijck, José van, *The Transparent Body: A Cultural Analysis of Medical Imaging* (Seattle: University of Washington Press, 2005).
Foucault, Michel, *The Birth of the Clinic: An Archaeology of Medical Perception*, trans. Alan Sheridan Smith (London: Routledge, 2003).
Kittler, Friedrich, 'Computergraphik. Eine halbtechnische Einführung', in *Paradigma Fotografie, Fotokritik am Ende des fotografischen Zeitalters, Bd. 1.*, ed. Herta Wolf (Frankfurt am Main: Suhrkamp, 2002), pp. 178–94.
Prentice, Rachel, *Bodies in Formation: An Ethnography of Anatomy and Surgery Education* (Durham, NC: Duke University Press, 2013).
Radman, Andrej and Stavros Kousoulas, eds, *3C: International Conference Proceedings* (Delft: Architecture Theory Chair in partnership with Jap Sam Books, 2015).
Rüting, Torsten, 'History and Significance of Jakob von Uexküll and of his Institute in Hamburg', *Sign Systems Studies*, Vol. 32.1, Issue 2 (2004), pp. 35–72.
Schulkin, Jay, *Rethinking Homeostasis: Allostatic Regulation in Physiology and Pathophysiology* (Cambridge, MA: MIT Press, 2003).
Thacker, Eugene, 'Vermittlung Und Antivermittlung', in *Die Technologische Bedingung: Beiträge Zur Beschreibung Der Technischen Welt*, ed. Erich Hörl (Berlin: Suhrkamp, 2011), pp. 306–27.
Tönnis, Marcus, *Augmented Reality: Einblicke in die Erweiterte Realität* (Berlin; Heidelberg: Springer, 2010).
Uexküll, Jakob von, *Theoretische Biologie* (Frankfurt am Main: Suhrkamp, 1973).
Uexküll, Jakob von, *A Foray into the Worlds of Animals and Humans: With a Theory of Meaning*, trans. Joseph D. O'Neil (Minneapolis: University of Minnesota Press, 2010).

CHAPTER 8

Swimming in the Joint

Rachel Prentice

With whose blood were my eyes crafted? (Donna Haraway, 1990)

A retired gynaecologist I know, citing nineteenth-century surgeon William Halsted, said that anything he can see, he can operate on. The statement appears to be self-evident. 'Exposure' is a surgeon's term for interventions that make injury or pathology available to vision and action. But what happens when new technologies reconfigure the relationship of hands, eyes, tools and patient body? This chapter examines the technical and perceptual skills that surgeons deploy as they work to see and to act upon patients' bodies. The surgeries recounted in this chapter exemplify moments when the relationships among seeing, acting and embodiment come into view, revealing how technology can lead to new perceptual experiences, but also how those experiences emerge from the broad cultivation of surgical embodiment.

Learning to see in surgery involves crafting more than eyes. Surgical sight emerges from a link between vision and action that is so tight that it urges reconsideration of the representational language of a medical gaze or disembodied cognition. Rather, sight comes into being with the embodied work that surgeons do. Sight and touch are intertwined; that is, they 'belong to the same world' in each individual's body and 'yet they do not merge into one'.[1] Put more simply, most people can sense what something will feel like when they see it and most can sense what something will look like when they touch it (for example, most of us can sense the roughness of a tree's bark before we touch it). All senses come into play during sensory interactions in ways that are typically taken for granted.

My work builds on eighteen months of fieldwork at three academic medical centres in North America, where I observed surgical procedures, interviewed surgeons and worked with a group building simulation technologies for teaching surgery. I became aware of the extent of physicians' perceptual training during an anatomy course that I took at the start of my ethnographic fieldwork in 2001. Often described as strictly visual, anatomical training actually is visual and spatial; trainees learn to identify and name parts by locating them in the body's three-dimensional volume. While observing surgeries in 2001 and again in 2006, I saw surgeons and trainees put their three-dimensional spatial sense to work in traditional surgeries and in minimally invasive surgeries. The language of representation is inadequate for describing such interactions. Neither an open operative site nor an operative site depicted on a monitor is an image the surgeon views. Rather, it is a three-dimensional space that the surgeon inhabits.

Accumulating Skill

The first lessons in the operating room – including scrubbing, maintaining sterility and obeying the staff – defamiliarise trainees with their own bodies, encouraging them to build a new, surgical stance toward the patient. Repetition of the small actions of surgery, such as retracting and stitching, aggregate and condense to become bodily habits. Years of cultivation of surgical habits leads to surgical skill, a term that surgeons use unflatteringly when qualified as technical proficiency alone, but that becomes high praise when accompanied by judgement and knowledge. Maurice Merleau-Ponty says that 'habit' is a 'rearrangement and renewal of the body image' through which the body becomes 'mediator of a world'.[2] 'Skill' connotes the effects of intentional training in ways 'habit' does not and can be defined as purposeful habituation. Building from this, 'craft' becomes 'skill, presence of mind and habit combined'.[3] The skilled body thus becomes the body habituated to intentional action through practice.

The accumulation of craft practices makes the body into a temporal joint that articulates past practices and present conditions. 'Our body comprises as it were two distinct layers, the habit-body and that of the body at this moment.'[4] Drawing on the experience of amputees, Merleau-Ponty illustrates this with the example of a phantom limb, which joins a past body habituated to life with the limb to a present body marked by its loss. In contrast, the habit-body that develops through practice in surgery becomes joined to a present in which variations in the milieu, such as new tools or unusual anatomy or pathology, generate improvisations that draw upon the general abilities of the surgeon's habit-body. The crafting of the surgeon's body also has a moral component: skill, judgement and accumulated procedural techniques all qualify the surgeon to undertake this high-stakes activity.

Merleau-Ponty's habit-body corresponds roughly to Pierre Bourdieu's description of how bodily habits can reflect and create culturally conditioned dispositions to act in particular ways.[5] But Bourdieu's 'habitus' is a more recognisably social concept: dispositions develop through practice in situations where symbolic, spatial and social structures instill particular ways of perceiving and acting.[6] Bourdieu does not address how the dispositions of the habitus adapt and change as worlds evolve. Saba Mahmood argues for practice as the embodying force, arguing that practices instill dispositions not only to act but also to believe in particular ways.[7] Treating the body as joining condensed social and physical practice with improvised action in the present allows us to consider continuity and change in embodiment. Changes in surgeons' institutional and technical worlds that become absorbed into the larger stream of surgical practice may impact their craft profoundly as they become incorporated into surgeons' embodied repertoires of skills.

When Sight Fails

I begin with two examples of open surgeries that disrupt any notion that surgical seeing is purely a product of highly trained eyes, revealing the broader embodiment at work. In these cases, surgeons see with their bodies. These two moments occurred during the same surgery. The patient was a middle-aged man with a Klatskin's tumour at the top of his bile duct. Dr Marcos Alexander, the hepato-biliary fellow, and Dr Jill English, the chief resident, made a long incision across the abdomen and had retracted ribs and reflected muscles and intestines to reveal the liver. I stood behind the anaesthesiologist's drape and looked over at the operative site. While Marcos and Jill worked to expose the tumour, the patient started to bleed heavily into his abdomen. The team kept working silently, looking for the source of the bleeding. The surgeons had nicked the patient's vena cava, the largest vein in the body, which returns all blood from the body to the heart. Dr Nick Perrotta, the attending surgeon, told the anaesthesiologist to call his chief, saying with typical surgical understatement, 'We've got a little bit of a problem.'

The vena cava runs deep, at the back of a curved abdominal space that cradles the liver. The upper abdomen was rapidly filling with blood. The surgeons could not see the vessel's path to the heart. Nick reached into the cavity, felt around, and then removed his hand and showed Marcos how he believed the vena cava ran. He held his palm upward and pushed his curved hand up and to the left, as though following the vessel's path as it ascended into the chest. He instructed Marcos to reach in and feel it. Marcos mimicked him, also tracing a curve to the left. The gesture both communicated the surgeons' understanding of the necessary action and rehearsed the proper movement. They repeated the movement of the palm until they agreed that they had identified the vessel's curvature. Nick repeated that the curved instrument would have to move to the left and not straight upward, where it would cause damage. Having traced the vena cava virtually, Marcos blindly slipped an instrument under the liver to lift it. Nick stitched the opening and, a few moments later, they closed the hole.

These surgeons could not see exactly what they were doing in this space. They could feel the path the retractor would have to take and they could demonstrate using hand gestures that they had felt it. The demonstration had two purposes: to confirm the venous anatomy's path and to rehearse the movements needed to slip the instrument under the liver. With this gestural practice, they easily lifted the liver and stitched the cut vessel. The gesture involved imagining and practicing, learning with their hands. In this case, knowing was based on accumulated practice and tactile signals. Both surgeons understood the vena cava's path with their bodies. But they could not, in the strictest sense, see it.

The second moment came during the same surgery. Late in this long operation, the blood pressure readout plunged. Blood pressure and other anaesthesia monitors extend the patient's body by making it emit signs that speak for the patient.[8] Monitors tell surgeons and anaesthesiologists alike whether the patient's body has destabilised. If the pressure drops, the surgeons must step away to give anaesthesiologists time to raise it. Glancing at the monitor, Jill, the chief resident, asked, 'Is this a real number?' The anaesthesiologist insisted that the numbers reflected a problem with their machines, not with the patient's body. The team proceeded. Nick asked the anaesthesiologists repeatedly if everything was OK. Each time he asked, he placed his hand inside the abdominal cavity and lifted his eyes to the monitors. The anaesthesiologists insisted that everything was fine, even while they rushed around trying to get their machines to work. They used a manual backup to ensure that blood pressure was adequate, but the surgeons could not see the numbers. The team completed the last steps of the resection, as well as the rest of the operation, without benefit of a monitor the surgeons could see.

After the operation, I asked Nick for his account of what happened. 'I could feel the aorta beating under my hand,' he said. When he placed his hand inside the abdomen, the strong pulse from the aorta told him that the patient's heart was pumping blood through his body adequately, defying the numbers on the screen. Machine and aorta told him different things. 'I would have preferred the numbers,' Nick said, wanting precision that his hand could not provide. In this case, though, Nick chose to trust his hand and its ability to understand the powerful pulse beating through the aorta, rather than the information provided by the machine. Touch, not vision, told him what he needed to know to continue the operation.

Clearly, vision was not the most significant sense in either moment: touch replaced sight in the first case and overrode it in the second. In the first moment, knowing meant virtually tracing the cava's path with one's hand. In the second, knowing meant trusting one's body despite the evidence given by one's eyes. Blood pressure provides important information in the operating room and there are many 'practices of engagement' that indicate blood pressure.[9] In this case, Nick chose to ignore the representation of blood pressure on the monitor in favour of information provided by his hand.

From Foucault onward, writers about medicine often have discussed the medical 'gaze', an amalgamation of sensory cues and an organisation of medical spaces, logics and apparatuses of knowing that could tell physicians what they would see if they could open the patient up at autopsy.[10] Foucault's concept of the gaze (*'le regard'*) represents a broad, socio-historical construction of perception that is quite close to the sense in which I discuss learning to see and act in surgery. But the gaze too easily comes to represent the slippage from vision to thought common to Western philosophical trends since Descartes and Locke.[11] The physicians and many technology designers I encountered while doing fieldwork tended to sublimate bodily knowledge under the cognitivist label of 'mental model'. Trainees' ability to locate venous anatomy in the body is an acquired skill that typically begins in the anatomy laboratory and becomes much more nuanced with surgical practice. However, shifting too rapidly into visual or cognitive language for this type of understanding easily elides other aspects of embodied knowing.

Inhabiting Minimally Invasive Space

Removal of a Klatskin's tumour is an open surgery: surgeons access the tumour through a large incision. The remaining surgeries I examine all were done using minimally invasive surgical techniques. Minimally invasive techniques involve threading a camera into a natural or artificial hole in the patient's body and performing the work while watching instruments on a monitor. The techniques that surgeons use today began to develop in the 1970s. Unlike traditional open surgeries, the technology moves the surgeon's eyes and hands away from the operative site as the surgeon uses instruments with long handles threaded through openings in the patient body while watching the action on a monitor that is typically located adjacent to and above the patient's body. Putting the action on a monitor is the first, critical move toward surgical simulation, robotic surgery and other remote surgical practices.[12]

The perceptual skills needed to work in minimally invasive space differ from those required during open surgery, leading to a new form of virtual embodiment. Surgeons have no direct manual contact with the insides of the patient's body. They also have a less direct kinaesthetic 'feel' for the body as transmitted through the instrument. Further, they cannot use their hands as probes, as Nick did when feeling the patient's aorta. They must continually extrapolate from a two-dimensional image to an interaction of bodies and instruments in three dimensions, sometimes making the screen space navigable for beginners by verbally identifying anatomical structures as they appear on the screen, giving them a stronger sense of movement through the body's volume.

The differences in embodied skills of beginners, competent practitioners and experts become very clear, very quickly. Amal Nassif was an earnest first-year resident, just beginning his second month of residency. While the team scrubbed for a second operation, a gall bladder removal, Tom asked Amal if he would like to hold the camera. He eagerly said yes. Once the team had anaesthetised and prepped the patient, Dr Cory Nguyen, the surgical fellow, inserted the ports into the patient's abdomen. She inserted the camera and handed it to Amal, saying, 'You keep the buttons up.' Cory began to dissect the gall bladder's connective tissues. She told Amal several times to rotate the camera or pull it back to keep steam from clogging the lens. After a half hour, Amal said, 'So "up" is looking down?' 'Yes', Cory said. Tom added, 'From the top down. That's what it means.' Amal's phrase is revealing: 'up' meant moving the camera's base upward, so the camera tip inside the patient's body pointed downward: seeing down required a counter-intuitive movement of hands and instrument. Amal had practiced with a simulator, but was just becoming aware of the effect of working over a fulcrum, which is just one way in which minimally invasive surgeries differ perceptually from open surgeries. He has not yet learned how to coordinate eyes, hands and instruments in this space. As will become clear from the examples that follow, experienced surgeons do more than look at the image depicted on a monitor. They treat these bodyscapes as three-dimensional spaces they work in, rather than as pictures they look at.

'Operating on Images'

Dr Harry Beauregard, a retired gynaecologist, began doing laparoscopic abdominal surgeries while looking through a microscopic eyepiece, an earlier generation of the technology. He found the late-career transition to the monitor alienating. During an interview, Harry said:

> It was the focus change from the patient to the monitor. That's where the action was and it was something I had to take into account. I mean I had to go there to do the work that the camera illustrated, allowed me to visualise ... And so I was leaving the patient and looking up to a monitor where there I could do stuff. And with the tools of the minimal access, if I looked at the patient, I couldn't do anything. You see how absolute it was? And I could look at the handles, but couldn't see on the inside. It was totally useless. I had to go to the monitor to operate. And that's why I started saying I was operating on images, not on patients.

Laparoscopy removed Harry's hands from the operative site. His hands were outside the patient and the operative site was hidden inside. To see what he was doing, Harry had to look at the monitor. 'I had to go there to work,' he said, because 'that's where the action was.' He described the operative site as though it moved to the monitor. The position of his hands did not change much from the microscopic system. But he talked as though his entire body moved with his eyes. He experienced himself as no longer working on patients, but on images. Vision and action came together on the screen.

I asked Dr Claire Franklin, a hand surgeon, to watch videotapes of a shoulder arthroscopy she had performed and to explain to me what she was doing. Unlike Harry, who was retired, Claire had performed minimally invasive surgeries since residency. The tapes depicted the scope's view, the view from inside the patient's body. The patient had torn his biceps tendon years before, destabilising his shoulder joint, wearing down protective cartilage and encouraging arthritic bone growth. Unlike the view of the abdomen through the scope, the view of the joint's interior was entirely unlike any anatomical view I have seen: the body looked abstract, like looking through a porthole at a red and white undersea floor with white tendrils undulating in the current. The view made me mildly seasick.

As the camera advanced into the shoulder, Claire said, 'This is somebody with a terrible shoulder, a terrible shoulder.' She described how she was running fluid through the joint, hence the sea-floor effect.

Claire: That's the outflow. It's also the cannula for instrumentation in the front. This is not a good first one for you to look at. This is his humoral head and there's just a lot of arthritis, a lot of fibrillation.
RP: So arthritis, it's not like a neat bone growth, it's this messy crap?
Claire: It's messy crap. It's just bare, bare bone. I'm coming from behind him and the glenoid is on our left and the big ball is on our right. So the camera is with me. It's kind of my view from chest level. So here I'm probing. I'm proving that he's got an arthritic shoulder. This is the remnant of his biceps tendon. This will definitely make you dizzy.
RP: You feel that or you see that?
Claire: I put through the cannula in front. I take the probe. That's my finger extender.

Claire had several ways of opening the operative site. The first was navigational. She named anatomical structures, such as the ball of the humerus and the glenoid, or shoulder socket, as they came into view. Verbal navigation can be particularly important with minimally invasive surgery because the two-dimensional view can be deceptive and requires skill to read. She also called out anatomical landmarks in the operating room with residents and hand surgery fellows. Every surgeon I have watched does this with minimally invasive procedures. But I have never seen surgeons do as much narration during an open surgery. Claire said she names the anatomy in part to establish common ground with the surgical team.

The second method of establishing the operative site was through probing. Claire tells her medical students that instruments are extensions of her body. She used the probe as an extension of her finger as she explored the arthritis. Probing the arthritis was the important action, not holding the probe. Maurice Merleau-Ponty uses the example of a blind man's cane to show how we use instruments to extend our senses – that is, our bodies and ourselves – towards objects in the world.[13] He argues that the cane extends the blind man's bodily consciousness into space. What remains unstated is that, like a surgeon, the blind man, too, needs practice to navigate in his world. Thus, the cane has structuring effects on the holder's experience.[14] Further, tapping does not resemble navigating either by sight or by unmediated touch. Both cane and probe have structuring effects on their respective users' embodiment.

Claire said something that suggests broader surgical embodiment: the image was her view 'from chest level'. This was an odd statement. We do not have eyes in our chests and, thus, have no view from chest level. But Claire located her body in relation to the patient's body. 'The camera is with me,' she said. She was standing behind the patient's right shoulder. Shoulder and scope were level with her chest. Just as the probe became her finger extender, the scope became her eye extender. Claire extended this technological eye from her body's position in space to the patient's body. Action began with her body and extended from there. It became a view from chest level. I asked for more detail:

RP: So then you feel the arthritis or you see it?
Claire: Yeah, both. It's very much a proprioceptive thing.

Claire proved that the man had an arthritic shoulder by sight and by feel. She verified the arthritis by probing the tissue's hardness. She said her identification of the arthritis was proprioceptive. Proprioception is our sense of our body in space, the sense that allows most of us to know, for example, where our left foot is without looking at it. Claire's statement that the visual and tactile confirmation of arthritis was proprioceptive appears to conflate vision with proprioception and seems to be incorrect. But I believe that something subtler was at work. Claire oriented her body and instruments so she could best see and operate on the patient's body. In this instance, the connection between sight and action was so tight that vision and proprioception merged. Claire saw and probed the arthritis through her body. She extended multiple senses to the operative site to make the diagnosis.

On the video screen, another right shoulder appeared. Claire used a probe to gently flick a small knob of flesh on the shoulder. She described this as 'physical doodling'. In other words, she was thinking about what she could do with this injury.

> Actually what I'm doing also is, I'm externally rotating the shoulder to see the tension of the muscle. [She points.] That's the middle gleno-humeral ligament. You can see the glenoid here. And you can see the humeral head over here. This is also a right shoulder. There's a lot of fibrillation coming down and actually this is probably going to be some of the rotator cuff falling in our face. The fibrillation is that gunky stuff. This is the rotator cuff tear.

Claire located us in relation to the anatomy and diagnosed the injury. She identified messy white tissue descending into the frame as the rotator cuff tear, describing it as 'falling in our face'. This odd grammatical construction suggests several aspects of the embodiment at work. Tissue waving against the camera lens showed up on the monitor as tissue blocking our view. Claire, who sat next to me in a computer lab, placed our faces at the meeting place of tissue with camera, oddly merging both our faces with the technological interface. This suggests that the apparatus of camera and monitor structured her perception: the monitor has just one 'face', the camera lens inside the joint. Multiple people can peer through it, however, so it became 'our face'. In other words, Claire located our faces at the interface of the camera with the interior of the patient's shoulder joint. I have heard Claire make statements like this several times, but only while doing arthroscopy, never while doing open surgery. Unlike Harry, she did not experience this as alienating. She described herself as part of a joint when she does arthroscopy. Instead of separating operative site from patient, the scope became part of Claire's surgical embodiment as they moved together inside the patient's body. The scope exerted its own agency in shaping her perception. Claire described her view as inside the patient's body and on the same scale as the magnified view of the shoulder's interior. As with the description of her 'view' from chest level, Claire and technology created a new form of embodiment. Put differently, camera and instruments have become incorporated as part of the embodied assemblage that allows her to work in minimally invasive space.[15] Eyes and instruments merged at the operative site.[16]

I discussed a similar relationship with Dr Ramesh Chanda, another hand surgeon, who also trained using both open and minimally invasive techniques. What he said is worth quoting at length:

> You have an image on the monitor. You have this thing in front of you, which is the actual patient, the patient's joint or whatever. In addition to this, there is also a third image, and that's the image, which is in your head. And it's a combination of the two, the patient's image that you see and the stuff you see on the monitor, but also takes into consideration some cognitive aspects, some other issues, the haptic feedback you are getting from your hands. It's a mental model or image or whatever and what I have felt as I have gone through my training is that I have tended to use that third image more and more, which in some ways draws upon what I am seeing on the screen, draws upon what I am faced with in front of me and am touching and holding and manipulating. So I am almost like, I almost imagine myself, almost routinely if I am doing an arthroscopy, sitting inside the joint. And I say, Oh, OK, I am looking up and I see the scaphoid or whatever, if I am in the wrist joint, for example. And of course the images on the screen are very important [for] guiding, in fact probably the most important. You certainly cannot do without that. But there are other pieces of information and that, in the end, becomes a guide.

Ramesh creates a composite mental image of the patient's joint. This image unites the on-screen visual, the kinaesthetic and tactile information coming from instruments and the patient's body, as well as his experience and knowledge of anatomy. Claire and Ramesh both struggled to describe their experience. Claire said, 'It's kind of my view from chest level.' Ramesh said, 'It's a mental model or image or whatever.' Both are unusually articulate people, but these moments of imprecision reveal how perplexing some of these perceptual issues are. They wrestled to describe what they know with their bodies. Claire's body merged with the scope; Ramesh dispensed with the technology and its limitations. He inhabited the patient's body when he operated. He said he did not have this experience when doing open surgery. I encouraged him to say more:

RP: It resonates very strongly with something [Claire] said, which is that when she is operating on the shoulder, she is part of the shoulder.

Ramesh: Yes, that's exactly how I feel.

RP: If you're thinking of yourself as inside the joint, do you actually position yourself, like my eyes are sitting on this piece of anatomy looking at whatever?

Ramesh: Yeah, and actually I would say I am sitting on that piece of anatomy, or rather that you are floating around, swimming around in the synovial fluid, so you can move around, look up, look down, look right, left. And actually the other thing is that you can also, in that mental model, come out of the joint. You can go in and out very easily, so you can visualise it from the outside. You can visualise it from the inside.

Ramesh imagined his whole body inside the patient's body. One could think of this as the disembodied gaze promised by writers about virtual reality.[17] But examining what Ramesh does while swimming in the joint reveals that this formulation is misleading. He looks at a monitor and, often, rotates the joint from the outside while using a probe to examine the internal effects of rotating, engaging with the joint visually and physically. Ramesh's actions condition his ability to mentally place himself in synovial space. He described his 'mental model' as a composite of the view on the monitor with other sensory information, especially kinaesthetic information. And he gave himself imaginary abilities that were technologically unavailable, such as the ability to move out of and back into the joint at unusual angles.

Multiple sensations are in play in minimally invasive surgery, including what the surgeon sees on the monitor, the tactile and kinaesthetic sensations transmitted from the instruments to the surgeon's hands and the surgeon's proprioceptive sense of their body in space. The real-time sensations of surgical action come together with the embodied skills developed in practice, as well as with knowledge of human anatomy and surgical procedure. Harry, perhaps because he did not begin his training using minimally invasive technology, experienced the patient as split in two: the image on the monitor and the patient's body. He left the patient's body to work on images. The two hand surgeons, in contrast, did not consider the monitor as such. The scale of their bodies was radically reduced, focused at the meeting place of scope and joint.

During a discussion over coffee, I asked the three surgeons to speculate about the differences. Harry gave two possibilities. The first related to when in their careers they began working with a monitor. He started late in his professional life. Harry also suggested that this difference could relate to the size of joints versus abdomens. He described the abdomen as a large room with darkened corners. It does not feel confined. Claire picked up his metaphor and began to play with it. 'A shoulder is like a closet', she said. 'Only it's like a California closet where everything should be neatly tucked away.' The joint-as-closet analogy creates a strong sense of the joint's confined spaces and the disorder that pathology creates. The patient's body contributes to these perceptual effects. Ramesh agreed with Harry and Claire and added that surgeons who work in the abdomen do not manipulate the body from the outside while viewing it from the inside. They manipulate instruments, but do not rotate limbs to see what changes inside the way that orthopaedists do. This suggests that arthroscopy – minimally invasive surgery in joints – more tightly links the surgeon's body and the patient's body.

Much later, after he read an early draft of this chapter, Harry found another explanation for the differences. He said that he is a man who worked on women's bodies. These bodies were unlike his own. The intimacy the hand surgeons experience would feel inappropriate, he said. As a gynaecologist, he spent his career with hands, instruments and eyes in intimate contact with women's bodies. But somehow, inhabiting a woman's body would have felt transgressive, suggesting a difference between putting his hands or eyes into the patient's body and putting himself into the patient's body. Thus, these perceptual relationships can be shaped by gender and cultural experiences. In other words, he objectified his own body as it came into intimate contact with the patient's body.

Conclusion

What do we make of these seemingly bizarre perceptual relationships? If Harry experienced his work site as the image, is the image 'just' a representation? And do these examples represent a complete departure from the bodily relations of open surgery?

I use surgical embodiment to explain the alterations in a surgeon's location or scale that can occur with minimally invasive surgery. As Harry said, surgical seeing and acting are inextricable. I want to take this further to consider how a 'paradigm of embodiment' as developed by Thomas Csordas can help explain sensory experiences as they are distributed by remote technologies.[18] Csordas describes embodiment as the existential ground of culture, showing how the body's actions and perceptions are culturally informed.[19] Our culturally informed bodies shape perceptions before they turn into ideas, concepts, representations and other abstract objects.

Surgeons cultivate specific habits of perception and thought during years of training. Their visual and tactile perceptions are highly trained. They must learn to identify anatomy in indistinct flesh. They must also distinguish normal from pathological tissues. And, unlike anatomists, they must learn to repair the patient's body while doing as little extraneous damage to it as possible. While surgeons learn their craft in the operating room, the symbolic, spatial and social structures of the operating room instill particular ways of perceiving and practicing upon the patient's body. New residents mince around the operating room and shrink away from the patient's body, afraid to touch anything. They rapidly gain confidence, but even fellows, who are fully qualified surgeons, have much to learn from more experienced surgeons. The differences in confidence, economy of motion and skill among surgeons with different levels of experience became very obvious to me after observing just a few surgeries.

Surgeons learn the body even as they create the body they want to learn. That is, a surgeon learns by repeated sculpting of anatomy in a patient's body. Practitioner's body and patient's body mutually articulate each other.[20] When blood, tissue, anatomy or technological failure disrupt sight, the surgeon uses other senses to envision the patient's body. When technology distributes the patient's insides and outsides, the equally distributed surgeon reunites it through the circuit of his or her own body. Harry, Ramesh and Claire demonstrated this. Harry talked as though his hands followed his eyes to the monitor. Claire talked about arthritis falling in our collective face.

And Ramesh described himself swimming in the joint. These surgeons' entire bodies were focused on the operative site, where seeing and acting come together.

The body's sense of space is brought into being through action. This is not a spatiality of position, but a spatiality of situation.[21] When spatial perception is radically altered by technology, the body creates a perceptually altered 'virtual body' that, according to Merleau-Ponty, 'ousts the real one to such an extent that the subject no longer has the feeling of being in the world where he actually is . . . [H]e inhabits the spectacle.'[22] The operative site, whether experienced on a screen or in the flesh, becomes a space that the surgeon inhabits.

The situation of surgery is to see and to act at the operative site. From an external, objective point of view, the surgeons' perceptions of themselves operating on images, or waving arthritic tissues out of their faces or swimming inside the joint appear bizarre. The surgeons themselves, if asked about the actual positions of their bodies in the operating room, would objectify their actions, describing themselves as an observer would. But these statements become clearer if we imagine surgical embodiment as developing from lengthy residence in a surgical culture dedicated to the art of seeing to intervene and intervening to see. The distributed bodies, instruments, sensations and knowledges focus on a single event: opening the operative site so the surgeon can work there. The work of surgery unites the surgeon's body, technologies for seeing and acting and the space of the patient's body. Harry located himself at the monitor so he could see enough to work. But he experienced his attention as divided between patient and monitor. Claire and Ramesh took a more radical step. They located their bodies in the one place where a person could see and operate without being divided – at the actual operative site, inside the joint.

Notes

This chapter is abridged from Rachel Prentice, *Bodies in Formation: An Ethnography of Anatomy and Surgery Education* (Durham, NC: Duke University Press, 2012).

1. Merleau-Ponty, *Visible and Invisible*, p. 134.
2. See also Merleau-Ponty, *Phenomenology of Perception*, pp. 142–3.
3. Mauss, 'Techniques of the Body', p. 58.
4. Merleau-Ponty, *Phenomenology of Perception*, p. 82.
5. Csordas, 'Embodiment as a Paradigm for Anthropology' and Csordas, *Embodiment and Experience*.

6. Bourdieu, *Outline of a Theory of Practice*.
7. Mahmood, *Politics of Piety*.
8. Hirschauer, 'Manufacture of Bodies in Surgery', p. 290.
9. Barad, *Meeting the Universe Halfway*, p. 133.
10. Foucault, *Birth of the Clinic*.
11. Rorty, *Philosophy and the Mirror of Nature*, pp. ix–xiv.
12. Satava, 'Virtual Reality for Medical Application', pp. 19–20.
13. Merleau-Ponty, *Phenomenology of Perception*, p. 143.
14. The blind man's cane as philosophical example has a long tradition beginning with Descartes's Fourth Discourse and continuing through Heidegger, Merleau-Ponty, Polanyi and others. Philosophers have used the cane to discuss the relations of objects to cane to human. Here, I am most interested both in the structuring effects of the technology (cane or minimally invasive surgery – MIS – camera) on the experience of the viewer (the phenomenological perspective), as well as the structuring effects that the technology has on that perception, an aspect of the relations of objects, bodies and apparatuses also discussed by Karen Barad, pp. 157–61.
15. See Barad, p. 157; Prentice, *Bodies in Formation*, pp. 56–7.
16. For an extraordinary discussion of perceptual changes that occur with technological mediation, see Reeves and Nass, *The Media Equation*.
17. Balsamo, *Technologies of the Gendered Body*; Gibson, *Neuromancer*.
18. Csordas, 'Embodiment as a Paradigm for Anthropology'.
19. Rajchman, 'Foucault's Art of Seeing'.
20. See Prentice, 'The Anatomy of a Surgical Simulation', and Prentice, *Bodies in Formation*.
21. Merleau-Ponty, *Phenomenology of Perception*, p. 100.
22. Merleau-Ponty, *Phenomenology of Perception*, p. 250.

Bibliography

Balsamo, Anne, *Technologies of the Gendered Body: Reading Cyborg Women* (Durham, NC: Duke University Press, 1996).

Barad, Karen, *Meeting the Universe Halfway: Quantum Physics and the Entanglement of Matter and Meaning* (Durham, NC: Duke University Press, 2007).

Bourdieu, Pierre, *Outline of a Theory of Practice*, trans. Richard Nice (Cambridge: Cambridge University Press, 1977).

Csordas, Thomas J., 'Embodiment as a Paradigm for Anthropology', *Ethos*, Vol. 18 (1990), pp. 5–47.

Csordas, Thomas J., *Embodiment and Experience: The Existential Ground of Culture* (Cambridge: Cambridge University Press, 1995).

Foucault, Michel, *The Birth of the Clinic: An Archaeology of Medical Perception*, trans. Alan Sheridan (New York: Vintage Books, 1973).

Gibson, William, *Neuromancer* (New York: Ace Books, 1986).

Haraway, Donna, 'Situated Knowledges: The Science Question in Feminism and the Privilege of Partial Perspective', in *Simians, Cyborgs, and Women: The Reinvention of Nature* (New York: Routledge, 1990), p. 192.

Hirschauer, Stefan, 'The Manufacture of Bodies in Surgery', *Social Studies of Science*, Vol. 21 (1991), 279–319.

Mahmood, Saba, *The Politics of Piety: The Islamic Revival and the Feminist Subject* (Princeton: Princeton University Press, 2005).

Mauss, Marcel, 'Techniques of the Body', in *Beyond the Body Proper: Reading the Anthropology of Material Life*, ed. Judith Farquhar and Margaret Lock (Durham, NC: Duke University Press, 2007), pp. 50–68.

Merleau-Ponty, Maurice, *Phenomenology of Perception*, trans. Colin Smith (London: Routledge & Kegan Paul, 1962).

Merleau-Ponty, Maurice, *The Visible and the Invisible*, trans. Alphonso Lingis (Evanston: Northwestern University Press, 1969).

Prentice, Rachel, 'The Anatomy of a Surgical Simulation', *Social Studies of Science*, Vol. 35 (2005), pp. 837–66.

Prentice, Rachel, *Bodies in Formation: An Ethnography of Anatomy and Surgery Education* (Durham, NC: Duke University Press, 2012).

Rajchman, John, 'Foucault's Art of Seeing', *October*, Vol. 44 (1988), pp. 88–117.

Reeves, Byron and Clifford Nass, *The Media Equation: How People Treat Computers and New Media Like Real People and Places* (Chicago: University of Chicago Press, 1996).

Rorty, Richard, *Philosophy and the Mirror of Nature* (Princeton: Princeton University Press, 1979).

Satava, Richard M., 'Virtual Reality for Medical Application', *Information Technology Applications in Biomedicine, 1997. ITAB '97. Proceedings of the IEEE Engineering in Medicine and Biology Society Region 8 International Conference*.

CHAPTER 9

Key-Hole Surgery: Minimally Invasive Technology

Jenny Dankelman

Introduction

Imagine that it would be possible to treat any patient in an early phase of a disease at any location in the body using tiny instruments that cause no trauma to healthy tissue. Our challenge is to develop a new generation of devices with highly advanced functionality at the tip for instantaneous diagnosis and targeted treatment, allowing diagnosis and therapy in one single procedure. Examples are thin steerable needles, catheters with multifunctional tips that can be manoeuvred in branching blood vessel systems and water jet cutting to treat cartilage in narrow joints. Tailored, slender and multi-steerable instruments, with interior optical fibres, can be used to transmit diagnostic information and to perform local treatment during currently challenging oncological and cardiovascular procedures.

There are many benefits for the patient and the costs for staying in the hospital are lower compared to open surgery. However, the minimally invasive approach is more challenging for the surgeon and interventionist. When slender instruments are inserted through tiny incisions in the skin or through tiny openings, the Degrees of Freedom (DOF) are reduced, depth perception is limited due to the 2D view on a monitor and haptic feedback is inhibiting by friction in instruments.[1] Therefore, psychomotor skills training is required. To support training, we develop training systems to enable residents to learn outside patients. These training systems are based on fundamental research into the field of eye–hand coordination, haptic feedback and objective assessment methods for psychomotor skills.

With the introduction of new technology in the operating room (OR), the environment has become technologically complex and prone to errors.[2] The complexity of the perioperative process, with its involvement of many people from different disciplines, makes structured and meaningful extraction of information about the workflow highly challenging. Therefore, improving the safety and efficiency of the perioperative process requires reliable and objective systems to monitor the workflow throughout all phases of the care pathway. To improve the monitoring of the workflow in the OR, we started the Digital Operating Room Assistant (DORA) project.[3] In this project we aim to automatically monitor activities such as instrument and equipment use, and give feedback when hazards occur related to instrument and equipment use. Furthermore, we aim to automatically determine the workflow in the OR for better OR planning and to determine when processes deviate from standard.

MISIT group: Within the Minimally Invasive Surgery and Interventional Technology (MISIT) group we work in a multidisciplinary team to improve minimally invasive techniques (www.misit.nl). The group consists of mechanical, biomedical and electrical engineers, industrial designers and clinicians. The MISIT group works on the development of novel instruments, performs research on the interaction between instrument and tissue and, finally, develops methods to improve the interaction between instruments and the users in the operating room environment.

In this chapter, the advantages and limitations of minimally invasive techniques will be described first. Then the link between minimally invasive techniques and clinical cartographies, making a representation (map) of the patients' anatomy, will be made. An important aspect during the application of minimally invasive procedures is the indirect way that visual and haptic information are fed back to the user. Aspects such as misorientation (an incorrect orientation), change of coordinate system and uncanny valley will be discussed. Minimally invasive applications: there are many different applications for treating patients in a minimally invasively manner. During minimally invasive surgical procedures (Fig. 35), also known as key-hole surgery, tissue is treated through a few tiny incisions in the skin. During key-hole surgery in the abdomen, to remove, for example, the gall bladder after identification of gall bladder stones, first the abdomen is inflated with carbon dioxide to create space for the intervention and then tiny hollow cannulas are inserted and placed in the skin through which long slender instruments, such as graspers and scissors, can be inserted. An endoscopic camera is also inserted through one of the cannulas to provide a view of the operating field to the surgeon. Other examples of minimally invasive procedures are arthroscopic meniscus procedures, needle interventions

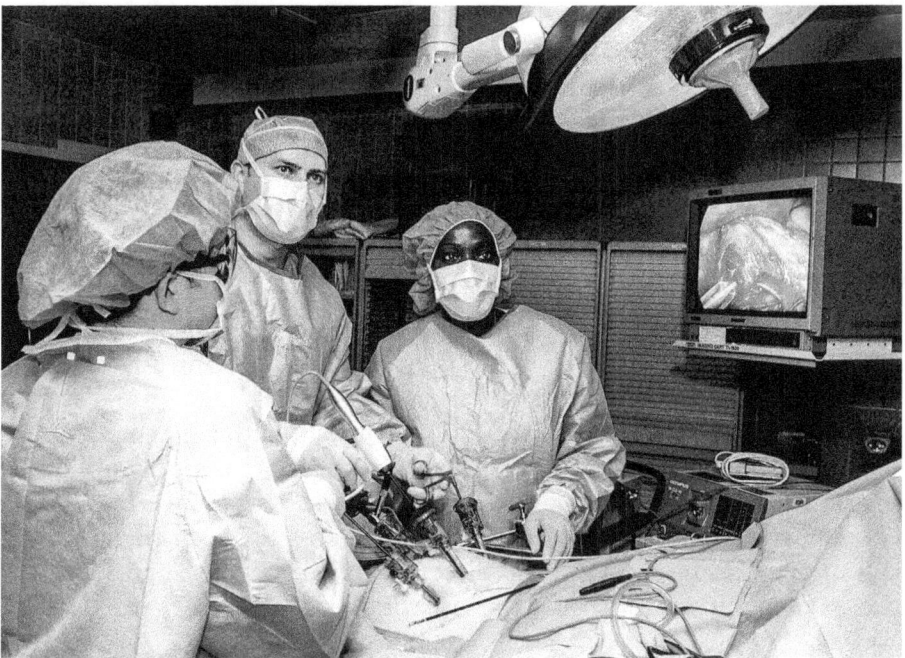

Figure 35 Minimally Invasive Surgery.
Source: author.

to take a biopsy, catheter interventions such as Dotter procedures to open up narrowed vessels with a balloon, hysteroscopic procedure to remove myoma from the uterus via the vagina and scull-base surgery to perform brain surgery via the nose.

Advantages for the patients: The number of minimally invasive interventions has increased tremendously during the past years. An important reason for this increase is the improved instruments, making minimally invasive procedures possible and yielding advantages to patients. During a minimally invasive procedure there is less damage to healthy tissue, reduced risk of wound infection, improved cosmetic outcome, shorter recovery time, reduced postoperative pain and a shorter hospital stay, which results in a faster resumption of regular activities. The latter result also reduces working hours of the nursing staff. Although the cost of instruments used during the procedure is higher, a minimally invasive procedure is mostly cost effective because hospitalisation is shorter. A minimally invasive approach may in some cases also be the only opportunity to treat very ill patients. An open procedure may be too invasive and a minimally invasive procedure can be the only possibility that the patient can handle. Recently, a new aortic valve was developed so that very ill patients can receive a new valve percutaneously with the use of a catheter.

Limitations for the surgeons and interventionists: Compared to open procedures, the minimally invasive surgical procedure implies many limitations for the surgeon and interventionist. Long instruments are inserted through small incisions, limiting the movement possibilities induced by the so-called fulcrum effect.[4] The degree of freedom to move the instrument is reduced (the number of DOF changes from six to four). Movements of the tip depend not only on the movement of the handle, but also on how far the instrument is inserted, so a scaling effect occurs. An endoscopic camera is inserted presenting a 2D image on the screen. Due to a misalignment of the natural line of sight with the camera orientation, misorientation can occur, limiting the ease of manipulation of the instruments and inducing hand–eye-coordination problems (Fig. 36).

Due to the interposition of long instruments between the hands of the surgeon and the tissue, the direct coupling between hand and tissue is lost.[5] This results in reduced haptic feedback, so that feeling of

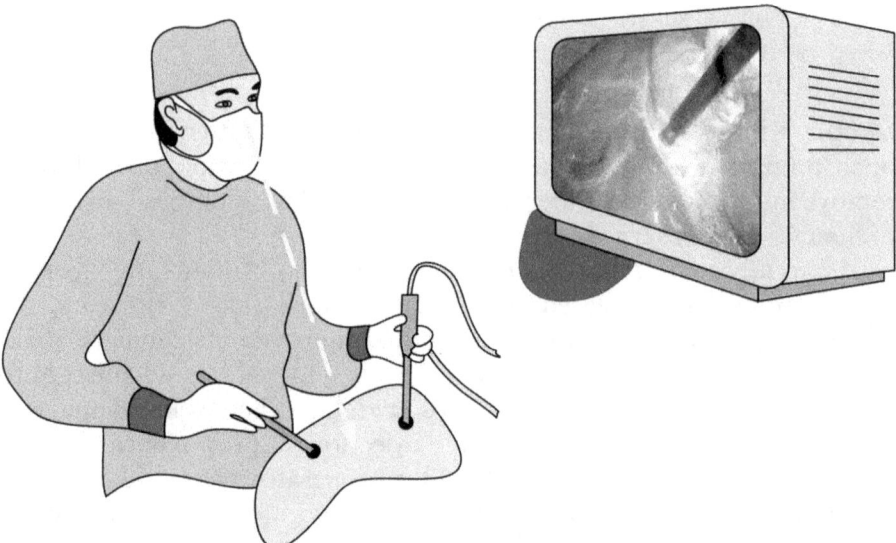

Figure 36 Misorientation due to misalignment of natural line of sight with the camera orientation; adapted from Mark Wentink, 'Hand-eye Coordination in Minimally Invasive Surgery: Theory, Surgical Practice & Training' (PhD Thesis, TU Delft, 2003).

Source: author.

tissue properties, and information of the forces applied to the tissue is disturbed, and this may lead to unintended tissue damage.[6] Finally, since holding the long slender instruments for a long time in a less than ergonomic position, performing minimally invasive surgical procedures frequently causes pain in the surgeon's fingers, wrist and shoulder, but also in the long term, hernias in lower back and neck.

Role of Technology/Instruments

As indicated above, there are many limitations for the surgeon when performing a minimally invasive procedure. To overcome these limitations several new instruments are under development to give the instruments more DOF and sensors to detect tissue properties. Breedveld worked on the development of tailored slender multi-steerable instruments (comparable, for example, to the bodily configuration of a snake)[7] (Fig. 37). Loeve et al. worked on instruments with adaptable stiffness (flexible during steering, and stiff as a platform for tissue manipulation) and with transmission of force information using new sensors, such as biocompatible force sensors.[8] Additionally, new technology is being developed to support instrument navigation, because during, for example, needle interventions, a platform is required for integrating positioning determination (global positioning system (GPS)) or image-guided positioning in relation to the anatomy of the patient. To achieve this, we integrated optical sensors (Fiber Bragg Gratings in optical fibers) in

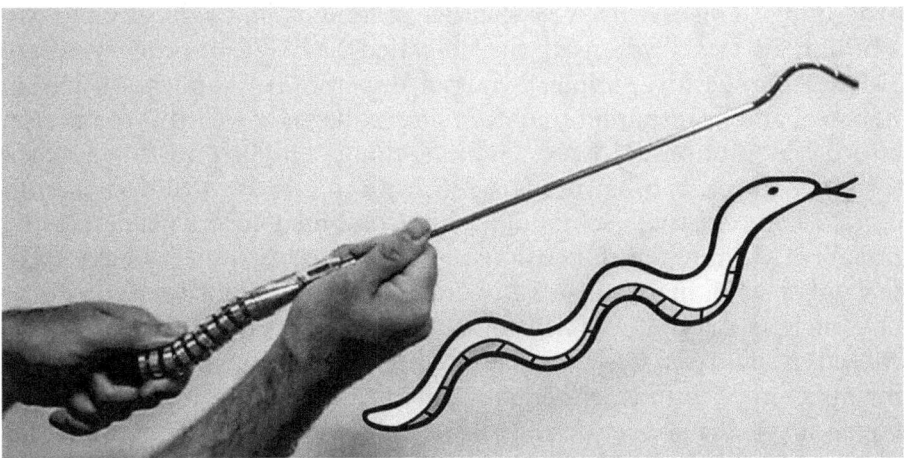

Figure 37 Multi-steerable instrument.
Source: Paul Breedveld.

steerable needles enabling accurate position determination without the use of other imaging systems such as X-rays.[9]

Robotics is a popular field within this area. Several systems have been developed to overcome the limitation related to the fulcrum and misorientation. One system, the Da Vinci system, is commercially available and broadly used for difficult minimally invasive procedures. During robotic procedure, the movements of the handles held by the surgeon are translated to movements of the instrument tip within the patient's body. The limitations caused by fulcrum are omitted and a 3D view is presented. The main limitations of currently used robotic systems, which may cause unsafe situations, are the reduced haptic feedback because force feedback is lost, the bulkiness of the system and the fact that the surgeon sits behind a console in their own environment removed from activities around the operating table,.[10] These limitations and the high costs motivate many groups in the world to work on alternative (robotic) systems such as handheld devices.

SMART Approach

Robotic solutions in a sterile environment with professionals without an engineering background result in a technologically complex setting, which is not always safe and the outcome for the patient is not always improved.[11] Therefore, we think that robotics is not always the preferred solution to overcome the limitations of minimally invasive procedures. This is why we, when we started as an engineering group working in this field, used the so-called SMART approach, where SMART stands for: Simple, Minimal dimensions, Application based, Reliable, and Transparent to use. S, simple, because we want to develop affordable solutions. However, making things simple is not an easy task for an engineer. It often needs a much more creative solution than in cases where existing technologies are combined to create new functionality. M, minimal dimensions, because the instruments have to fit through a small incision, so automatically the solutions for extra function such as steering and sensing should have minimal dimensions as well. A, application based: when looking for solutions that can be used for many applications, either the instrument does not work optimally for each specific application, or becomes too complex. Therefore, it is better to optimise the instrument for a specific task. R, the instruments used should be reliable and surgeons should be able to use them in a safe way. T, transparent to use means that it should be ergonomic and the design should be such that it is immediately clear how it should be used. This SMART approach does not fit very well with robotic solutions. That is why researchers from the MISIT group in Delft started to look into ways to come up with more simple mechanical solutions. For steering, Breedveld came up with a very creative idea inspired by nature.[12] He investigated how squid can move their tentacles. In a tentacle a ring of muscles oriented in the longitudinal direction is encircled by other muscle cells. Breedveld translated this idea into a mechanical solution for tiny long instruments. By using a ring of wires enclosed between springs, a much simpler (thus SMART) solution was discovered than the more traditionally used steerable mechanisms using four cables needing a complex construction to keep them in place. Other examples of SMART solutions are the use of water jets to make holes in bones in joints (to promote recovery of cartilage), and the use of isolators to make surgical procedures outside a sterile environment possible.[13]

Cartographies and Minimally Invasive Surgery

During surgery it is important to have a good representation (map) of reality. This is not simply for visual information, but also for information about other tissue properties, such as, for example, stiffness of tissue structures. The special setting of operating through tiny incisions introduces some clinical cartographical challenges.

Visualisation of the Operating Field

Breedveld[14] wrote a good overview of the specific observation aspects that comes with the minimally invasive approach. Visual feedback is provided by a camera (endoscope) and the surgeon looks at a monitor resulting in an indirect way of observing tissue. This is not the only limiting factor. The limited depth perception due to the absence of shadows, stereovision, movement parallax and the unnatural line of sight of the endoscope disrupt the surgeon's hand–eye coordination.

Limited Degree of Freedom

Manipulations are limited due to the incision. Breedveld et al.[15] offer an insightful overview of these aspects. The limitations of manipulation concern both spatial and grasping movements. The differences from moving instruments in an open environment to having them work through a tiny incision could be seen as a change of coordinate systems. We are normally used to manipulating our instruments in a 3D free space with x,y,z movements and three rotations – a total of six DOF. As soon as our long instruments have to move through an incision, the freedom of movement is limited to four: in–out movement, axial rotation and two angular movements in the polar coordinate system (Fig. 38). Moreover, the movements of the hand are mirrored and scaled in relation to the movements of the instrument tip. This is often called the 'fulcrum' effect and hampers natural hand–eye coordination.

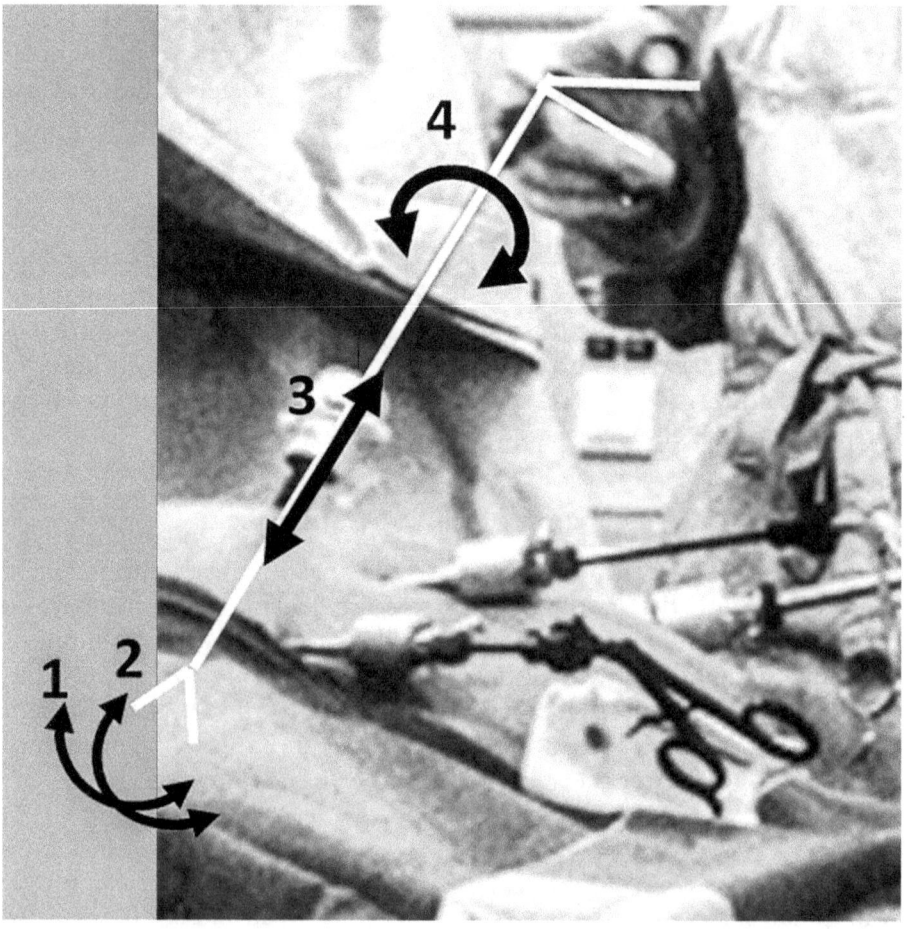

Figure 38 Limited degrees of freedom. The numbers indicate the 4 Degrees of Freedom (DOFs).
Source: author.

Misorientation and Hand–eye Coordination

During minimally invasive surgery the line of sight from the surgeon's eye on to the operating field is often not aligned with the orientation of the endoscopic camera. Consequently, hand movements do not correspond with the movements of the instrument tip on the monitor. Furthermore, the monitor is not located near the operating field. The difference in perspective results in a sense of misorientation, an incorrect or inappropriate orientation. Wentink showed that camera rotation and adding shaft as reference help to improve hand–eye coordination and reduce misorientation, but these tools did not work well in practice.[16] By rotating the camera, a natural line of sight can be obtained. This adaptation worked well in an environment with abstract tasks but for the surgeon rotation of the camera caused rotation of the anatomy and caused difficulties in detection of the anatomy of the patient. The use of deflectable instruments, such as used in some robotic systems, and using a steerable endoscope may help to improve the natural line of sight and reduce hand–eye coordination problems.[17] With deflectable instruments, such as the Da Vinci's endowrist,[18] the coupling between hand and instrument movements is restored.[19] The surgeon moves the instrument handle at the console (the master) while the instrument is controlled by a device (the slave) at the operating table. This often provides the possibility to improve the kinematic relation between instrument tip movements. Unfortunately, this system lacks haptic feedback, and placing the surgeon behind a console in their own virtual world may have a negative influence on safety.[20]

Haptics

Haptics refer to all the physical sensors that provide a sense of touch at the skin level in combination with position, movement and force feedback information coming from muscles and joints.[21] Haptics is closely linked to proprioception, which is the information of the position of body parts, obtained through sensory receptors in muscles, tendons and joints. Not being in direct contact with tissue due to the interposition of instruments results in difficulties to control the grasp force. Friction in instruments, friction between the cannula and instrument and abdominal wall properties all influence the transfer of force information.[22] The relation between pinching and pulling forces at the tip does not resemble the forces in the hand. It is shown that insufficient control of pinch-and-pull force results in tissue slippage or the use of excessive force. Both can lead to tissue damage.[23]

Training Surgeons

The limitations mentioned above, caused by the reduced degree of freedom, hand–eye coordination problems and reduced haptic feedback, necessitates extensive training in order to obtain the required psychomotor skills to perform a minimally invasive procedure safely.

Minimally invasive surgery training outside patients is a major item with regard to the ethical aspect of using patients as learning objects and the governmental policy of discouraging the use of animal models. In the past years many training systems have been developed,[24] including physical trainers (box trainers), virtual reality (VR) trainers with and without haptic feedback and hybrid ones. Box trainers have natural haptic feedback but lack the automatic assessment of task performance and are therefore not often used. Getting performance feedback via a score is a large motivator for training. VR trainers are often able to provide automatic assessment since they use standardised tasks. However, the rendering of the complex (tissue) environment is difficult and not enough data on forces and structures/properties of tissue is available to provide realistic haptic feedback. Recently, Horeman developed a force sensing system able to use in box trainers. This gives the opportunity to train to apply the right amount of force during tissue manipulation in physical box trainers with natural haptic feedback (Fig. 39).[25]

In surgery, the 'see one, do one, teach one' concept was for a long time the standard training method. This method can be effective for simple tasks; however, for more complex motor and cognitive tasks other training methods are required.[26] According to Dankelman et al., a large variety of skills is required to be able to perform a minimally invasive procedure. Examples are: technical skills, motor skills, quick acting, surgical judgment, cognitive knowledge, team work and so on.[27] To master these skills, intensive training is needed. Several theories obtained from other disciplines are discussed in relation to minimally invasive surgical skills.[28] Examples are the three different stages at which learning skills can be identified (cognitive, associative and autonomous),[29] different learning styles and the adapted level of behaviour (skill-based, rule-based or knowledge-based).[30]

Coordinate System Change

When using training simulators, it is essential that training is effective, that is the objectives are met and that training is efficient, which means that costs, and thus time, be minimised.[31] The most important aspects

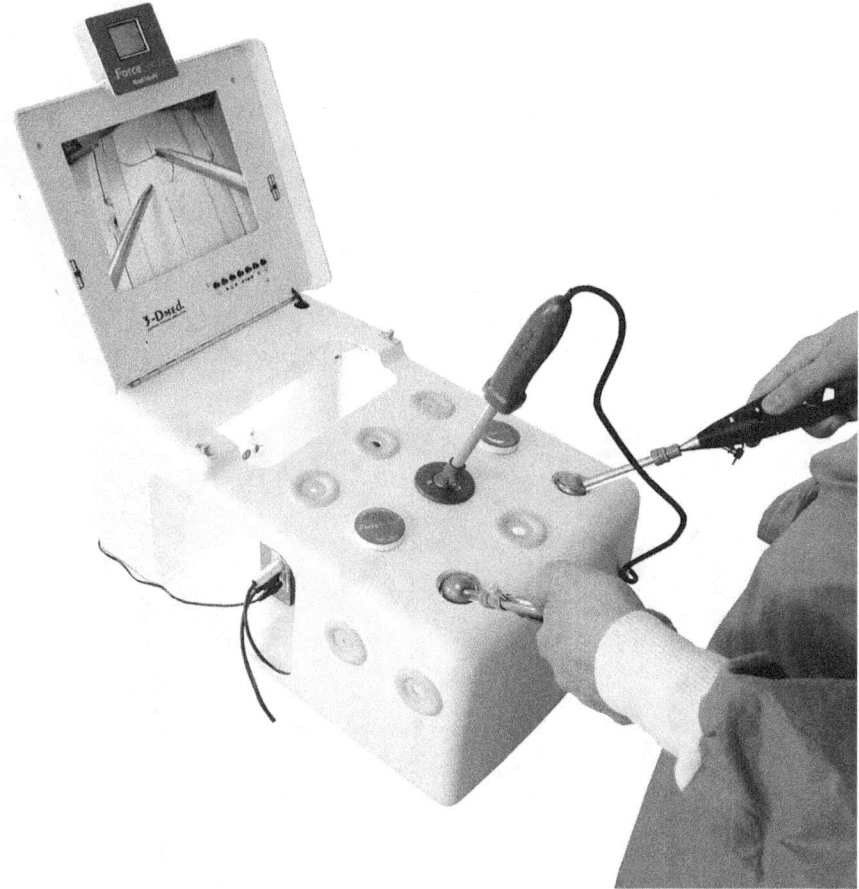

Figure 39 The ForceSense box trainer with sensors for tracking instrument path and measuring tissue manipulation forces.

Source: Tim Horeman, Jenny Dankelman, Frank Willem Jansen and John J. van den Dobbelsteen, 'Assessment of Laparoscopic Skills Based on Force and Motion Parameters', *IEEE Transactions on Biomedical Engineering*, Vol. 61 (2014), pp. 805–13.

learned in these simulators is to get used to the limited DOF, hand–eye coordination problems and limited depth perception. To get used to the limitation due to the fulcrum effect, it may be that a 'switch' in our brain is needed from Cartesian to polar coordinates. This is different from a switch from 3D to 2D. It is known that some trainees have tremendous difficulty getting used to this minimally invasive setting where instruments go through an incision point. An interesting question arises whether these trainees have difficulties making the switch from the Cartesian to polar coordinate system.

Optical illusion/gaze change: If the switch from the open to the minimally invasive setting can be seen as switch, the question arises as to whether a switch between coordinate systems also needs a switch in the brain when moving the instruments. Could this switch in coordinate system be compared with the switch seen in optical illusions? We all know the picture of the old and young lady[32] (Fig. 40). At a certain moment one lady can be observed, often causing difficulties in recognising the other one. Hence, it is not easy to switch between the two settings and it is also impossible to see both pictures at the same time. It may be that a necessary part of a surgeon's training is to smoothe out the switch between Cartesian and polar coordinate systems.

Figure 40 Optical illusion: The old and young lady. Old German postcard, 1988.
Source: <https://en.wikipedia.org/wiki/My_Wife_and_My_Mother-in-Law> (accessed 19 August 2015).

Uncanny Valley

In 1970 Masahiro Mori hypothesised that a person's response to a human-like robot would shift from empathy to revulsion as it approached, but failed to attain, a lifelike appearance. It is described as a response of revulsion to robots that look and move almost, but not exactly, like healthy humans.[33] The *uncanny valley* refers to the dip on a graph of the observer's comfort level as subjects move toward a healthy, natural likeness, described as a function of a subject's aesthetic acceptability (Fig. 41). The effect is mainly described in the fields of robotics and 3D computer graphics. The question arises as to whether this reaction also occurs in *clinical cartographics*.

For example, as explained above, during the training of surgeons a VR environment can be used. For training it is important that the trainee has a good internal representation of the system, the task and the environment in which he or she is interacting. During surgical simulation, the trainee needs to develop a 'map' representing the environment in which they have to perform a task.

There has been discussion about how accurately the virtual environment should represent real surgical tissue. Real tissue is difficult

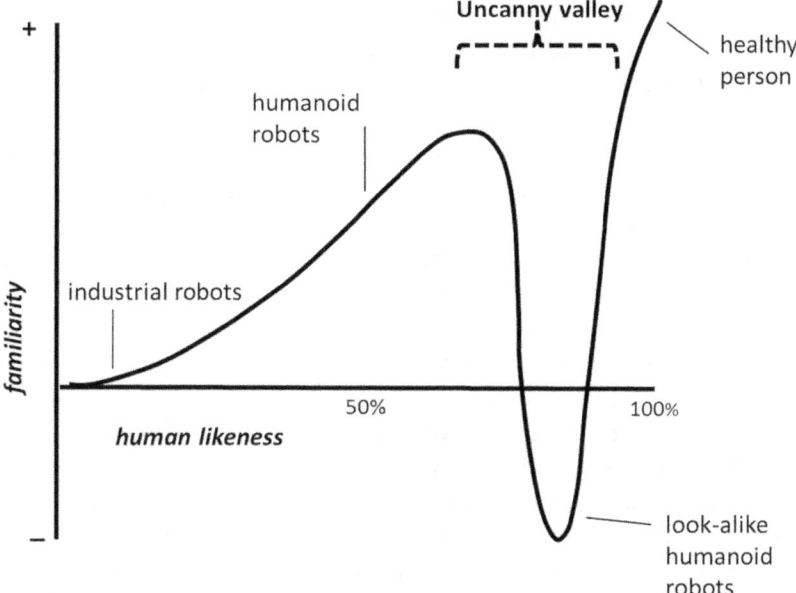

Figure 41 Uncanny Valley; adapted from Mori, 'Uncanny Valley', Uncanny Valley, <https://en.wikipedia.org/wiki/Uncanny_valley> (accessed 19 August 2015).

to simulate, due to its complex properties, especially when the tissue's response to manipulation has to be included. Rendering the response of inhomogeneous tissue material to, for example, cutting and stretching is extremely difficult. Hence, during tissue manipulation the visual information obtained from the operating field does not exactly match the real situation. Not only does visual information deviate from reality, but the haptic information also deviates. Hence, what the trainee feels does not exactly correspond with the situation when using real instruments and tissue. That may also be one of the reasons they prefer training in the OR on patients to training on a device. Providing realistic haptic feedback is complicated, because the reaction forces on movements are very fast and rendering it is too time consuming. Hence, during VR simulation both visual and haptic information are hampered and the effect may be that the trainee ends up in the dip of the uncanny valley.

In a training setting we performed research into the effect of providing realistic haptic feedback during training laparoscopic tasks.[34] We discovered that when haptic feedback was not accurate, the performance measured afterwards had decreased. It was better to leave out tasks in which force application was needed. Hence, it may be better to use abstract tasks during VR training and to focus only on hand–eye-coordination and depth perception tasks. With the same reasoning it may sometimes be better to leave out properties when the details cannot be presented accurately enough. Thus, during the development of training tasks, it may sometimes be preferable to omit detail in order to avoid negative effects. Although this is not exactly the same as the uncanny valley effect, it shows that when it is not possible to reach the exact representation it is better to deviate further away from the real situation and, in the case of surgical training, to provide an abstract environment. It is not unlikely that this effect also occurs in other VR situations.

Architectural uncanny valley: Literature on whether the uncanny valley response can also occur in architecture could not be found. However, the phrase 'architectural uncanny valley' showed up on the Internet in relation to games.[35] It describes that the detail of the way a surface responds to, for example, weather conditions, the depiction of shadows and textures, and other factors that contribute to a realistic looking structure, are important factors for presenting architecture in games. A game can pose its own uncanny valley hypothesis, when something about the gravity in the virtual world does not match up with human experience.[36]

When making a map of reality it is important to understand the effect that adding highly detailed precision may have on the perceiver. Too much detail may bring the perceiver into a situation of the dip of the uncanny valley. Really precise maps are well understood, but it is also good for cartographers to realise that the uncanny valley effect may occur when using very precise maps. The effect of moving/changing objects may even increase the effect. Hence, it could be interesting to explore whether achieving a highly detailed representation in the field of cartographics may also influence the perceiver. With current computer technology, the drawings of buildings look more and more realistic, while more abstract drawings may in some situations be better perceived. The uncanny valley may occur when the observer is unsure about the level of reality of the map and the most exact representation may cause mistrust and revulsion.

The Digital Operating Room Assistant (DORA)

The operating room is a place where the technological complexity has increased over the past few years. In an observation study, on average two incidents related to equipment could be identified occurring during a gallbladder procedure. Examples of these incidents were: equipment that was not present in the OR, monitor not working well and connections not right. Most of these problems could be solved in a few minutes, however it was observed that these incidents disrupt the normal workflow during the per-operative process and also reduce efficiency. Moreover, due to this highly technical complex setting, patient safety issues may emerge. To improve instrument and equipment use and to improve efficiency and safety, the DORA project was started: the Digital Operating Room Assistant (Fig. 42). DORA aims at the development of an autonomous system, which monitors the surgical progress with all its critical steps and presents this information in a structured and understandable way. This system can be used for personnel in the OR to have better insight in the current process, but also to optimise OR planning and to further optimise the whole OR process.

Several DORA research pilot projects are currently running, focusing on the pathway of patients in the hospital,[37] on the status of equipment and instrument nets[38] and on the identification of surgical tasks. To monitor the status of equipment, all the devices were tagged with a radio-frequency (RFID) tag and sensors were placed in stock areas and ORs. During the start of the procedure DORA automatically checks whether all the equipment is present in the OR, checks safety status and

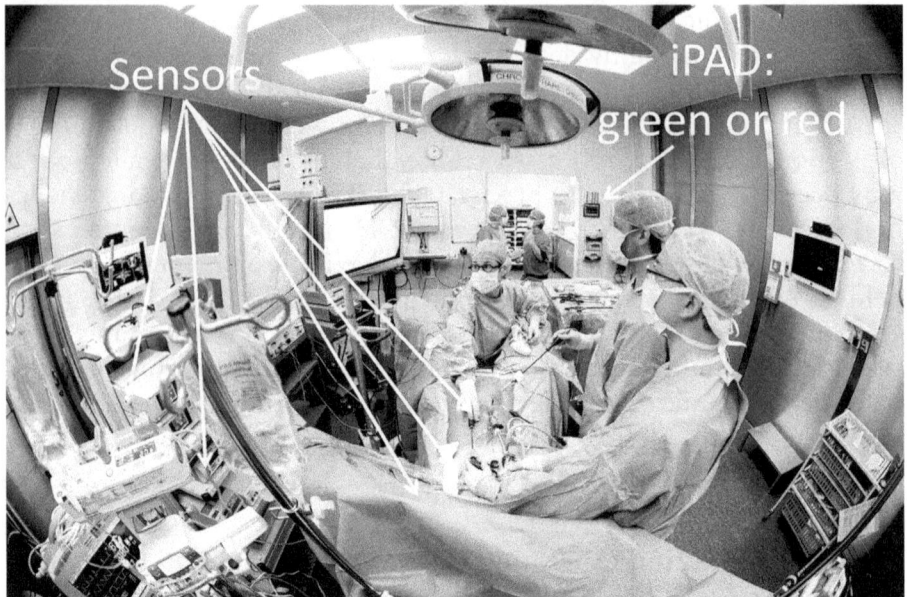

Figure 42 DORA system with sensors to monitor the surgical process.
Source: author.

maintenance, and it reports whether problems had occurred in the past that have not yet been solved.

In another DORA project, sound is used to detect OR activities and to determine the phase of the OR in real time. This information can be used to improve OR planning and scheduling. OR planning and the time patients need to be in the recovery room can be further optimised by combining patient characteristics such as age and gender, using signal processing and subsequent data mining[39] and pattern analysis methodologies of the fused sensor data. Furthermore, through statistical analysis of large numbers of patient and operation records,[40] DORA can discover possible means of increasing efficiency. In the end it is DORA's aim to provide an overview, a 'map' of OR safety status. This safety map can then be used to adapt the surgical process and to improve the surgical treatment of patients.

Conclusion

In this chapter an overview has been given of minimally invasive technologies and its advantages and disadvantages. The result of the indirect approach of working via tiny incisions resulting in limitation related to instrument movements and vision are explained. Due to these limitations surgeons need systems to be able to train to manipulate tissue in a safe environment outside patients. Aspects such as misorientation, hand–eye coordination and uncanny valley and the effect of providing DORA's safety map need to be understood to improve the minimally invasive surgical process.

Notes

1. Stassen et al., 1999. 'Open versus Minimally Invasive Surgery'; Breedveld et al., 'Manipulation in Laparoscopic Surgery'; Breedveld et al., 'Observation in Laparoscopic Surgery'.
2. Verdaasdonk et al., 'Problems with Technical Equipment'.
3. Guédon et al., 'Safety Status System'.
4. Breedveld, 'Manipulation in Laparoscopic Surgery' and 'Observation in Laparoscopic Surgery'.
5. Westebring-Van der Putten et al., 'Haptics in Minimally Invasive Surgery'; Westebring-Van der Putten et al., 'Force Feedback Requirements'; Van den Dobbelsteen et al., 'Friction Dynamics of Trocars'; Van den Dobbelsteen et al., 'Indirect Measurement of Pinch and Pull Forces'.
6. Heijnsdijk et al., 'Influence of Force Feedback'; Rodrigues et al., 'Suturing Intraabdominal Organs'.
7. Breedveld et al., 'New, Easily Miniaturised Steerable Endoscope'.
8. Loeve et al., 'Endoscope Shaft-Rigidity Control Mechanism'.
9. Henken et al., 'Error Analysis'.
10. Szold et al., 'European Association of Endoscopic Surgeons'.
11. Ibid.
12. Breedveld et al., 'New, Easily Miniaturised Steerable Endoscope'.
13. Horeman et al., 'Isolator System'.
14. Breedveld et al., 'Observation in Laparoscopic Surgery'.
15. Ibid.
16. Wentink et al., 'Endoscopic Camera Rotation'
17. Wentink, 'Hand-eye Coordination'.
18. Szold et al., 'European Association of Endoscopic Surgeons'.
19. Wentink, 'Hand-eye Coordination'.
20. Szold et al., 'European Association of Endoscopic Surgeons'.
21. Westebring-Van der Putten et al., 'Haptics in Minimally Invasive Surgery'.
22. Westebring-Van der Putten et al., 'Force Feedback Requirements'; Van den Dobbelsteen, 'Friction Dynamics of Trocars'.
23. Heijnsdijk et al., 'Slip and Damage Properties'.
24. Verdaasdonk et al., 'Transfer Validity of Laparoscopic Knot-tying Training'; Chmarra et al., 'Systems for Tracking Minimally Invasive Surgical Instruments'.
25. Horeman et al., 'Assessment of Laparoscopic Skills'.
26. Dankelman et al., 'Fundamental Aspects of Learning'.
27. Ibid.

28. Ibid.
29. Fitts and Posner, *Learning and Skilled Performance*.
30. Rasmussen, 'Skills, Rules and Knowledge'.
31. Dankelman et al., 'Fundamental Aspects of Learning'.
32. Optical illusion <https://en.wikipedia.org/wiki/Optical_illusion https://en.wikipedia.org/wiki/My_Wife_and_My_Mother-in-Law> (accessed 19 August 2015).
33. Mori, 'Uncanny Valley'. Uncanny valley, <https://en.wikipedia.org/wiki/Uncanny_valley> (accessed 19 August 2015).
34. Chmarra et al., 'Force Feedback'.
35. McNeil, 'Architectural Uncanny Valley'.
36. Ibid.
37. Wauben et al., 'Tracking Surgical Day Care Patients'.
38. Guédon et al., 'A RFID Specific Participatory Design Approach'.
39. Bouarfa and Dankelman, 'Workflow Mining and Outlier Detection'.
40. Bouarfa et al., 'Prediction of Intraoperative Complexity'.

Bibliography

Bouarfa, Loubna, and Jenny Dankelman, 'Workflow Mining and Outlier Detection from Clinical activity logs', *Journal of Biomedical Informatics*, Vol. 45 (2012), pp. 1,185–90.

Bouarfa, Loubna, Armin Schneider, Hubertus Feussner, Nassir Navab, Heinz U. Lemke, Pieter. P. Jonker and Jenny Dankelman, 'Prediction of Intraoperative Complexity from Preoperative Patient Data for Laparoscopic Cholecystectomy', *Artificial Intelligence in Medicine*, Vol. 52 (2011), pp. 169–76.

Breedveld, Paul, Jules S. Scheltes, Esther M. Blom and Johanna E. I. Verheij, 'A New, Easily Miniaturised Steerable Endoscope', *IEEE Engineering in Medicine and Biology Magazine*, Vol. 24 (2005), pp. 40–7.

Breedveld, Paul, Henk G. Stassen, Dirk W. Meijer and Jack J. Jakimowicz, 'Manipulation in Laparoscopic Surgery: Overview of Impeding Effects and Supporting Aids', *Journal of Laparoendoscopic & Advanced Surgical Techniques*, Part A, Vol. 9 (1999), pp. 469–80.

Breedveld, Paul, Henk G. Stassen, Dirk W. Meijer and Jack J. Jakimowicz, 'Observation in Laparoscopic Surgery: Overview of Impeding Effects and Supporting Aids', *Journal of Laparoendoscopic & Advanced Surgical Techniques*, Part A, Vol. 10 (2000), pp. 231–41.

Chmarra, Magdalena K., Kees (C.) A. Grimbergen and Jenny Dankelman, 'Systems for Tracking Minimally Invasive Surgical Instruments', *Minimally Invasive Therapy & Allied Technologies*, Vol. 16 (2007), pp. 328–40.

Chmarra, Magdalena K., Jenny Dankelman, John J. van den Dobbelsteen and Frank Willem Jansen, 'Force Feedback and Basic Laparoscopic Skills', *Surgical Endoscopy and Other Interventional Techniques*, Vol. 22 (2008), pp. 2,140–8.

Dankelman, Jenny, Magdalena K. Chmarra, Emiel G. G. Verdaasdonk, Laurents P. S. Stassen and Kees (C.) A. Grimbergen, 'Fundamental Aspects of

Learning Minimally Invasive Surgical Skills', *Minimally Invasive Therapy & Allied Technologies*, Vol. 14 (2005), pp. 247–56.

Fitts, Paul Morris and Michael I. Posner, *Learning and Skilled Performance in Human Performance* (Belmont, CA: Brock-Cole, 1967).

Guédon, Annetje C. P., Linda S. G. L. Wauben, Dirk F. de Korne, Marlies Overvelde, Jenny Dankelman and John J. van den Dobbelsteen, 'A RFID Specific Participatory Design Approach to Support Design and Implementation of Real-Time Location Systems in the Operating Room', *Journal of Medical Systems*, Vol. 39 (2015), pp. 168–78.

Guédon, Annetje C. P., Linda S. G. L. Wauben, Marlies Overvelde, Joleen H. Blok, Maarten van der Elst, Jenny Dankelman and John J. van den Dobbelsteen, 'Safety Status System for Operating Room Devices', *Technology and Health Care*, Vol. 22 (2014), pp. 795–803.

Heijnsdijk, Eveline A. M., Hans de Visser, Jenny Dankelman and Dirk J. Gouma, 'Slip and Damage Properties of Jaws of Laparoscopic Graspers', *Surgical Endoscopy and Other Interventional Techniques*, Vol. 18 (2004), pp. 974–9.

Heijnsdijk, Eveline A. M., A. Pasdeloup, Arjan J. van der Pijl, Jenny Dankelman and Dirk J. Gouma, 'The Influence of Force Feedback and Visual Feedback in Grasping Tissue Laparoscopically', *Surgical Endoscopy and Other Interventional Techniques*, Vol. 18 (2004), pp. 980–5.

Henken, Kirsten R., Jenny Dankelman, John J. van den Dobbelsteen, Lun K. Cheng and Maurits S. van der Heiden, 'Error Analysis of FBG-Based Shape Sensors for Medical Needle Tracking', *IEEE-Asme Transactions on Mechatronics*, Vol. 19 (2014), pp. 1,523–31.

Horeman, Tim, Jenny Dankelman, Frank Willem Jansen and John J. van den Dobbelsteen, 'Assessment of Laparoscopic Skills Based on Force and Motion Parameters', *IEEE Transactions on Biomedical Engineering*, Vol. 61 (2014), pp. 805–13.

Horeman, Tim, Frank Willem Jansen and Jenny Dankelman, 'An Isolator System for Minimally Invasive Surgery: The New Design', *Surgical Endoscopy and Other Interventional Techniques*, Vol. 24 (2010), pp. 1,929–36.

Loeve, Arjo J., Dick H. Plettenburg, Paul Breedveld and Jenny Dankelman, 'Endoscope Shaft-Rigidity Control Mechanism: "FORGUIDE"', *IEEE Transactions on Biomedical Engineering*, Vol. 59 (2012), pp. 542–51.

McNeil, Joanne 'Architectural Uncanny Valley' <http://rhizome.org/editorial/2011/nov/14/steam-postcards/> (accessed 9 October 2015).

Mori, Masahiro 'The Uncanny Valley', trans. Karl F. MacDorman and Takashi Minato, *Energy*, Vol. 7, Issue 4 (1970), pp. 33–5.

Mori, 'Uncanny Valley', Uncanny Valley, <https://en.wikipedia.org/wiki/Uncanny_valley> (accessed 19 August 2015).

'Optical illusion' <https://en.wikipedia.org/wiki/Optical_illusion https://en.wikipedia.org/wiki/My_Wife_and_My_Mother-in-Law> (accessed 19 August 2015).

Rasmussen, Jens, 'Skills, Rules and Knowledge: Signals, Signs and Symbols, and Other Distinctions in Human Performance Models', *IEEE Transactions on Systems, Man, and Cybernetics*, Vol. 13 (1983), pp. 257–66.

Rodrigues, Sharon P., Tim Horeman, Jenny Dankelman, John J. van den Dobbelsteen and Frank Willem Jansen, 'Suturing Intraabdominal Organs: When Do We Cause Tissue Damage?', *Surgical Endoscopy and Other Interventional Techniques*, Vol. 26 (2012), pp. 1,005–9.

Stassen, Henk G., Jenny Dankelman and Kees (C.) A. Grimbergen, 'Open Versus Minimally Invasive Surgery: A Man-machine System Approach', *Transactions of the Institute of Measurement and Control*, Vol. 21 (1999), pp. 151–62.

Szold, Amir, Roberto Bergamaschi, Ivo Broeders, Jenny Dankelman, Antonello Forgione, Thomas Lango, Andreas Melzer, Yoav Mintz, Salvador Morales-Conde, Michael Rhodes, Rick Satava, Chung Ngai Tang and Ramon Vilallonga, 'European Association of Endoscopic Surgeons (EAES) Consensus Statement on the Use of Robotics in General Surgery', *Surgical Endoscopy and Other Interventional Techniques*, Vol. 29 (2015), pp. 253–88.

Van den Dobbelsteen, John J., Ruben A. Lee, Maarten van Noorden and Jenny Dankelman, 'Indirect Measurement of Pinch and Pull Forces at the Shaft of Laparoscopic Graspers', *Medical & Biological Engineering & Computing*, Vol. 50 (2012), pp. 215–21.

Van den Dobbelsteen, John J., Abel Schooleman and Jenny Dankelman, 'Friction Dynamics of Trocars', *Surgical Endoscopy and Other Interventional Techniques*, Vol. 21 (2007), pp. 1,338–43.

Verdaasdonk, Emiel G. G., Jenny Dankelman, Johan F. Lange and Laurents P. S. Stassen, 'Transfer Validity of Laparoscopic Knot-tying Training on a VR Simulator to a Realistic Environment: A Randomised Controlled Trial', *Surgical Endoscopy and Other Interventional Techniques*, Vol. 22 (2008), pp. 1,636–42.

Verdaasdonk, Emiel G. G., Laurents P. S. Stassen, Maarten van der Elst, Tom M. Karsten, and Jenny Dankelman, 'Problems with Technical Equipment during Laparoscopic Surgery – an Observational Study', *Surgical Endoscopy and Other Interventional Techniques*, Vol. 21 (2007), pp. 275–9.

Wauben, Linda S. G. L., Annetje C. P. Guédon, Dirk F. de Korne, and John J. van den Dobbelsteen, 'Tracking Surgical Day Care Patients using RFID Technology', *BMJ Innovations*, Vol. 1 (2015), pp. 59–66.

Wentink, Mark, 'Hand-eye Coordination in Minimally Invasive Surgery: Theory, Surgical Practice & Training' (PhD Thesis, TU Delft, 2003).

Wentink, Mark, Paul Breedveld, Dirk W. Meijer and Henk G. Stassen, 'Endoscopic Camera Rotation: A Conceptual Solution to Improve Hand-eye Coordination in Minimally-invasive Surgery', *Minimally Invasive Therapy & Allied Technologies*, Vol. 9 (2000), pp. 125–31.

Westebring-Van der Putten, Eleanora P., Richard H. M. Goossens, Jack J. Jakimowicz and Jenny Dankelman, 'Haptics in Minimally Invasive Surgery – A Review', *Minimally Invasive Therapy & Allied Technologies*, Vol. 17 (2008), pp. 3–16.

Westebring-Van der Putten, Eleanora P., John J. van den Dobbelsteen, Richard H. M. Goossens, Jack J. Jakimowicz and Jenny Dankelman, 'Force Feedback Requirements for Efficient Laparoscopic Grasp Control', *Ergonomics*, 52 (2009), pp. 1,055–66.

PART IV
Philosophy

CHAPTER 10

Elasticity and Plasticity: Anthropo-design and the Crisis of Repetition

Sjoerd van Tuinen

Today, a key concept at the intersection between medicine, design and politics is that of plasticity. It is the centrepiece of the neurosciences, where it is used to describe the way in which the brain moulds itself, developing its own history and historicity beyond genetic predetermination. But it also has a long philosophical history, from its invention by Goethe to Hegel, Nietzsche and Freud, in which it constitutes the transformative continuity between nature and history and between the neuronal and the mental. For Hegel, plasticity characterises human thought itself in the radically immanent and necessary way it develops as well as retains its past. Pure dialectics, his name for the self-movement of thought motivated by the power of the negative, is the way in which the mind comes from the body that disappears in what it becomes. It is the conceptual mediation of their ongoing conflict, the internal resistance that makes forming and unforming coincide in the continuous self-transformation of reason or spirit.[1] But according to what criteria and ends does reason produce itself, and thus also design and compose its brain?

This question lies at the heart of humanism and post-humanism, as their modes of thought are bound to the actual becoming of human life. In what follows, I bring together the work of two contemporary philosophers of human enhancement, Peter Sloterdijk and Catherine Malabou, in order to restage this intrinsic relation between thought and life. I will first introduce the German philosophical anthropologist Sloterdijk, whose oeuvre contains one of the most important critical and clinical cartographies of the present. While he is known for his writings in media theory, philosophy of technology and the life sciences, his work makes up a more general philosophy of life framed in terms of a generalised immunology.[2] After an exposition of his recent theory of 'anthropotechnics' based on habit and repetition, I will further develop his notion of the autoplasticity of man through Malabou's much better known concept of plasticity. If, with and against Sloterdijk, we nonetheless diverge from Malabou, this is because plasticity, while bearing an enormous potential for imaginative thought, ultimately does not suffice as its model or image. From a medical or immunological perspective rather than merely a biological one, it provides too weak or non-binding a link between thought and life.[3] Instead, we must respond to plasticity in a more recalcitrant or elastic way, that is, with a non-modern image of thought understood as learning to protect, and take care of what we – including all those 'others' who are affected by our decisions and those others of whom we are composed – will become.

In the discourse around plasticity, elasticity is usually equated with infinite flexibility and uncritical adaptivity. By contrast, I propose the concept of elasticity to invert the modernist relation between, on the one hand, the infinite plasticity of modes of living and, on the other, those finite subjectivities that do not live up to the standards of science and capitalism, and that the thrust of modernity actively wants to leave behind in the name of progress. I argue that the concept of elasticity, contrary to that of plasticity, could have the power to reorient thought beyond the modern division of labour between knowledge and action, towards the production of that future continuity between past and present that today is everywhere so fatefully lacking. Sloterdijk shares the criticism of metaphysical humanism by Karl Marx and Martin Heidegger, only to be repeated more recently by Giorgio Agamben or Bruno Latour: man is not initially alienated from himself only to find himself again as the future outcome of history, rather, he is that Promethean something capable of generating himself from the start.[4] Human history, in other words, is not the story of the negation of the negation of man, but the prospect (*avenir*, as opposed to *futur*) man appropriates and assumes for himself.[5] We are not after finitude, but before finitude, even if it is an unlimited finitude. Following phenomenologist Hermann Schmitz, Sloterdijk refers to this conversion in our relation to history as a 're-embedding' of the subject and a 'turn (*Kehre*)' towards the 'total care (*Sorge*)' of the world.[6] This conversion should not be understood in the sense of the unconditioned freedom of existentialism, to be sure, since the individual subject is only the fold or form of an actual becoming (elasticity) and not its agency, which, like the brain, is immediately and infinitely divided over a network of networks (plasticity). Yet to think is to answer clinically to the critical challenge of how to take

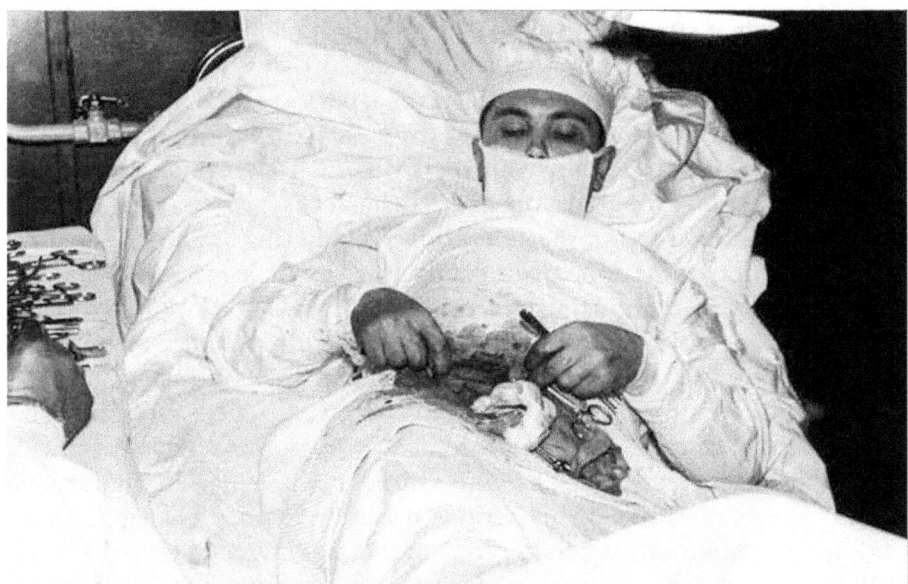

Figure 43 The Russian surgeon Leonid Rogozov, stranded in Antarctica with the Sixth Soviet Antarctic Expedition, in 1961 performs a self-operation: under local anaesthesia, the twenty-seven-year-old removes his own appendix.

Source: Rare Historical Photos, Leonid Rogozov, <http://rarehistoricalphotos.com/leonid-rogozov-appendix-1961/> (accessed 18 October 2016).

responsibility for the practical effects of the facticity of being-there, that is, how to participate in and slow down the events that happen to us yet of which we are never only a passive part.

Operable Man

A key figure in Sloterdijk's recent works is that of 'operable man', the human whose condition is characterised by the fact that he operates on himself while simultaneously letting himself be operated on.[7] Modern man increasingly finds himself in an 'auto-operative curvature' that puts him in constant relation to his own passivity, not in the form of resignation or submission, but in the form of a free and active cultivation or care.[8] In the most extreme case, the auto-operative curvature becomes a circle and we operate directly on our individual selves. Sloterdijk gives three examples, all cases of self-surgery. The first is that of Leonid Ivanovich Rogozov, a Soviet general practitioner who, during his stay at a research station in the Antarctic, was forced to perform an appendectomy on himself. The second is that of American Alpinist Aron Ralston, who after being stuck for five days after an accident in the mountains, decided to break his own forearm and cut off the flesh with a blunt pocket knife. And the third is the British performance artist Heather Perry, who used a local anaesthetic and a special drill for the trepanation on her own skull, apparently in order to fight her chronic fatigue and attain a higher level of consciousness.[9] In each case, we are dealing with an immense capacity for the toleration of suffering made possible by an extreme determination to act. With Martin Heidegger, we could see this capacity of simultaneously acting and being acted upon as typical for the steeled subjectivity of modernity implied by modern technology. The more everyday subjectivity of operable man, by contrast, is less self-centred and more mediocre. Not without irony, it is closer to what Heidegger calls *Gelassenheit* or 'releasement' (or more religiously, grace): the decentred subjectivity of he who affirms his reticular entanglements by enlarging the radius of his actions precisely by also enlarging those of others over himself. When I switch on the TV or when I take the train, I choose for my own profit to let others do something with me. The more networked our world becomes, the more my passivity is implied by my activity: I have to make myself passive in order to be able to become active. In terms of design, this implies not only that we have to learn anew to distinguish what is within our power and what lies beyond, but also the realisation that we are able to make ever less ever better. In medicine, this self-relativisation means that the figure of the client slowly but surely replaces that of the patient. On a discursive level, but also on a juridical and operational level, we claim more and more competence vis-à-vis our own ever-expanding relations of dependence at the same time that we claim the right to powerlessness. In this way the setting of an operating room or a dialysis centre has become exemplary of the human condition in general. In fact, Sloterdijk will argue, individual humans are not thrown (*Geworfenheit*), but born (*Getragenheit*). We have never been bound to the human condition, since we are always already products of pre-human and trans-human processes of 'air conditioning': starting with the womb, we are always already embedded in atmospherical interiorities in which we do not coincide with ourselves, but find ourselves in 'ecstatic immanence' with our surroundings.[11] Living is nothing but the assumption of ourselves as intervening in the design of that that has already begun without us. *Dasein* is Design,[12] *geworfener Entwurf*.[13]

Gelassenheit, of course, is far removed from the self-understanding of the moderns, who maintain a deep mistrust for almost any form of passivity, or indeed for the past itself. In temporal terms, *Gelassenheit* implies that we establish a continuity between the past we passively inherit and the future we actively co-construct. Thus, in medical progress, a balance is usually kept between passive patience about what is scientifically achievable (realism) and active impatience about what is still to be done (optimism). Perhaps in this respect we should say that medicine has never been modern. But in almost any other modern domain from science to political economy, by contrast, there is the ceaseless activity of a permanent revolution. It is for this reason that modernity has led to what Alfred North Whitehead has called a 'bifurcation of nature': a strict separation and purification of what lies in the power of humans (Culture) and what does not (Nature), to the detriment of the latter. As Jean-Paul Sartre once put it, to actualise our freedom is to make something out of what we have been made into. It would be bad faith to remain an object among other objects; the only tolerable thing to do is to resist all passivity and replace it with total mobilisation. Similarly, when Karl Marx defined man as a species-being (*Gattungswesen*), an animal capable of (re)producing himself, the point was that this reproduction is not bound to our biological needs, but can happen according to any standard whatever (even that of beauty). What we are coincides with what we make and how we make it.[14] History is the transformation of nature by man, just as productive work would be the nature of man (*homo faber*), as is reflected by Marx's famous idealisation of the course of a day in a future communist society from *The German Ideology* (1945): 'to hunt in the morning, fish in the afternoon, rear cattle in the evening, criticise after dinner'. This ceaseless activity is reflected by the functionalism of modernist architecture, in which permanent revolution converges with permanent visibility. Its panoptic structures are adapted not to the needs of the human species, but, as Michel Foucault has shown, to the relentless self-production and self-transformation of man through the physical environment he inhabits. The moulding of the new man proceeds by a perpetual cutting away of everything that is old. It is the production of the producer himself.

The question is of course whether this architectonic principle of operability on a grand scale is really the sign of our freedom, or whether it does not rather reveal the unprecedented moralisation of public space under the sign of mechanical work, effectuated by means of a rigorous penetration of everyday life by disciplinary designs. Or more generally speaking, does not the modern fear of repeating the past inspire a new kind of repetition, the most graceless and servile of all, namely the monotonous repetition of the present actuality without a future? Here we get a first glimpse of what Sloterdijk calls a crisis of repetition.

Modernity

Sloterdijk agrees with Marx that man himself is a product of repetition. But work is only one kind of repetition and everything depends on the capacity to make a difference between different modes of repetition. At a biological or animal level, after all, the idea that man constantly reproduces himself is not exactly new. Even if the technology and science of human reproduction are in constant development, humanity as such appears to be quite satisfied with the tried-and-tested mechanisms of organic reproduction and affiliative variation that are co-evolutionary with itself. However, the self-reproduction of man exceeds the level of what has evolved in nature. Already at the level of evolution, nothing is fixed and species are adrift. But if there is such a thing as culture or will, this is because the power of repetition also constitutes a labour of man on himself. As self-generating power of 'exercise', repetition is coextensive with culture at large.[15] 'To be human is to exist in an operatively curved space, in which actions retroact on the actor, works on the worker, communications on the communicator, thoughts on the thinker, and feelings on the feeler.'[16] Sloterdijk famously speaks of a 'human park' or a 'human incubator': a habitat in-between nature and culture in which man produces and reproduces himself by means of domesticating rituals, ideas, practices, gestures, techniques, texts and all sorts of newer media. This is where Sloterdijk is closest to the empiricist tradition from Félix Ravaisson and William James to Gilles Deleuze and Pierre Bourdieu. Accordingly, 'habit' reveals 'the paradox of repetition': by constituting an intermediary milieu or elastic reversibility between passivity and activity, it draws something new – the very difference between nature and will – from repetition.[17] Habit is plastic and, to some degree, 'autoplastic': it is not just a passive response to a stimulus, but also inventive in its own containing way. As a consequence, it is not just a cultural thing. Nature and culture, impression and expression are not opposed but two sides of the same tendency from receptivity to spontaneity. Habitual repetition is what makes possible the hybrid determination of humans as artificial or prosthetic by nature, as well as the new functional aggregates of man and machine we celebrate today.

One of the central concepts in Sloterdijk's more recent books based on this concept of habitual repetition is that of 'anthropotechnics' (or *anthropo-urgy*, 'the work(ing) of man').[18] While repetition is active by definition, we initially and for the larger part of our history and prehistory undergo the process of repetition passively – both habit (the present) and memory (the past), Deleuze writes, are 'passive syntheses'.[19] Every tradition is the product of man's prehistoric labour on himself, and begins as the unquestionable imposition of a collective power of command. Plato called the process by which culture transmits itself to next generations *paideia*, the '*art* on the child'; humanists prefer to call it *Bildung* or formation. In order to take away all moralistic prejudices about the finality of this primordial repetition of culture, Friedrich Nietzsche regarded anthropotechnics as the 'morality of customs' (*Sittlichkeit der Sitte*): the inscription of a capacity to remember (moral conscience) in the bodies of human animals through painful and violent training (a 'mnemotechnics') that remains without a moral aim in itself.[20] As a dressage in hardship, culture is both the practice and the result of its own repetitive moulding of the nervous systems of its children. It follows that all morality is first of all slave morality: all of culture is initially a matter of forced inheritance.[21]

Just as anthropotechnics goes beyond natural reproduction, however, it also exceeds slave culture. Freedom initially manifests itself when cultural repetition is turned against itself and self-determined individuals manage to distance themselves from biological and local bonds of filiation and alliance.[22] This becomes possible as soon as, by means of exercises on themselves, men learn to actively intervene in the passive repetitions of which they are the result. Ascetics, from monastic rituals to the protracted training of athletes and the strained exercises of musicians, are circular drills that create self-referential relations that commit individuals to cooperate with their own subjectivation and thus switch to the active side of repetition.[23] The patriarchal transmissions of Antiquity and the apostolic transmissions of monotheist religions are based on such secessionist repetitions.[24] They also form the core of modern humanism, to the extent that it marks the transition from a logic of reproduction to a logic of auto-domesticative optimisation, that is, a logic of anthropo-design based on techniques of self-drilling and self-enhancement through which the human condition is modified and thus kept in shape.[25]

A crisis of repetition occurs, finally, when the constancy and duration of a culture is threatened by copy mistakes, that is, when the repetitions spill over into non-intended consequences that interrupt and turn against the tradition. If modernity is the 'age of side effects', a 'Copernican mobilisation' resulting in 'global cultural entropy', this is because it produces innovative effects that cannot be integrated in the line of cultural filiation – speaking in very contemporary terms, we could say it is obsessed with 'disruptive innovation'.[26] Like a nuclear reactor, modernity undermines its own sustainability by producing too many 'glorious bastards' or *enfants terribles*, that is, figures of increasing asymmetry between past and future: the mystic, the protestant, the entrepreneur, the *nouveau riche*, the discoverer, the *virtuosi* who develop unexpected capacities, the planter, the inventor, the parvenu, the self-made man, the proletariat, the artist-genius, the intellectual, the revolutionary, the manager, the populist politician.[27] Living the fantasy of a life without presuppositions, without past, without original sin and without origin *tout court*, they are figures for whom action prevails over passion. As in the auto-operative circle, they either cut their relations to the world or take them in and reinvent them as their own. In the latter case, they are the *hommes du monde*, the heroes of classical modernity. Yet to the extent that, instead of assuming their origin or even themselves, they seek permanent mobilisation, permanent insurrection, permanent innovation, permanent conversion, they also contribute, as Sloterdijk argues in his last book *Die schrechlicke Kinder der Neuzeit. Über das anti-genealogische Experiment der Moderne*, to the shock of an insisting and persisting crisis of repetition that increasingly threatens to drain our souls, exhaust our bodies and destroy the earth.[28]

Having arrived at this diagnosis of the state of modern life as essentially corrupt or decadent, as a time out of joint, one wonders what is its sense, that is, its critical meaning and clinical evaluation. Sloterdijk refers to his political orientation as conservative and 'conserving' (*Konservatorisch*), or more precisely, as an 'elastic conservatism'.[29] The resilient stretching of a culture in time depends on its immunisation against novelty as much as on its capacity for integrating changes in its conditions, such that Sloterdijk understands Chinese culture, Jewish culture or the Catholic church as 'success stories of strictly controlled replications'.[30] Accordingly, the essence of a culture or civilisation would be the stable repetition of the same, such that cause and effect or subject and object of repetition more or less coincide.[31] As we saw with Hegel, the same goes for modernity. If man is by definition an autogenic effect, coproduced by the repercussions of his actions on himself, then its specificity is the anthropotechnical attempt of man to become the exclusive subject of his own reproduction. But this is impossible, for as we learned from the genealogy of morals, in reality the subject of repetition never coincides with its object. In principle, every identity is individuated in an open-ended process and appears as such only at the end of a series, not at the beginning. Worse still, every attempt to make the beginning coincide with the end can lead to 'malign repetitions', in which selfish systems – Sloterdijk discusses the systems of modern penal camps, modern schools and contemporary art – lose their elasticity and come to revolve only around themselves.[32] The crisis of humanism was therefore already announced by Nietzsche as the advent of Last Man, that childless end product of humanist individualism: 'The entire West has lost those instincts out of which institutions grow, out of which the future grows . . . One lives for today, one lives very fast – one lives very irresponsibly: it is precisely this which one calls "freedom".'[33]

If we continue to invoke Nietzsche, this is to emphasise the ambiguity and ambivalence of this crisis. The modern individual may have been the historical object of repetition for several centuries, but the subjective side of repetition is much harder to identify as it involves all sorts of non-human constituents including natural and unconscious processes as well as socio-economic developments and biopolitical technologies. What is necessary is a non-modern perspective on modernity. For even if all production is reproduction, this is never just the exact repetition of the same, but always the production of the similar 'with non-resembling means'.[34] It doesn't follow that all continuity in time is merely a fleeting illusion, but rather that whatever subsists in time is plastic in nature. Whether we are dealing with a thing, an institution or a civilization, it is always both the memory of the forces of repetition that have inscribed themselves in it and the capacity for the relative dissolution and metamorphosis of their traces. Plasticity enables the absorption of an exhausted form of life into a neighbouring form, not as passage from one well-delineated, total form to the other (transformation), but as a 'transdifferentiation' between forms (deformation).[36] Fatigue is thus the objective limit where the past must be forgotten if it is not to become the gravedigger of the present and its future.[37] It is integral to the modification of habits or the formation of new habits. What appears as a historical discontinuity of identity is in fact a more liquid continuity of becoming across differences and distances. It is not the many obvious but disparate repetitions of breathing, heartbeat and cell renewal that give organic unity to living bodies, for example, but their rhythmic continuity, in which new tensions are prepared between them as a result of the resolution of old ones.[38] Habit unifies the duration between, and extending through, repetitions, such that permanence itself is a pattern of changes, a communication of 'events' in a radically discontinuous time.[39] We are so much obsessed with cultural transmission and survival that, as Claude Lévi-Strauss famously said, we want to make history the driving force of its own development.[40] But perhaps the real danger for the survival of a (social) body lies precisely in the auto-immunitary attempts to conceive of repetition and transmission only in the image of what is already given and at the cost of the return of its potential of becoming. This is why Nietzsche's ultimate concept of repetition is the negentropic idea of the eternal return, 'the being of becoming'.[41] Plasticity implies a reverse ontology: we are always already caught up in complex feedback loops of which the effects have the potential to retroact on the causality of repetition and thus to produce a third passive synthesis that makes the present and the past coexist in the future. There is no repetition without excess, but 'if repetition makes us ill,' Deleuze writes, 'it also heals us'.[42] In this sense of an immanent alterity, of repetition as a power of the false, where continuity and discontinuity become indiscernible in the thickness of a becoming, the critical difference between the transcendental subject of repetition and its empirical object not only constitutes a post-human condition of man – operable man as a plastic multiplicity, or *plastes et fictor*, as Pico della Mirandola famously put it – it also necessitates, as I intend to demonstrate, a more elastic conception of conservation than Sloterdijk himself, despite his manifest Nietzschean inspiration, usually provides.

Immunology

Regardless of whether we evaluate the modern crisis of repetition as a breakdown of the old or as a breakthrough of the new, its message is clear: *Du mußt dein Leben ändern*; you must change your life. Indeed, as a crisis of the coherence and consistency of our habits and their traditional authority, it may well be the only authority whose imperative we can accept. It tells us that we must re-implicate ourselves in processes that exceed us on all sides. Whether this concerns a new relation between the local and the global in network culture, a renegotiation between rich and poor in political economy, a redefinition of the relationship between man and the biosphere in political ecology or indeed of the relation between soul and body in a new form of embodiment, operable man appears to himself under new, immunological premises, according to which he must learn to actively take responsibility for what he passively undergoes.

This re-implication of our own passivity is inseparable from what Sloterdijk calls the explicitation of the immunological paradigm. The problem of immunity is that of translating the human back into nature and incorporating the non-human.[43] As such it remains the Outside of all classical humanisms, but by becoming explicit, it also becomes its fate, or as we refer to destiny today: complexity, risk or uncertainty. Once it has been unfolded, it will never return to its previous folded state of implicitness. But precisely for that reason, immunological knowledge is more than just knowledge: it also changes the way we think and relate to our passivity. It can function as a prosthesis of trust that re-intimates us within nature or, what comes down to the same, within our bodies, now no longer understood as that alien prison of the soul but as our own potential or latent disposition. Sloterdijk refers to this trust as a 'second naivety', an anthropotechnical naivety in the same sense as when Deleuze emphasises that we do not yet know what a body can do.[44] It is here that both politics and science could regain a critical and clinical sense. In addition to providing insight in what we are and predicting what we can do, they cannot withdraw themselves from the affirmative interpretation and selective evaluation of how we are passively becoming. The modern bifurcation of nature has led to a culture of experts and specialists, figures of anomalous origin and mutual animosity for whom thought, whether in the guise of science or politics, has become reduced to knowledge and reflection about, but also disconnected from, life. In order to avoid this cultural stagnation and general stultification and make itself relevant for the future, the immunological paradigm inevitably reconnects thought to life itself, recovering a living or 'natural relationship' to its past and its culture, just as it turns philosophy into what Sloterdijk calls *Biosophie*. If what we are is constituted by habitual series, and indeed life itself is a vital continuum of repetitive behaviour, then knowledge too becomes an immunological 'act ... conducted in such a way that its current execution co-conditions its later executions'.[45] If life is a homeostasis brought about through self-organising repetition, then immunological thought explicitly wants to be a continuation of life, a consciously conducted auto-affective experiment with its own conditions. The question it raises is indeed how, building on our passivity, knowledge can get to the productive side of repetition, intervene in it and thus truly become a thought of plasticity, instead of just its indifferent reflection in knowledge. For this reason, we may agree with Sloterdijk that 'just as the nineteenth century under the cognitive sign of production, and the twentieth century under the sign of reflexivity, in the same way the future should present itself under the sign of the exercise'.[46]

One way of further exploring this reunion of thought and being is through Catherine Malabou's elaboration of the concept of plasticity. As we have seen, plasticity, like habitual repetition, is both active and passive, meaning 'simultaneously the capacity to take a form (concrete or clay for example are designated as "plastic" materials) and to give a form (as in art or plastic surgery)'.[47] As a consequence, it is both preformation and transformability, shifting the thresholds between the organic and the inorganic, between the innate and the acquired, or indeed between the human body as an organism and machine technologies. Composed of folds, fields and layers, the brain is not a fixed entity, but an infinite series of modifications of the nervous system, of the nerves, of the neuronal and of our synapses. Perhaps we should say that the brain is a cascade of repetitions, of repeated determinations but also of their frictions and interstitial indeterminations, such that its recursivity not only fine-tunes existing brain pathways, but also generates new connections within brain regions, for example in the case of irreversible brain damage. With its twin conditions of receptivity and change, in other words, plasticity is the path-dependent, future potential of the brain.

The problem is that in the descriptive approach of science, this potential remains unthought. While neurobiology has explicitated an enormous amount of knowledge on the plasticity of the brain, Malabou argues that we still need to 'implicate' a 'consciousness' or 'representation' of plasticity in the brain itself.[48] If today it makes sense to say that 'we are our brain', this still sounds misleadingly familiar. As Alva Noë has pointed out, it would be much more astonishing if it turned out that we are not our brain. The real question is: what is being and how does being a brain in the process of being made present itself to us?[49] Reductionist materialists would like to see strong ties, or even a coincidence between consciousness and the brain, while idealists would like to cut all ties between them. But what if they were weak ties, such that, even if thought and life are connected, they do not mirror each other (that is, 'adequation without correspondence')? For Malabou, the concept of plasticity forms the condition of the very coherence between what we know of the brain and how we relate to the brain.[50] Before we have explicit knowledge of how the brain conditions us (or indeed: itself), we already implicitly mould the brain under the influence of our cognitive experience. What a brain can do is therefore not just an epistemological question but an ontological question. Contrary to all the contemporary tendencies towards its naturalisation, the brain is never only a given, because it must always be made in order to exist. More than the object of the laws of neuroscience, the brain is first of all the *milieu* of thought, its matter, just as thought is not *what* the brain is, but what it does, the *how* or manner of the brain, its very performance. The brain, says Malabou, is a 'future producing machine' and plasticity is the 'eventual dimension of the machinic'.[51] In this sense, her question 'what to do with our brain?' – also the title of one of her books – strongly resonates with Sloterdijk's imperative 'you have to change your life'. The explicitation of the synthetic nature of the brain cannot but imply a minimum of care that replaces the freedom to stay unconscious about the limits of our sovereignty: 'We are not only asking the question of repetition; repetition has become the question, what questions us.'[52] By means of the plasticity in repetition, we do not just design our own lives, we also design our own brain. Precisely because we now know that it is not finished and never will be, but also that our current habits of moulding it are unsustainable, we must ask what to do with the brain, how to work and modify it.

Learning What To Do and Why?

Design, ever since the word was coined, has always meant a combination of thought and action. If today we are lacking a consciousness of plasticity and merely hear the cry of its imperative, this is because our capacity for thought, as Hannah Arendt famously observed, no longer keeps up with our capacity for knowledge and action.[53] The discrepancy between the neurological promises and possibilities, for example, and the political, philosophical and cultural space for action on these promises and possibilities could not be bigger. We make our own brain no less than we make social order, but we certainly do not realise it in the same degree. As a consequence, thought and action do not communicate. If, for example, we are confronted with an 'epidemy of depression',[54] a widely spread disconnect or weakening of neuronal connections, our sense of plasticity is to take antidepressants that stimulate the neurochemical transference in order to repair and protect the plastic capacities of the brain.[55] But as Malabou points out, plasticity in these examples is reduced to the capacity to work and 'function well', in other words, to flexibility. In physical terms, being flexible means to be able to bend, to give way and take form, but not the capacity to produce form itself. In psychopolitical terms, it means impotent suffering, obedience and consilience, the opposite of the capacity of resistance. As modern self-entrepreneurs, we like to think of our lives as works of art, even if we are generally unmoved by the various options of styling them differently. With the brain things are even worse. We celebrate adaptivity and creativity in the form of temporary contracts, part-time work and increasing mobility, all while we generally accept the brain as a natural given, as a closed-loop system at the intersection of social science and bio-engineering ('finally we can measure'). In this way, our consciousness of the brain coincides with the new spirit of capitalism.

We reduce the plastic potential of the brain to the alienated and displaced image of the world – the *Kopfkino* of our ostentative precarity – and fail to see that it is also a biopolitical construction. Everything happens, as Malabou puts it in a very Nietzschean way, 'as if we knew more about what we can bear than about what we can create'.[56] But like the social, the brain is not just a faculty of passive toleration; it is a field of activity. It is history in the making. And while we are exposed to history more than we make it, in this case the question is indeed how, building on our passivity, we can become at least one of the subjects of this history? How can we think what we are doing? The answer – and here, too, Sloterdijk and Malabou converge – implies a 'cybernetic' conception of freedom: by learning.

Everybody knows that learning is not just a cognitive affair. It means not only to think differently, but also to live and feel differently; it is a matter of conversion, as opposed to the mere accumulation of information and knowledge. There is no learning, no futurity, without a relation to the past and to history, including, perhaps today first of all, that of the brain. According to Malabou, we must therefore 'respond in a plastic way to the plasticity of the brain'.[57] Unlike flexibility, which fixates the brain between biological determinism and economical multiple usability, plasticity delocalises the brain by producing transformational effects. As Deleuze and Guattari say, the brain is more like grass than like a tree.[58] Being neither inside nor outside, it is an interstitial whole, not an integrated whole. To learn from the brain is thus to cultivate its interstices beyond biological determinism and make new circuits. As in *AI*, the essence of intelligence is in the capacity of linear processes to interrupt their automaticity and produce a residual interference. It implies an experiment with a cerebral interactive network whose fragmentary organisation is determined, not by some 'administrative centre', but on and through its immanent outside.

In this experiment/experience of learning, it is not caution that Malabou lays the emphasis on. Without an immunity crisis, without a breakdown of our habits and automatisms, the brain is doomed to remain a caricature of the world. For this reason, she distinguishes two types of plasticity: positive and destructive. In positive plasticity, a continual balance is kept between the capacity for change and the aptitude for remaining the same, in other words, between future and memory, between giving and receiving form. Plasticity is an ongoing process in which some destruction is necessary, yet this does not contradict a given form, but makes it possible. Like the Ship of Theseus or

Otto Neurath's bootstrap, our brains and bodies must be constantly but gradually reconstructed, such that they retain their essence in the form of a complex continuity in a sea of discontinuities.[59] Destructive plasticity, by contrast, is a kind of plasticity that does not repair and in which the smallest accident suffices for the biggest possible deformation. It is the type of destruction wreaked by cerebral lesions, but also by a sudden burst of anger. Instead of the repetition of the same, plasticity here becomes the repetition of difference, the production of the singular. Malabou is primarily interested in this second type of plasticity, because it forces the brain to reinvent itself and discover its freedom in relation to the traces from the past. Only when the continuum of repetition that puts neuronal functioning in mutual functional dependency with the normal functioning of the world is interrupted, then does the brain become capable of turning itself into an event and in this way of de- and reprogramming itself.[60] When the pressure of flexible polymorphy exceeds our limits, there is a rupture, a point where we do not bend, but find our own form. An 'explosion' (Malabou speaks with Bergson of a 'reversal of the law of the conservation of energy') forces the brain to break with flexible polymorphy and to renegotiate its relation to the world in a non-pathological and non-obedient way. The alternative in plasticity, according to Malabou, is therefore not terror or fixed identity, destruction or impression of form. Rather, the immanence of explosion and generation (or: the partiality of death) is the condition of possibility of formal resilience. Plasticity has to be critical in order to become clinical, that is, in order to make a difference.

But is the concept of plasticity enough to learn from the brain? This is the immunological question I would like to raise. A plastic rupture, after all, is not something you 'want' for the sake of it. If it happens, it happens behind your back, unintentionally, in the interstices between your own reasons and the body-brain you inhabit, and thus at the risk of leaving you unprotected against the chaos of an uninhabitable form of embodiment, something that approximates what Deleuze and Guattari have described as a 'black hole'.[61] Plasticity means that our life can go on without us.[62] It may pose no threat to the continuity of the brain, but it may well lead to a radical discontinuity in consciousness and thought. As such, it is a change that is potentially without measure or resistance. When we become what we are, our brains do not necessarily become who we are, such that, in the worst case, the one we will become doesn't care that he is what he has become. Jairus Grove has therefore argued that plasticity is not so much a matter of hope as it is of horror. Reading the history of cybernetics as a trial run of neuroplasticity, his point is that plasticity turns out to be less the ability to learn than the ability to control and predict outside the humanist horizon. As a consequence, it may be interesting as a speculative problem, but it is limited in practical scope: 'the challenge of neuroplasticity is necessary but insufficient to formulate a politics or an ethics'.[63]

In fact, we should wonder whether, in practice, destructive plasticity and flexibility are not equally indifferent to thought.[64] In both cases, there is a neutralisation of subjectivity. While it is clear that the plasticity of the brain implies a resingularisation and could provoke reflexivity, it is not at all clear how thought matters in the becoming of the brain. It is as if the plasticity of the brain doesn't really need thought to organise itself, but merely makes our narcoleptic 'selves' suffer from one disposition to the next – it is, after all, a change of unconscious habits without our conscious intervention. But also, as if the answer to the question 'Que faire de notre cerveau?' is already known and leaves little to be learned from the risks of disaffection and disattachment to which the brain is exposed. As Sloterdijk says, the very form of the question, which dates back to Lenin and the avantgardes, expresses a kind of ontological energy, an extreme certainty that doing something is still possible even when all existing possibilities seem exhausted.[65] While risk is the price of progress, moreover, it becomes increasingly hard to see how anyone is likely to profit from the leaps taken by the entrepreneurs of neurocapitalism. By contrast, what we have previously defined as a crisis of repetition therefore means precisely that we do not know what to do, not even if we can do anything at all. This is also Grove's point: yes, plasticity means that our thought, freedom and life are contingent. But as a consequence, knowledge immediately acquires an all the more acute practical and political – or in Sloterdijk's words, immunological – component. In a situation of radical disorientation, there are either too little or too many motives to act.[66] We have to change our lives, but we cannot spontaneously act on the basis of what we know is possible to do with our brain. On the contrary, the knowledge of plasticity seems to induce in us the same radical disinterestedness and irresponsibility as does the experience of plasticity itself. Hence, it is doubtful that our way of relating to plasticity, its thought or reflection, must itself be plastic. Rather, it seems that the notion of plasticity is not enough to establish more than just a theoretical coherence between thought and brain, because in practice they do not communicate in the same way. As a consequence, Malabou's attempt to solve the ambivalence of plasticity by dissecting it into the two contraries of flexibility and resistance remains abstract as well, with freedom or creative thought only appearing in the blind transgression of the limits to flexibility.

Precisely because plasticity offers us no promise of return, the image of elasticity, understood as a different way of relating reflectively to plasticity, is a better, and perhaps the only human(ist) way to address the tension between obedience and (self-) creativity. Insofar as it is grounded on the eternal return of difference instead of the same, it could make for an important supplement to Malabou's concept of plasticity. For Malabou, elasticity is the same as flexibility, the natural limit of variably present forms that excludes the plastic labour of the negative.[67] But while flexibility is the ideological form of plasticity, elasticity is the capacity of returning, if not *to* an original form, then at least *of* something in the original form across a difference. Unlike the sheer consilience of flexibility, then, elasticity also constitutes the *re*-silience that allows for the forming of a processual self with and against its destruction. If positive plasticity relies on formal continuity, the reason for this constancy is found in the elastic manner in which this continuity, as soon as it has come into being, is (re)produced in discontinuous matter. In destructive plasticity, by contrast, there appears to be neither reason nor cause.[68] Yet no explosion is total and plasticity is always partial, a dialectics between emergence and destruction of form. Here, too, only the elasticity of the deformed form can give sense to the appropriation of an explosion and develop it into a new degree of freedom. Once there is 'homeostasis' (that, seen from the perspective of genesis instead of structure, is always a principle of 'allostasis', that is, of heterogenesis and meta-stability), self-preservation proceeds as constant, internally directed self-creation in and on the outside. Precisely because plasticity is limitless and will ultimately kill you, it must be dissociated from itself in order to be made viable. Stasis is crucial. Even if it is born from passivity, as a slip in(to) the environment that camouflages it, elasticity gradually becomes a cause in itself, a force of internalisation. It is thus the conative capacity (the 'passibility'[69]) to deal with the tension between the genesis of form and its ongoing deflagration. Structurally open but operationally closed, it is the *vis elastica* that integrates within oneself the world as it exceeds one's strength.[70] Since the body and the brain are never just natural givens but must be constructed in their continuity and variation by the very ways we inhabit them, the question is how do we learn to make ourselves at home in them. It is not sufficient to say what the body or the brain is or can do; we have to make a difference by making them ours, that is, to repeat them and to produce ourselves through repeating them. To supplement plasticity with elasticity is therefore to supplement the question 'what to do?' with the question that has to be answered by any living entity, that is any mode of being that has a stake in the continuity in its vital conditions: 'why here now?'[71] While the former question belongs to knowledge and action, possibility and actuality, only the latter is capable of orienting thought in the presence of its ground and thus of taking us on a learning curve, even if it ultimately takes us beyond itself in repetition. Today, the plastic brain is the image and etiology of bare life; it is the cornerstone of biopolitical management and social engineering of probabilities and possibilities. But thought is precisely the elastic imagination by which we claim a stake in how we 'become what we are' and thus in that that we will become. If the brain is the plastic power or potential of thought's becoming, thought is our elastic capacity to learn and grow with its interstices.

Vertical Tension

Far from the sign of a sentimental humanism that subordinates the body to the mind or separates intentionality from its embodiment, elasticity is the anthropotechnical power of incorporation, of rising to the occasion. Instead of reducing cerebral life to our image and seeking total dominance over it, we make ourselves relevant to the adventure of the immanent discontinuity of its future development. The elasticity of habit is generative of both person and brain, rendering them mutually inclusive in their singular becoming. If what the brain 'is' will be decided by the manner in which it is exercised, all we can do is experiment and follow the brain's restless plunge into chaos.[72] Born from difference (natal difference), the immunological challenge for us is to orient ourselves in this groundless ground of existence as in a field of distances, neighbourhoods, moods and vectors. In the question of the ends of man, as Jacques Derrida has famously pointed out in his deconstruction of humanism, there can be no transcendent answer, because the ultimate *end* of man is precisely the end of *man*.[73] Or in the words of Malabou, the plasticity of repetition is the raw material of our lives in which we return even when our essence is dissolved: 'the human is sculpting a certain relationship to repetition and ... in return, this relationship is sculpting it'.[74] But from the immunological point of view, a point of view that is no longer modern, the way we respond to a crisis of repetition must be elastic. Given a certain historical situation, we cannot but protect our mode of living. Only the fragility of habit can open a space of expectation and desire, and is therefore a prerequisite for belief in the future. Thought is this immanent orientation process that produces its criteria as it operates, becoming sensitive to the validity and viability of differences, demonstrating its aptitude for return and for engendering a heritage or tradition.

For Sloterdijk, the guiding intuition in this orientation is that of the self. Just as Ravaisson's concept of habit must be understood as a disposing disposition, a virtue (*hexis*) of actualisation acquired through and composed of previous acts by which a being has a hold on itself and its future,[75] Sloterdijk defines habit not by what we are but by what we possess while being possessed, which leads him to a 'rehabilitation of egoism' as primordial virtue.[76] He quotes Ernst Bloch: 'I am. But I do not have myself. That is why we first become.'[77] A habit is a kind of self-addiction: we come in possession of ourselves through relations of being-possessed. (Again, this explains why the humanist project of man to become fully in possession of itself is doomed to fail. New habits imply new compulsive repetitions.) To contract a habit (derived from *habitus* (*habere/ekhein*): to have/hold) is to integrate the elements and repetitions from which we come in a new way. I always begin as 'guest of the other', but this beginning is simultaneously the appropriation of the difference out of which this relation with the other comes.[78] The stone becomes a tool in my hands at the same time that I become a stonecutter. Habit is thus the self-referential and imaginative manner in which a being makes the pass from an unstable *methexis* (participation in the other) to a stable individual. For Sloterdijk, it provides a basic orientation for immunological thought, namely the *Standortvorteil Ich*. The human individual, he argues, is not a given but the political project of the ongoing conditioning of an elastic atmosphere or immune system: 'The individual is a futile passion, but a passion it should nonetheless remain.'[79]

Sloterdijk further specifies this orientation in terms of a 'vertical tension (*Vertikalspannung*)': the affirmation of resistance in an autogenetic gesture by which we make a difference. Only by seeking-producing friction do we secede from and get to the 'other side of habit', for example from the repeated repetitions of religion and mass media to the repeating repetition of art or sports. It is true that an elastic that returns to its initial form is not interesting. But just as elasticity is an essential feature of plasticity, elasticity without the rigidity and resistance of plasticity is only the lowest degree of elasticity. In reality, there can be no reproduction without turning repetition against itself in a creative self-intensification generated from within a tortuous but virtuous circle. Whereas plasticity gives us a real experience of thought's synthetic groundlessness ('How do you internalise a cerebral lesion?'),[80] this outside must be folded into thought and synthesised relative to thought's grounding movement. The elastic human, for Sloterdijk, therefore holds the middle between an athlete and an acrobat. He practices a 'subversion from above', a 'supraversion' of actual existence.[81] Thought or design, Sloterdijk argues, is this transcendental ascensional pull in the production of a *Halt im Haltlosen*, a *gekonnte Abwicklung des Nichtgekonnten*.[82] It is a kind of parallel action, the ritualistic and imaginary overcoming of the gap between what we are able to do and what we are not able to do. The more complex the machine, the slicker its interface must look. Design is the power of impotence; it is pure gesture, pure self-referentiality, pure self-potentialisation.[83]

When Sloterdijk speaks of elasticity as vertical tension, he therefore develops it as an aristocratic posture. 'At any time you should act in such a way that you personally anticipate the better world in the worse.'[84] This 'ethical difference', which he traces back to Heraclitus, is in fact already implied by the concept of habit, since 'habit can only be surpassed by habit' (Thomas à Kempis). Habits not only habituate but also habilitate. The elasticity of habit makes us pass from optimism to meliorism: becoming is better than being. Immunological thought is not a transcendent movement back-and-forth between the possible and the real, but a constant and immanent focus on 'the best (*eris*)'.[85] If plasticity is a cold and indifferent change without subjectivity, then elasticity is the capacity to synthesise and subjectivise change into self-enhancement and self-amplification.[86] Just as in sports, we are ultimately not in competition with others but with ourselves, working on and with an outside that is necessarily more intimate than any relative inside, because it is the very potential of our becoming better. Every habit, every tendency, can be thought of as duplicitous and incomplete, that is, as transductive relation in which we must take care for the best, which may contain the worst as much as be contained in it.[87] *You* have to change *your life*.

Notes

1. Hegel, 'Preface to the Second Edition', p. 19.
2. For a more encompassing discussion of autoplasticity in terms of new media of self-writing, see Van Tuinen, 'Transgenous Philosophy'.
3. Joining a coalition of feminist theory, science and technology studies and environmental movements, Sloterdijk gives an immunological redefinition of the finite/infinite relation and warns of the bad infinity, or metaphysical 'infinitism' (Sloterdijk, *Sphären II*, pp. 410–11, footnote 173) that lives in denial of its immunological premises.
4. Agamben, *Infancy and History*, pp. 107–9, 113–14.
5. Latour, 'A Cautious Prometheus?'
6. Sloterdijk, *Du mußt dein Leben ändern*, pp. 493–518, esp. 510, and 691–8, esp. 697.
7. Sloterdijk, *Nicht gerettet*, pp. 212–34.
8. Sloterdijk, *Du mußt dein Leben ändern*, p. 590; Sloterdijk, *Nicht gerettet*, pp. 69–81.
9. Sloterdijk, *Nicht gerettet*, pp. 595–7.
10. Sloterdijk, *Zeilen und Tage*, p. 54.
11. For a discussion of Sloterdijk's concept of design as air conditioning, see Van Tuinen, 'Airconditioning Spaceship Earth'.
12. Oosterling, 'Dasein as Design'.
13. Heidegger, *Beiträge zur Philosophie*, p. 56.
14. See Fischbach, *La production des Hommes*.
15. 'I define as exercise every operation by means of which the actor acquires or improves his qualification for the next execution of the same operation, regardless of whether this operation is declared an exercise or not.' Sloterdijk, *Du mußt dein Leben ändern*, p. 14.
16. Sloterdijk, *Du mußt dein Leben ändern*, pp. 174–5.
17. Deleuze, *Difference and Repetition*, p. 70.
18. Sloterdijk, *Du mußt dein Leben ändern*, p. 629.
19. Deleuze, *Difference and Repetition*, p. 71.
20. Nietzsche, *On the Genealogy of Morals*, essay II, par. 2.
21. Sloterdijk, *Die schrecklichen Kinder der Neuzeit*, pp. 245–55.
22. Sloterdijk, *Du mußt dein Leben ändern*, p. 639.
23. Ibid. pp. 175, 301–7.
24. Sloterdijk, *Die schrecklichen Kinder der Neuzeit*, pp. 229–311.
25. Ibid. p. 358.
26. Ibid. p. 92.
27. Ibid. pp. 312–28, 54, 485.
28. Ibid. p. 23; Sloterdijk, *Im Weltinnenraum des Kapitals*, pp. 241–2.
29. Sloterdijk, *Du mußt dein Leben ändern*, p. 17; Sloterdijk, *Heilige und Hochstapler*, p. 8.

30. Sloterdijk, *Die schrecklichen Kinder der Neuzeit*, p. 234.
31. James, *Habit*, pp. 51–2.
32. Sloterdijk, *Du mußt dein Leben ändern*, pp. 683–4.
33. Nietzsche, 'Twilight of the Idols', p. 493.
34. Gilles Deleuze distinguishes between resemblance as producer and resemblance as produced. Deleuze, *Francis Bacon*, p. 98.
35. Barad, *Meeting the Universe Halfway*, p. 151. Cf. Deleuze and Guattari, *A Thousand Plateaus*, p. 498.
36. Malabou, *What Should We Do with Our Brain?*, p. 16.
37. Nietzsche, *Sämtliche Werke*, 7.251.
38. Langer, *Feeling and Form*, p. 126.
39. Nietzsche, *Sämtliche Werke*, 7.579.
40. I borrow this reference to Lévi-Strauss's concept of 'warm societies' from Georges Didi-Huberman's discussion of plasticity as material force of becoming between survival (*Nachleben*) and rebirth, memory and metamorphosis, after-effect (*Nachwirkung*) and effect, and body and style in *Das Nachleben der Bilder*, p. 177.
41. Deleuze, *Nietzsche & Philosophy*, p. 24.
42. Deleuze, *Difference and Repetition*, p. 21.
43. Lemm, 'Nietzsche, *Einverleibung*'.
44. Sloterdijk, *Sphären. Plurale Sphärologie: Band III: Schäume*, p. 202.
45. Sloterdijk, *Scheintod im Denken*, p. 19.
46. Sloterdijk, *Du mußt dein Leben ändern*, p. 14.
47. Malabou, *What Should We Do with Our Brain?*, p. 5.
48. Ibid. p. 11.
49. Ibid. p. 69.
50. Ibid. pp. 3–4.
51. Ibid. p. 38.
52. Malabou, 'From the Overman to the Posthuman', p. 71.
53. Arendt, *The Human Condition*, p. 6.
54. Pignarre, *Comment la dépression est devenue une épidémie*.
55. Cf. Heather Perry's auto-trepanation to combat her physical exhaustion.
56. Malabou, *What Should We Do with Our Brain?*, p. 13. Because of this structural analogy between neuro-degenerative illness and social handicaps, Malabou argues that an unemployed person without financial support is exactly like an Alzheimer's patient, that is someone with no 'network' (Ibid. pp. 40–54).
57. Ibid. p. 30.
58. Deleuze and Guattari, *A Thousand Plateaus*, p. 17.
59. Malabou, *What Should We Do with Our Brain?*, p. 56.
60. Ibid. p. 73.
61. Deleuze and Guattari, *A Thousand Plateaus*, pp. 333–4.
62. 'Plasticity is not habit, but it is a condition of habit.' (Carlisle, 'The Question of Habit', p. 2.)

63. Grove, 'Something Darkly This Way Comes', p. 250.
64. In the same sense, Deleuze has repeatedly warned that schizophrenia is a potential of thought, but not its model or image. It is only at this latter level that learning becomes indispensable: 'In *our universal schizophrenia*, we need reasons to believe in this world' (Deleuze, *Cinema 2*, p. 172; emphasis in the original).
65. Sloterdijk, *Du mußt dein Leben ändern*, p. 614; Sloterdijk, *Die schrecklichen Kinder der Neuzeit*, pp. 54–74.
66. Sloterdijk, *Die schrecklichen Kinder der Neuzeit*, p. 85.
67. Malabou, 'Plasticity and Elasticity'.
68. Malabou, *Ontology of the Accident*, pp. 59–61.
69. As Ed Cohen has argued, instead of the self-environment opposition a 'natural elasticity' or natural healing propensity of organisms (*vis medicatrix naturae*) lies at the basis of pre-modern (pre-mid-nineteenth century) medicine. Cohen, *A Body Worth Defending*, p. 4.
70. Sloterdijk speaks of 'the discovery of the world in man' (Sloterdijk, *Du mußt dein Leben ändern*, pp. 507–11). On the *vis elastica*, see Deleuze, *Francis Bacon*, p. 41.
71. Stengers, 'Introductory Notes on an Ecology of Practices'.
72. Sloterdijk, *Du mußt dein Leben ändern*, p. 255. As Deleuze and Guattari write, the brain 'plunges into and confronts chaos' (Deleuze and Guattari, *What is Philosophy*, p. 210).
73. Derrida, *Margins of Philosophy*, pp. 109–36.
74. Malabou, 'From the Overman to the Posthuman', p. 62.
75. Ravaisson, *Of Habit*, pp. 49, 77.
76. Sloterdijk, *Du mußt dein Leben ändern*, pp. 376–8, 391; Sloterdijk, *Nicht gerettet*, p. 32.
77. Sloterdijk, *Zeilen und Tage*, p. 89.
78. Nancy, *Hegel*, p. 57.
79. Sloterdijk, *Ausgewählte Übertreibungen. Gespräche und Interviews 1993–2012*, p. 444.
80. Malabou, *Ontology of the Accident*, p. 29.
81. Sloterdijk, *Stress und Freiheit*, p. 135. Deleuze would speak of a 'counter-effectuation', although Sloterdijk blames Deleuze for putting too much emphasis on the event: 'He ascribes to the event what belongs to exercise' (Sloterdijk, *Zeilen und Tage*, pp. 151, 164).
82. Sloterdijk, 'Das Zeug zum Design'.
83. As Sloterdijk argues, there is no freedom without resistance or stress, just as there is no difference between positive and negative freedom, because it is only here that we break with 'tyranny of the possible' and seek to find our worthiness in the immediate realisation of 'the best'. Modern decadence, by contrast, would be due to the fact that the project of Enlightenment has raised our sensibility far beyond the reality principle, thus fatally disconnecting thought from the activity of the world (Sloterdijk,

Stress und Freiheit, pp. 29, 47, 57–8). 'Enlightenment,' however, 'as it first took form in ancient sophistry, is above all a prophylaxis of helplessness' (Sloterdijk, *Ausgewählte Übertreibungen*, p. 293). This prophylactic quality is also the essence of a manner or attitude (*Haltung*): when habitual tricks and easy answers fail, we find agency (*existentielles Können*) precisely in *amechania* (Ibid. pp. 394–5).
84. Sloterdijk, *Du mußt dein Leben ändern*, p. 506.
85. Ibid. pp. 304–9. On the upgrading of being as the latent metaphysics of late modernity, see Sloterdijk, *Ausgewählte Übertreibungen*, pp. 156, 163.
86. Malabou, *Ontology of the Accident*, p. 11.
87. Emphasizing its duplicitous motive, Bernard Stiegler describes *eris* or artistocratic culture as emulative competition (as opposed to levelling, imitative competition) or 'the elevation towards an always possible best, *artiston*' (Stiegler, *Decadence of Industrial Democracies*, p. 55).

Bibliography

Agamben, Giorgio, *Infancy and History: On the Destruction of Experience*, trans. Liz Heron (London; New York: Verso, 2007).

Arendt, Hannah, *The Human Condition* (Chicago: The University of Chicago Press, 1958).

Barad, Karen, *Meeting the Universe Halfway* (Durham, NC: Duke University Press, 2007).

Carlisle, Clare, 'The Question of Habit in Theology and Philosophy: From Hexis to Plasticity', *Body & Society*, Vol. 19, Issue 2–3 (2013), pp. 1–28.

Cohen, Ed, *A Body Worth Defending: Immunity, Biopolitics, and the Apotheosis of the Modern Body* (Durham; London: Duke University Press, 2009).

Deleuze, Gilles, *Cinema 2: The Time Image*, trans. Hugh Tomlinson and Robert Galeta (Minneapolis: University of Minnesota Press, 1989).

Deleuze, Gilles, *Difference and Repetition*, trans. Paul Patton (New York: Columbia University Press, 1994).

Deleuze, Gilles, *Francis Bacon: The Logic of Sensation*, trans. Daniel W. Smith (London; New York: Continuum, 2004).

Deleuze, Gilles, *Nietzsche & Philosophy*, trans. Hugh Tomlinson (New York: Columbia University Press, 2006).

Deleuze, Gilles and Félix Guattari, *A Thousand Plateaus: Capitalism and Schizofrenia*, trans. Brian Massumi (Minneapolis: University of Minnesota Press, 1987).

Deleuze, Gilles and Félix Guattari, *What is Philosophy*, trans. Hugh Tomlinson and Graham Burchell (London; New York: Verso, 1994).

Derrida, Jacques, *Margins of Philosophy*, trans. Alan Bass (Chicago; Sussex: The University of Chicago Press, 1982).

Didi-Huberman, Georges, *Das Nachleben der Bilder. Kunstgeschichte und Phantomzeit nach Aby Warburg*, trans. Michael Bisschoff (Berlin: Surhkamp Verlag, 2010).

Fischbach, Franck, *La production des Hommes. Marx avec Spinoza* (Paris: VRIN, 2014).

Grove, Jairus, 'Something Darkly This Way Comes: The Horror of Plasticity in an Age of Control', in *Plastic Materialities: Politics, Legality, and Metamorphosis in the Work of Catherine Malabou*, ed. Brenna Bhandar and Jonathan Goldberg-Hiller (Durham, NC: Duke University Press, 2015), pp. 233–63.

Heidegger, Martin, *Beiträge zur Philosophie*. Gesamtausgabe 65 (Frankfurt: Klostermann, 1989).

Hegel, Georg Wilhelm Friedrich, 'Preface to the Second Edition', in *The Science of Logic*, trans. George di Giovanni (Cambridge: Cambridge University Press, 2010), paras 29–30.

James, William, *Habit* (New York: Henry Holt and Company, 1914).

Langer, Susanne, *Feeling and Form* (New York: Charles Scribner's Sons, 1953).

Latour, Bruno, 'A Cautious Prometheus? A Few Steps Toward a Philosophy of Design with Special Attention to Peter Sloterdijk', in *Measuring the Monstrous: Peter Sloterdijk's Jovial Modernity*, ed. Sjoerd van Tuinen and Koenraad Hemelsoet (Brussels: KVAB, 2009), pp. 61–71.

Lemm, Vanessa, 'Nietzsche, *Einverleibung* and the Politics of Immunity', *International Journal of Philosophical Studies*, Vol. 21, Issue 1 (2013), pp. 3–19.

Malabou, Catherine, 'Plasticity and Elasticity in Freud's *Beyond the Pleasure Principle*', *Diacritics* (Winter 2007), pp. 78–85.

Malabou, Catherine, *What Should We Do with Our Brain?* (New York: Fordham University Press, 2008).

Malabou, Catherine, *Ontology of the Accident. An Essay on Destructive Plasticity*, trans. Carolyn Shread (Cambridge: Polity, 2012).

Malabou, Catherine, 'From the Overman to the Posthuman: How Many Ends?', in *Plastic Materialities: Politics, Legality, and Metamorphosis in the Work of Catherine Malabou*, ed. Brenna Bhandar and Jonathan Goldberg-Hiller (Durham, NC: Duke University Press, 2015), pp. 61–72.

Nancy, Jean-Luc, *Hegel: The Restlessness of the Negative*, trans. Jason Smith and Steven Miller (Minneapolis; London: University of Minnesota Press, 2002).

Nietzsche, Friedrich, 'Twilight of the Idols', in *The Portable Nietzsche*, trans. Walter Kaufman (New York: Viking Press, 1954), pp. 463–564.

Nietzsche, Friedrich, *Sämtliche Werke*. Kritische Studienausgabe in 15 Bänden, ed. G. Colli and M. Montinari (München: De Gruyter/dtv, 1999).

Nietzsche, Friedrich, *On the Genealogy of Morals: A Polemic*, trans. Michael A. Scarpitti (London: Penguin Books, 2013).

Oosterling, Henk, 'Dasein as Design. Or: Must Design Save the World?', in *From Mad Dutch Disease to Born to Adorno. The Premsela Lectures 2004–2010*, trans. Laura Martz (Amsterdam: Premsela, 2010), pp. 115–40.

Pignarre, Philippe, *Comment la dépression est devenue une épidémie* (Paris: La Découverte, 2012).

Rare Historical Photos, Leonid Rogozov, <http://rarehistoricalphotos.com/leonid-rogozov-appendix-1961/> (accessed 18 October 2016).
Ravaisson, Félix, *Of Habit*, trans. Clare Carlisle and Mark Sinclair (London; New York: Continuum, [1838] 2008).
Sloterdijk, Peter, *Sphären II: Globen. Makrosphärologie* (Frankfurt am Main: Suhrkamp Verlag, 1999).
Sloterdijk, Peter, *Nicht gerettet: Versuche nach Heidegger* (Frankfurt am Main: Suhrkamp Verlag, 2001).
Sloterdijk, Peter, *Sphären. Plurale Sphärologie: Band III: Schäume* (Frankfurt am Main: Suhrkamp Verlag, 2004).
Sloterdijk, Peter, *Im Weltinnenraum des Kapitals* (Frankfurt am Main: Suhrkamp Verlag, 2005).
Sloterdijk, Peter, *Du mußt dein Leben ändern. Über Anthroropotechnik* (Frankfurt am Main: Suhrkamp Verlag, 2009).
Sloterdijk, Peter, *Scheintod im Denken: Von Philosophie und Wissenschaft als Übung* (Berlin: Suhrkamp Verlag, 2010).
Sloterdijk, Peter, *Stress und Freiheit* (Berlin: Suhrkamp Verlag, 2011).
Sloterdijk, Peter, *Zeilen und Tage. Notizen 2008–2011* (Berlin: Suhrkamp Verlag, 2012).
Sloterdijk, Peter, *Ausgewählte Übertreibungen. Gespräche und Interviews 1993–2012*, ed. Bernhard Klein (Berlin: Suhrkamp Verlag, 2013).
Sloterdijk, Peter, *Die schrecklichen Kinder der Neuzeit* (Berlin: Suhrkamp Verlag, 2014).
Sloterdijk, Peter, *Heilige und Hochstapler: Von der Krise der Wiederholung in der Moderne* (Berlin: Suhrkamp Verlag, 2015).
Sloterdijk, Peter, 'Das Zeug zum Design' <https://www.youtube.com/watch?v=zpgizaftqwI> (accessed 9 April 2016).
Stengers, Isabelle, 'Introductory Notes on an Ecology of Practices', *Cultural Studies Review*, Vol. 11, Issue 1 (2005), pp. 183–96.
Stiegler, Bernard, *The Decadence of Industrial Democracies: Disbelief and Discredit, Volume 1*, trans. Daniel Ross and Suzanne Arnold (Cambridge; Malden: Polity Press, 2011).
Van Tuinen, Sjoerd, 'Airconditioning Spaceship Earth: Peter Sloterdijk's Ethico-Aesthetic Paradigm', *Society and Space*, Vol. 27, Issue 1 (2009), pp. 105–18.
Van Tuinen, Sjoerd, 'Transgenous Philosophy: Post-Humanism, Anthropotechnics, and the Poetics of Natal Difference', in *In Medias Res: Peter Sloterdijk's Spherological Poetics of Being*, ed. Willem Schinkel and Liesbeth Noordegraaf-Eelens (Amsterdam: Amsterdam University Press, 2011), pp. 43–66.

CHAPTER 11

Automata, Man-machines and Embodiment: Deflating or Inflating Life?

Charles T. Wolfe

Introduction

In what follows I reflect on a conceptual pair that is not, at first sight, a piece of delightful symmetry: mechanism and embodiment. I first examine the complexity involved in early modern mechanistic approaches to organic life (contrary to some popular misconceptions), and then ask what they imply for our understanding of embodiment. Conversely, I examine in addition some of the core claims in contemporary embodiment discourse and subject it to some critical evaluation with regard to its potential anti-naturalism. What happens conceptually when we try to take account of the *reality* of actual mechanisms, but also of the *reality* of embodiment? Embodiment can be a challenge to mechanistic models, not just in a negative sense as in the usual 'mere machines or mechanical models of life cannot grasp X (fill in your favorite feature of embodied, personal, fleshly features of life)', but also in a positive sense, as a kind of explanatory challenge that, I will suggest, spurs on the elasticity and ambition of the mechanistic project, as we shall see below with the 'marveling' at Vaucanson's mechanical duck, but also in the irreducibly organic quality of La Mettrie's 'man-machine'.[1] To paraphrase a Deleuzo-Spinozian slogan, how much can mechanistic explanation do? If we look at the functional dimension of machines, which itself opens on to what we may call 'teleomechanism' (discussed below with respect to models of organism such as the 'animal economy', especially in eighteenth-century vitalism,[2] which are ultimately *structural*, in the sense that they study the properties of a system of interacting parts), the answer is: quite a bit. I conclude with some general reflections pointing to an 'affective' idea of the machine (shades of what Deleuze and Guattari, in the early 1970s, called 'desiring machines').[3]

One

Thirty years after Donna Haraway's ambitious, programmatic and somewhat polyphonous 'Cyborg Manifesto' – both a plea for hybridity and a kind of performative/enactive proof of its claims that points to the 'cyborgisation' of reality everywhere around us – it may not be easy to restore the habits of mind in which the machine and the body are opposed. But such habits are still firmly entrenched in embodiment discourse; in scholarship which targets, for example, an apparently dogmatic Cartesian mechanism that treats living bodies like machines ('As a machine, the body became objectified; a focus of intense curiosity, but entirely divorced from the world of the speaking and thinking subject');[4] and in many writings dealing with the theory of organism and organismic biology, which I have also worked on but that shall not be my concern here. These firmly entrenched habits reflect something of a failure to recognise or appreciate the flexibility, productivity and 'tolerance' (both at the ontological and at the explanatory levels) of mechanistic projects. In that sense, we need some clarification of what early modern mechanism could mean.

Major figures of mechanism in early modern medicine (or 'iatromechanism'), such as Boerhaave and sometimes Borelli, do speak of the mathematically specifiable mechanical properties of the bodies they study as laws of nature, since these ensure that the appeal, for example to the functioning of a pump or a sieve to explain a heart or a liver, is backed up by further guarantees. But what is their overall mechanistic commitment? This ranges from the idea of the world as a machine (clockwork, design) to a mechanistic ontology of the particles or components that compose the physical world, to – more interestingly for present purposes – an interest in the heuristic potential of mechanism, for example automata (from Descartes's fountains to 'living machines'). This heuristic potential is of particular interest because it is both adapted to and challenged by the specific reality of embodiment – the challenge of mechanical models faced with the living body.

Early modern mechanists do not necessarily deny or neglect the specific features of embodiment. Either because, like Borelli, they reflect on the 'shadowy similarity' between automata and living bodies (this is his own term: 'automata have a certain shadowy sameness (*umbratilem similitudinem*) to animals in that both are organic self-moving bodies').[5] Or because they seek to grasp the distinctively functional properties of bodies (as I discuss in the next two sections). From automata and man-machines to structural models of organism like the animal economy, there is a fascination with the 'challenge of Life'. Witness this description of Vaucanson's digesting duck by the Oxford literary scholar Joseph Spence, in 1741:

> If it were only an artificial duck that could walk and swim, that would not be so extraordinary: but this duck eats, drinks, digests and sh-ts. Its motions are extremely natural; you see it eager when they are going to give him his meat, he devours it with a good deal of appetite, drinks moderately after it, rejoices when he has done, then sets his plumes in order, is quiet for a little time, and then does what makes him quite easy.[6]

This is quite different from a picture we may have, of 'dead mechanism'. For indeed, at least as far back as Friedrich Engels[7] (leading to a commonplace in twentieth-century Marxist discourse but also, in a kind of development that is not aware of its origins, in recent theoretical moves such as 'new materialism'), there is an opposition between a misguided mechanistic standpoint and a better, organismic and/or humanist perspective (famously so in Sartre).

Hermann von Helmholtz, in 'On the Interaction of Natural Forces', makes a version of Engels's classic point, appealing less to the advances of nineteenth-century biochemistry and more to thermodynamics, except his target is not the materialists of the previous century, but its automata:

> To the builders of automata of the last century, men and animals appeared as clockwork which was never wound up, and created the force which they exerted out of nothing. They did not know how to establish a connexion between the nutriment consumed and the work generated. Since, however, we have learned to discern in the steam engine this origin of mechanical force, we must inquire whether something similar does not hold good with regard to men.[8]

Both Engels and Helmholtz are mistaken, however. I shall address their different versions of this 'enhanced life science' standpoint on the previous century, in turn.

Two

First, as regards materialism: it actually reacts quite often against the strictures of seventeenth-century mechanism, even if core vitalist concepts such as the 'animal economy' are *not strictly, or wholeheartedly, anti-mechanist*. For example, Denis Diderot insists on 'what a difference there is between a copper or silver watch, and a living watch', as he put it in his late, unfinished manuscript on the 'elements of physiology'.[9]

Elsewhere in the same text, he writes that an instrument made of wood or iron cannot feel, while an instrument made of flesh can feel:

> Difference between the clamp of a wooden or iron set of pliers, and that of pliers made of flesh or two fingers. The wooden clamp does not feel, the flesh clamp feels; the wooden clamp does not suffer, the flesh clamp suffers; the wooden clamp cannot be tickled.[10]

But what is this difference between a copper and a living watch, or a metal clamp and flesh-and-blood hand? It is not because of some kind of innate Aristotelian teleology in the flesh that is lacking in the iron or the wood. Recall Aristotle's influential argument for why a hand separated from the body is no longer a hand: the material structure of a part per se matters less than 'where' it is: 'Blood will not be blood, nor flesh flesh, in any and every state.' A hand can only be understood as a hand inasmuch as it belongs to an ensouled body, that is, matter animated by a form. Thus the material part, the hand, is derivative of the formal part, the soul. It is precisely this mere homonymy between a 'dead' hand and a 'live' hand that materialists miss, in Aristotle's view. If each animal and part would be defined by shape and colour, 'Democritus would be right' but 'the dead man has the same conformation of shape [as a man], but nevertheless is not a man'.[11]

Instead, the difference between copper and living watches, in Diderot, is twofold: (i) in types of arrangement and (ii) in the *type of matter itself* (including the difference between merely spatial contiguity and specifically organic continuity as Diderot presents in *Le Rêve de D'Alembert*, insisting that mere contiguity lacks 'network' properties, which a designer of neural networks may, or may not object to). The latter difference is also asserted by La Mettrie in *L'Homme-Machine*: 'The body is but a watch, whose watchmaker is the new chyle.' He adds:

> Nature's first care, when the chyle enters the blood, is to excite in it a kind of fever which the chemists, who dream only of retorts, must have taken for fermentation. This fever produces a greater filtration of spirits, which mechanically animate the muscles and the heart, as if they had been sent there by order of the will.[12]

Very summarily, La Mettrie is playing on the most classic mechanistic analogy (the watch or clock) and infusing the clockwork with living chemistry (chyle is the vital substance in organic chemical processes).

Do these differences between a copper watch and a living watch, or between an ordinary machine and a self-organising machine powered by chyle, amount to an *ontological difference*? In fact, early modern mechanists do not seem to insist on such an ontological difference (and nor do eighteenth-century vitalists in their focus on the organism, contrary to an equally common misconception), although an interesting passage in a 1640 letter from Descartes to Mersenne may run counter to this. Descartes himself toys with the opposition between machine and organism, or mechanical matter and living flesh, as precisely playing out at different levels: empirically (as in Helmholtz's comment on automata and bodies) and ontologically (as in Diderot's comment on flesh watches and wooden clamps, with its faint Aristotelian resonance). In Descartes's terminology, the different levels are called physical (or moral) and metaphysical:

> Speaking as a metaphysician, one might well build a machine that supports itself in the air like a bird, because birds, at least according to me, *are* such machines. But not speaking physically or morally, because it would take springs so subtle and overall so strong, that they could not be made by men.[13]

Rather, there is an insistence on complexity, structural and functional concepts, what I have called elsewhere, borrowing Timothy Lenoir's term but applying it to an earlier period with a subtly inflected meaning, 'teleomechanism'.[14] That is, on an explanation of systems (whether these be machines, automata or living bodies) that blend teleological features such as an appeal to purpose and function, and mechanistic features such as an account of their material properties and the interaction of their components.

As such, it is not just because later figures such as La Mettrie and Diderot seem, not just cognisant of but deeply concerned with 'the flesh', as Merleau-Pontyan phenomenologists may call it (relying on an opposition between *Leib*, the flesh as body possessed of subjectivity, and *Körper*, the mere physical body in space), that our historiography and our intellectual categories need improving on, especially compared with the picture painted by Engels and Helmholtz. It is also because *there is no such thing as pure, blind, cold mechanism*. And even when the body is treated mechanistically by the poster child of iatromechanism, Descartes, it is at the very least in a 'systemic' sense, as Barnaby Hutchins has emphasised: 'instead of reduction to corpuscular mechanics, Descartes explains the operation of the body through whole systems whose components exist at different levels (for at least some, central cases)'.[15] Hutchins is influenced by some of the recent literature on mechanisms, for example, Stuart Glennan:

> The complex-systems approach to mechanisms does not suppose that unification derives from unity of fundamental mechanisms. According to the complex-systems approach, mechanisms are collections of parts and parts are objects, but the objects that are parts of mechanisms may themselves be complex structures.[16]

Three

Even pure forms of mechanism, if they exist, have a *functional* dimension, visible in the increasing focus on (a) the structure (or 'fabric') and purpose of the body, (b) its description in purposive terms and (c) properly teleomechanist descriptions of 'the human machine' as an integrated system of mechanisms and higher purposes, the 'animal economy'. And my final point shall be that this ontology responds in ways we may not have expected to the challenge of embodiment.

In chapter 1 of Richard Lower's *Tractatus de Corde* (1669), we read:

> I must preface my account of (the movement of the blood) by some remarks on the Position and Structure of the heart. When these have been duly considered and collated, it will be easier to grasp how carefully both its Fabric and Position are adapted for movement, and how fittingly everything is arranged for the distribution of the blood to the organs of the body as a whole.[17]

The language of 'position', 'structure' and 'fabric' is quite striking here. These are not notions one can derive from basic atomic properties. Of course, the more the emphasis is on a kind of interconnection (called sympathy, consensus, cohesion and so on, often with chemical specifications such as the notions of action and reaction) rather than the nature of the components, the further away we are from a componential, mechanistic ontology, in the sense of, for example, this classic statement by Descartes:

> I think that all these bodies [sc. salt, sulphur, mercury and the four elements of the philosophers] are made of the same matter, and the only thing which makes a difference between them is that the tiny parts of this matter which constitute some of them do not have the same shape or arrangement as the parts which constitute others.[18]

Moving into the eighteenth century, we can see this in the attempts by a series of authors, including vitalist physicians associated with the Montpellier Medical Faculty but also Maupertuis, to model the structural, systemic and 'network' quality of the living organism (often, the human or animal body), using the language of the 'animal economy'. Interestingly, this modelling often employed the metaphor of a bee-swarm (that is, one organ is to the whole organism as an individual bee is to the bee-swarm).[19]

The more the emphasis is on interconnection, the further away we are from a componental, mechanistic ontology (the emphasis is more of *a relational property*).[20] Not necessarily because what we are seeing in the appeal to 'commixture', 'mutual influence', 'action and reaction' is a *rival ontology* (that is, organismic ontology as opposed to mechanistic ontology). But because it is *not an ontology*. Mechanism, whether in its pure or complex forms, has a functional and systemic dimension, and the more it emphasises the latter, the more it moves away from being a pure ontology. I hope this point is clear, but we should not lose sight of the other key feature I have mentioned: embodiment. Because, as in my response to Engels and Helmholtz, it is not true to claim that early modern materialists (including ones who admired Vaucanson, like La Mettrie and Diderot) were blind to the reality of embodiment. And these features are more related than we may think, because analogical and otherwise heuristic appeals for example to automata (but also the internal rhetoric of an inventor like Vaucanson about his digesting duck) were often ways of seeking to do justice to the unique properties of organic bodies.

Four

The relation between mechanical analogies and embodiment can also be presented as a response to a challenge, and it shows its complex relation to embodiment.[21] Borelli, like Descartes before him and Vaucanson after him, emphasises that part of the significance of artificially created mechanical objects (including but not restricted to automata: it can also be a clock or a pump) is that they enable a further theoretical but more generally cognitive engagement with the properties of natural objects. The machine here is functioning as a kind of go-between, enabling the interface between ontology and heuristics, within which actual machines can serve as 'matière à penser', so to speak. In Georges Canguilhem's elegant phrase: 'Essentially, a machine is a mediation or as mechanics say, a relay (*relais*).'[22]

Notice how far a machine as 'relay' or analogy is, from the stark opposition between 'a watch made of copper and a watch made of flesh', in Diderot's evocative image. The more analogical it is, the more it can serve as a heuristic: in Borelli's words, which I have quoted earlier, 'automata have a certain shadowy sameness (*umbratilem similitudinem*) to animals in that both are organic self-moving bodies'. Similarly, automata could be fascinating *both* because they were a form of mechanism, *and* because they called attention to the specifically 'vital' (teleological, purposive, intentional, homeostatic . . .) properties of living beings. As Riskin put it, discussing Vaucanson's digesting or defecating duck:

> The defecating Duck and its companions commanded such attention, at such a moment, because they dramatised two contradictory claims at once: that living creatures were essentially machines and that living creatures were the antithesis of machines.[23]

For La Mettrie, as we have seen, our machine is very much an *organic* machine, a flesh and blood system. It is a 'machine' in the sense that our drives, our urges, our instincts, our hormones, our blood sugar and, as the Ancients would have said, our youth or senescence (just the sort of thing the embodiment theorist would dwell on) don't allow us to function in a kind of absolute freedom. This machine is not one that boils down to a foundationalist ontology of the sort we today may call physicalism:

> Man is so complex (*composée*) a machine that it is impossible to get a clear idea of the machine beforehand, and hence impossible to define it. For this reason, all the investigations which the greatest philosophers have conducted a priori, using their intellects, have been vain.[24]

Similarly, however, much mechanist language is present in some of the vitalist accounts of the animal economy, the latter authors – the first to be clearly termed 'vitalists' – also stress, not mysterious vital forces, not a Stahlian soul powering the body, but a specifically (and 'real') *organic* structure: 'They [*the mechanists*] did not even pay attention to the organic structure of the human body which is the source of its main properties.'[25]

The fear of the 'dehumanising' force of the machine with regard to the warm world of the organic turns out to be misplaced, on the basis of an overview of some key texts spanning disciplines such as medicine, philosophy and, well, emerging robotics (the duck). But the embodiment theorist could respond that the real issue is not the opposition between the mechanical and the organic, but the lack of recognition of 'selfhood' in even these complexified, hybridised forms of mechanism: even a detailed account of 'organic structure' does not seem to yield the sense that 'someone is home'. And, often, embodiment is meant to connote a sense of first-personness, as opposed to the 'body-as-organism of biology', associated with externality. Thus, Karen Dale distinguishes between two ways of conceptualising the body:

> These are, first, the *historical body* – a body recognised as being constructed differently over time through social and cultural forces; and, second, the *phenomenologically lived body* – the body we experience in our everyday lives as the medium through which we 'know' our world. Taken together, these approaches to the body may be distinguished from the body-as-organism of biology by using the term 'embodiment'.[26]

There are several possible responses here. One is to insist that early modern materialism was in fact strongly concerned with embodiment and the nature of the flesh, in contradistinction to more austere forms of mechanism (and the mind–body dualism they could include).[27] Another is to point to the richness of these forms of mechanism and the conspicuous fact that, for example when it came to projects for automata such as Vaucanson's 'defecating duck', the specific nature of organic life was the challenge, not what was denied, as I have sought to make clear above. A third is to enquire into exactly what is being defended by embodiment theorists, aside from the by now rather obvious or 'granted' fact that bodies have a history and perhaps a historicity, and a degree of self-construction as well as their 'genetic' basis.[28] It is to this that I turn in closing.

Five

Consider the difference between these three claims for embodiment: (i) an American teenager and a medieval peasant, an obese person and a rail-thin chain smoker are not *in their body* in the same ways: both at the level of self-perception and of historicity. An X-ray or an MRI scan do not access what is in each case unique and personal to the lived body; (ii) a 'reluctance to conceive of cognition as computational and [an] emphasis on the significance of an organism's body in how and what the organism thinks';[29] here, the idea is that the properties of an organism's body limit or constrain the concepts an organism can acquire: 'an organism's body in interaction with its environment replaces the need for representational processes thought to have been at the core of cognition. Thus, cognition does not depend on algorithmic processes over symbolic representations', and 'the body or world plays a constitutive rather than merely causal role in cognitive processing';[30] (iii) the body is not in space like a physical object – a claim associated with 'embodied phenomenology'. There are different ways of identifying or dividing up these claims. I shall focus on one: the extent to which they are *naturalistic*, in the sense of being compatible with the set of analyses procured at any given time by the natural sciences (bearing in mind that naturalism, for instance in Spinoza or John Dewey, can be an impressively broad doctrine).

Claim (ii), which is a summary of the research programme known as 'embodied cognition', going back to early insights from authors such as Hubert Dreyfus and significantly expanded by Francisco Varela et al., and later Andy Clark in the 1990s until now, is fully naturalistic; it is meant to modify and/or amend work in the cognitive sciences. Claim (i) is not anti-naturalistic and is compatible with a weak naturalism, even if it focuses on dimensions of personhood or subjectivity that are not themselves accessible to scientific modelling or explanation. However, claim (iii), which is characteristic both of a radical strand of post-Husserlian phenomenology with Merleau-Ponty and of the current trend known as enactivism, rests on a fundamentally anti-naturalistic posit. The same point was made in more humoristic terms by the English cultural critic Eagleton (replace 'postmodernism' with 'embodiment theory'):

> Postmodernism is obsessed by the body and terrified of biology. The body is a wildly popular topic in US cultural studies – but this is the plastic, remouldable, socially constructed body, not the piece of matter that sickens and dies. The creature who emerges from postmodern thought is centreless, hedonistic, self-inventing, ceaselessly adaptive. He sounds more like a Los Angeles media executive than an Indonesian fisherman.[31]

In a way, these two caricatural figures, the media executive and the Indonesian fisherman, convey two aspects of recent embodiment theory, which Eagleton has cleverly turned into a kind of contradiction. On the one hand, embodiment theory is meant to free us from the tyranny of our biology (as exemplified by the media executive); on the other, in its fascination with the uniqueness of the flesh and its subjectivity, embodiment theory seems to hover around the Indonesian fisherman in his 'materiality'. But let me focus a bit more on the implicit tension between 'biology' and what some authors (including Dale as quoted earlier) call the 'phenomenological lived body' – that is, again, a tension involving the commitment to naturalism.

Theorists for whom the experience of the lived body is apart from natural science as a whole tend, perhaps unsurprisingly, to sound quite reactive. They want somehow to rescue (an entity? an intuition?), maintaining, with Merleau-Ponty, that the 'flesh' (as opposed to body in a merely spatial sense, like a brick or a glass of water) exists at least in part 'outside of physical space'.[32] Thus, the living body – indeed, any organism – 'is an individual in a sense which is not that of modern physics'.[33] This is anti-naturalistic, but curiously, it is a point frequently repeated in enactivist discourse, which seeks to be part of the discussion as to the nature of cognitive science. The enactivist leitmotiv is that the world of our experience is inseparably in interaction with our physiology, sensory system and the environment: focus on sensorimotor activity, life as movement; 'pigeons, for example, bob their heads up and down to recover depth information'.[34] But there is also a non-naturalistic commitment here. A phenomenology of the body is always a subjectivism,[35] for any such reflection on 'corporeity' treats our self-awareness as foundational. Differently put, the phenomenologist of the body cannot accept that cortical microstimulation, as has been done in experiments with macaque monkeys,[36] could produce a new phenomenology, a new set of memories and so on.

Lest I sound severe in judging 'embodied phenomenology', recall Merleau-Ponty's sacralisation of the living organism, which is, to be really specific, a metaphysics of transubstantiation, for he equates the sensation of an embodied being to a mystical communion with divine presence:

> Just as the sacrament not only symbolises . . . an operation of Grace, but is also the real presence of God . . . in the same way the sensible has not only a motor and vital significance but is a way of being in the world . . . sensation is literally a form of communion.[37]

Merleau-Ponty makes the same point without referring to the metaphysics of transubstantiation, but instead to the opposition between 'third person' and 'first person', in which of course he insists – in this more like a 'vitalist' than a 'phenomenologist' – that I am simply unable to understand the body if I think of it from an external standpoint, 'therefore the body is not an object'.[38] Catholic metaphysics aside, we should see what is involved in claiming that the organism is not in physical space. The ontological status, the uncaused causal role of selfhood here sounds much like early twentieth-century entelechies or vital forces. It plays hyper-interiority against spatiality.

Instead of safely distinguishing between types of embodiment discourse, and promoting one at the expense of the other, it may also be the case that a theoretical insistence on the fact that we are indeed creatures of the flesh means that *any form* of embodiment discourse can tend towards a mysticism of the flesh (the wisdom of the body, the body is not in space . . .), an out-of-control insistence on subjectivity, first personness and opposition between flesh and body. Yet, thinking back to my three types of embodiment claims, if (i) is less relevant to cognitive and biological discussions, and (iii) above – the phenomenological variant – is subjectivist and anti-naturalistic, a version of (ii) is most interesting, in Andy Clark's presentation. In fact, Clark further complicates my story because he distinguishes between two arguments for embodiment, *both of which* belong to (ii) in my distinction above:

> One of those strands depicts the body as special, and the fine details of a creature's embodiment as a major constraint on the nature of its mind: a kind of new-wave body-centrism. The other depicts the body as just one element in a kind of equal-partners dance between brain, body and world, with the nature of the mind fixed by the overall balance thus achieved: a kind of extended functionalism (now with an even broader canvas for multiple realisability than ever before).[39]

'Body-centrism' means that the theorist tends towards 'biochauvinism'.[40] This is perhaps one further aporia in addition to the others we have encountered . . .

Conclusion

If mechanistic theories of the body turned out to be less foreign to embodiment than some of the scholarship, a lot of the contemporary theory and perhaps our common intuitions would hold, conversely, embodiment theories turned out to often collapse into (or be beholden to) an anti-naturalistic mysticism of the flesh. As regards my desire to distinguish the latter theories in terms of their more or less strong, weak or non-existent commitments to naturalism, a final observation is that naturalistic versions of embodiment theory are compatible with the expanded forms of mechanism described above. The power of mechanistic analogies is not a 'denial of the flesh'. That is, the construct known as 'mechanistic materialism' is something of a conceptual monster, for most early modern materialists were deeply concerned with what we would call embodiment – pleasure, the flesh, sensation, organic sympathies, instincts, psychosomatic interconnection, the specific nature of the brain and so on. Embodiment theorists (and their close cousin, new materialists) often define themselves in contradistinction to 'mechanistic materialists' – but in most cases, materialism need not claim that 'only matter exists'. In the elegant terms of John Sutton and Lyn Tribble, it can be 'firmly pluralist' in its ontologies: 'even if all the things that exist supervene on or are realised in matter, the materialist can still ascribe full-blown reality to tables and trees and tendons and toenails and tangos and tendencies'. An account including the brain need not exclude 'memories, affects, beliefs, imaginings, dreams, decisions, and the whole array of psychological phenomena of interest to literary, cultural, and historical theorists'.[41] The same applies to the machine, to speak with La Mettrie:

> To be a machine, to feel, to think, to know how to distinguish good from evil, as well as blue from yellow, in a word, to be born with an intelligence and a sure moral instinct, and to be but an animal, is thus no more contradictory, than to be an ape or a parrot and to know how to give oneself pleasure.[42]

We are machines, to be sure, but we are necessarily organic and, furthermore, *affective* machines. To risk a curious *rapprochement*, I believe the early modern mechanist would agree with Félix Guattari, for whom,

> As opposed to a thinker such as Heidegger, I do not believe that the machine is something which turns us away from being. I think that the machinic phyla are agents productive of being. They make us enter into what I call an ontological heterogenesis. I am not making an opposition between the technological world (the ontic) and ontology. The whole question is one of knowing how the enunciators of technology, including biological, aesthetic, theoretical, machines, etc., are assembled, of refocusing the purpose of human activities on the production of subjectivity or collective assemblages of subjectivity.[43]

Notes

1. Vaucanson, *Le Mécanisme du Flûteur automate*.
2. See, for instance, Ménuret de Chambaud, 'Œconomie Animale (Médecine)', and Wolfe and Terada, 'Animal Economy as Object and Program'.
3. Some of my suggestions regarding mechanism and embodiment are not dissimilar to the more peremptory statements made by Donna Haraway in her famous 'Cyborg Manifesto': 'American radical feminists ... insist on the organic, opposing it to the technological ... But there are also great riches for feminists in explicitly embracing the possibilities inherent in the breakdown of clean distinctions between organism and machine and similar distinctions structuring the Western self. It is the simultaneity of breakdowns that cracks the matrices of domination and opens geometric possibilities' (Haraway, 'A Cyborg Manifesto', p. 174). Haraway celebrates the overcoming of boundaries (animal/human, machine/organism, and of course gender boundaries). But my concern is not with charting a new, biology-friendly course for North American

feminism, although this relates to my final remarks on embodiment and its mystification. For further discussion on this debate, see Davis, 'New Materialism and Feminism's Anti-Biologism'.
4. Sawday, *The Body Emblazoned*, p. 29.
5. Borelli, *De Motu Animalium*, Vol. II, prop. CXVI, p. 319.
6. Spence, *Letters from the Grand Tour*, p. 104.
7. 'The materialism of the past century was predominantly mechanistic, because at that time . . . only the science of mechanics . . . had reached any sort of completion . . . For the materialists of the eighteenth century, man was a machine. This exclusive application of the standards of mechanics to processes of a chemical and organic nature – in which the laws of mechanics are also valid, but are pushed into the background by other, higher laws – constitutes the specific (and at that time, inevitable) limitation of classical French materialism.' Engels, *Feuerbach und der Ausgang der klassischen deutschen Philosophie*, p. 278. For my criticisms of this point see: Wolfe, 'The Allure of the Flesh'.
8. Helmholtz, 'On the Interaction of Natural Forces', pp. 230–1.
9. Diderot, *Œuvres complètes*, vol. 17, p. 335.
10. Ibid. p. 499.
11. Aristotle, *Generation of Animals* I 18, 722b34; *Metaphysics* Z 11, 1036b32; *Parts of Animals* I 1, 640b29, b35, in *A New Aristotle Reader*.
12. La Mettrie, *L'Homme-Machine*, p. 105.
13. 'On peut bien faire une machine qui se soutienne en l'air comme un oiseau, *metaphysice loquendo* [metaphysically speaking], car les oiseaux mêmes, au moins selon moi, sont de telles machines, mais non pas *physice* ou *moraliter loquendo* [speaking physically or morally, that is in practice], pour ce qu'il faudrait des ressorts si subtils, et ensemble si forts, qu'ils ne sauraient être fabriqués par des hommes', in a letter to Mersenne, 30 August 1640, AT III, 163. I consulted the translation in Des Chene, *Spirits and Clocks*, p. 110, n. 9 (thanks also to Barnaby Hutchins); emphasis in the original.
14. See: Lenoir, 'Teleology without Regrets'. See also: Wolfe, 'Teleomechanism Redux'.
15. Hutchins, 'Descartes, Corpuscles and Reductionism', p. 671 refers to 'mechanism and multilevel systems'. He adds that 'Descartes cannot give an account of the heartbeat without also referring to and relying on everything involved in respiration and circulation. Each plays a necessary role in explaining how the heartbeat works: in the absence of circulation or respiration, there would be no heartbeat. And each plays its role within a specific organisation: if respiration did not precede the entry of blood into the left ventricle, the blood would be "too rare and too fine" for the process to continue; if circulation did not follow the active phase of the heartbeat, there would be no blood to re-enter the heart' (p. 676); this describes a system. A systemic explanation rather

than a pure mechanist-reductionist explanation, involves 'the entire composition that explains the effect (rather than the behaviour of individual corpuscles), where each component is taken from the level that is explanatorily relevant for that component' (p. 687).
16. Glennan, 'Rethinking Mechanistic Explanation', p. 352.
17. Lower, *Tractatus de Corde*, p. 2.
18. In a letter to the Marquess of Newcastle, 23 November 1646, AT IV, 568; emphasis added.
19. See: Ménuret de Chambaud, 'Observation (Gram. Physiq. Méd.)', pp. 318b–319a; Diderot, *Rêve de D'Alembert*, in *Œuvres complètes*, vol. XVII, pp. 121–3; Wolfe and Terada, 'Animal Economy as Object and Program'; and Sheehan and Wahrman, *Invisible Hands*.
20. This portrayal of mechanism as applied to the body, as increasingly structural and relational, could perhaps be fruitfully compared to Eric Schliesser's account of gravity as a relational property (Schliesser, 'Without God'. (Thanks to Dana Jalobeanu for this suggestion.) One difference is that in the latter case, gravity is relational, it is an ontology, whereas here there is a gradual move away from ontology in favour of description of systems.
21. I have devoted more attention to this specific issue elsewhere. See: Wolfe, 'Le mécanique face au vivant'.
22. See, for instance: Canguilhem, 'Aspects du vitalisme', 87; Clark, *Being There*, p. 87.
23. Riskin, 'The Defecating Duck', p. 612.
24. La Mettrie, *L'Homme-Machine*, p. 66.
25. Ménuret de Chambaud, 'Œconomie Animale (Médecine)', p. 364b.
26. Dale, 'Body and Organisation Studies', p. 11; emphasis in the original.
27. Wolfe, 'The Allure of the Flesh', pp. 83–106.
28. An elegant statement, although perhaps so broad as to no longer be a definition, was Caroline Walker Bynum's: 'There is no clear set of structures, behaviors, events, objects, experiences, words, and moments to which body currently refers. Rather, it seems to me, the term conjures up two sharply different groups of phenomena. Sometimes body, my body, or embodiedness seems to refer to limit or placement, whether biological or social. That is, it refers to natural, physical structures (such as organ systems or chromosomes), to environment or locatedness, boundary or definition, or to role (such as gender, race, class) as constraint. Sometimes – on the other hand – it seems to refer precisely to lack of limits, that is, to desire, potentiality, fertility, or sensuality/sexuality . . . or to person or identity as malleable representation or construct. Thus body can refer to the organs on which a physician operates or to the assumptions about race and gender implicit in a medical textbook, to the particular trajectory of one person's desire or to inheritance patterns and family structures.' Walker Bynum, 'Why All the Fuss about the Body?', p. 5.

29. See Shapiro, 'The Embodied Cognition Research Programme'. This approach to cognition overlaps with Malafouris and Renfrew's beautiful analyses of the 'cognitive life of things' I cannot go into here: 'Our vision of the cognitive life of things is inspired more by the hybrid image of the potter skilfully engaging the clay to produce a pot, than by the linear architecture of a Turing-machine.' 'Things have a cognitive life because intelligence exists primarily as an enactive relation between and among people and things, not as a within-intracranial representation.' Malafouris and Renfrew, 'Introduction', pp. 3–4. For further discussion, see Wheeler, 'Mind, Things and Materiality'.
30. Shapiro, *Embodied Cognition*, p. 4.
31. Eagleton, *After Theory*, p. 186.
32. Merleau-Ponty, *Structure of Behaviour*, p. 209.
33. Ibid. p. 154.
34. Shapiro, *Embodied Cognition*, p. 64.
35. See also the remarks on internalism and externalism in Malafouris and Renfrew's 'Introduction' in *Cognitive Life of Things*.
36. See Romo et al., 'Somatosensory Discrimination'.
37. Merleau-Ponty, *Phenomenology of Perception*, p. 212.
38. Ibid, p. 198. I discuss this in relation to Canguilhem's (non?-)response to Merleau-Ponty in Wolfe, 'Was Canguilhem a Biochauvinist?'
39. Andy Clark names Larry Shapiro as having a 'body-centric' view. The weaker view that Clark defends says that we don't have to be in human bodies to have the minds we do. Clark, 'Pressing the Flesh'. For an extension of Clark's viewpoint into actual 'embodied robotics', building on the early robotics work of Rodney Brooks, see: Symons and Calvo, 'Computing with Bodies'.
40. Di Paolo, 'Extended Life'.
41. Sutton and Tribble, 'Materialists are not Merchants of Vanishing'.
42. La Mettrie, *L'Homme-Machine*, p. 112.
43. Guattari, 'On Contemporary Art', p. 50.

Bibliography

Aristotle, *A New Aristotle Reader*, ed. John Lloyd Ackrill (Princeton: Princeton University Press, 1987).

Borelli, Giovanni Alfonso, *De Motu Animalium*, 2 vols, trans. Paul Maquet *On the Movement of Animals* (Berlin: Springer, 1989).

Canguilhem, Georges, 'Aspects du vitalisme', in *La connaissance de la vie*, revised edition (Paris: Vrin, [1952] 1965).

Clark, Andy, *Being There: Putting Brain, Body and World Back Together* (Cambridge, MA: MIT Press, 1997).

Clark, Andy, 'Pressing the Flesh: A Tension in the Study of the Embodied Embedded Mind?' *Philos. Phenomenol. Res.*, Vol. 76, Issue 1 (2008), pp. 37–59.

Dale, Karen, 'The Body and Organisation Studies', in *Anatomising Embodiment and Organisation Theory* (London: Palgrave Macmillan, 2001), pp. 8–31.

Davis, Noela, 'New Materialism and Feminism's Anti-Biologism: A Response to Sara Ahmed', *European Journal of Women's Studies*, Vol. 16 (2009), pp. 67–80.

De La Mettrie, Julien Offray, *L'Homme-Machine*, in *Œuvres Philosophiques*, volume 1, ed. Francine Markovits (Paris: Fayard-'Corpus', 1987).

Descartes, René, *Œuvres*, 11 volumes, ed. Charles Adam and Paul Tannery (Paris: Vrin, 1964–1974) (cited AT).

Des Chene, Dennis, *Spirits and Clocks: Machine & Organism in Descartes* (Ithaca, NY and London: Cornell University Press, 2001).

Di Paolo, Ezequiel, 'Extended Life', *Topoi*, Vol. 28 (2009), pp. 9–21.

Diderot, Denis, *Œuvres complètes*, ed. Herbert Dieckmann, Jacques Proust and Jean Varloot (Paris: Hermann, 1975–).

Diderot, Denis, *Le Rêve de D'Alembert*, in *Œuvres complètes*, ed. H. Dieckmann, J. Proust et al., vol. XVII, ed. J. Varloot et al. (Paris: Hermann, 1987).

Eagleton, Terry, *After Theory* (London: Allen Lane, 2003).

Engels, Friedrich, *Feuerbach und der Ausgang der klassischen deutschen Philosophie* (1888) in Marx, Engels, *Werke*, volume 21 (Berlin: Dietz Verlag, 1982).

Glennan, Stuart, 'Rethinking mechanistic explanation', *Philosophy of Science*, Vol. 69 (2002), S342–S353.

Guattari, Félix, 'On Contemporary Art', in *The Guattari Effect*, ed. Éric Alliez and Andrew Goffey (New York: Continuum, 2011), pp. 40–53.

Haraway, Donna, 'A Cyborg Manifesto: Science, Technology, and Socialist-Feminism in the Late Twentieth Century', *Socialist Review*, Vol. 80 (1985), pp. 65–108, reprinted in *Simians, Cyborgs and Women: The Reinvention of Nature* (New York: Routledge, 1991), pp. 149–81.

Helmholtz, Hermann von, 'On the Interaction of Natural Forces', cited in Minsoo Kang, *Sublime Dreams of Living Machines* (Cambridge, MA: Harvard University Press, 2011).

Hutchins, Barnaby, 'Descartes, Corpuscles and Reductionism: Mechanism and Systems in Descartes's Physiology', *The Philosophical Quarterly*, Vol. 65, Issue 261 (2015), pp. 669–89.

Lenoir, Timothy, 'Teleology without Regrets. The Transformation of Physiology in Germany, 1790–1847', *Studies in History and Philosophy of Science*, Vol. 12 (1981), pp. 293–354.

Lower, Richard, *Tractatus de Corde* (1669), in R. T. Gunther, *Early Science in Oxford*, volume IX (Oxford: Oxford University Press, 1932).

Malafouris, Lambros and Colin Renfrew, 'Introduction', in *The Cognitive Life of Things: Recasting the Boundaries of the Mind*, ed. Lambros Malafouris and Colin Renfrew (Cambridge: McDonald Institute for Archaeological Research Publications, 2008), pp. 1–12.

Ménuret de Chambaud, Jean-Joseph, 'Observation (Gram. Physiq. Méd.)', in *Encyclopédie ou Dictionnaire raisonné des sciences, des arts et des métiers*, ed. Denis Diderot and Jean le Rond D'Alembert, volume XI (Paris: Briasson, 1765), pp. 313–21.

Ménuret de Chambaud, Jean-Joseph, 'Œconomie Animale (Médecine)', in *Encyclopédie ou Dictionnaire raisonné des sciences, des arts et des métiers*, ed. Denis Diderot and Jean le Rond D'Alembert, volume XI (Paris: Briasson, 1765), pp. 360–6.

Merleau-Ponty, Maurice, *Phenomenology of Perception*, trans. Colin Smith (London: Routledge Kegan Paul, 1962).

Merleau-Ponty, Maurice, *The Structure of Behaviour*, trans. Alden L. Fisher (Boston: Beacon Press, 1963).

Riskin, Jessica, 'The Defecating Duck, or, the Ambiguous Origins of Artificial Life', *Critical Inquiry*, Vol. 29, Issue 4 (2003), pp. 599–633.

Romo, Ranulfo, Adrián Hernández, Anótonio Zainos and Emilio Salinas, 'Somatosensory Discrimination based on Cortical Microstimulation', *Nature*, Vol. 392, Issue 6,674 (1998), pp. 387–90.

Sawday, Jonathan, *The Body Emblazoned: Dissection and the Human Body in Renaissance Culture* (London: Routledge, 1995).

Schliesser, Eric, 'Without God: Gravity as a Relational Quality of Matter in Newton', in *Vanishing Matter and the Laws of Motion: Descartes and Beyond*, ed. Dana Jalobeanu and Peter Anstey (London: Routledge, 2011), pp. 80–102.

Shapiro, Lawrence, 'The Embodied Cognition Research Programme', *Philosophy Compass*, Vol. 2, Issue 2 (2007), pp. 338–46.

Shapiro, Lawrence, *Embodied Cognition* (London: Routledge, 2011).

Sheehan, Jonathan and Dror Wahrman, *Invisible Hands: Self-Organisation and the Eighteenth Century* (Chicago: University of Chicago Press, 2015).

Spence, Joseph, *Letters from the Grand Tour*, cited in Minsoo Kang, *Sublime Dreams of Living Machines* (Cambridge, MA: Harvard University Press, 2011).

Sutton, John and Evelyn Tribble, 'Materialists are not Merchants of Vanishing: Commentary on David Hawkes, "Against Materialism in Literary Theory"', *Early Modern Culture*, Vol. 9 (2011), 2011 <http://emc.eserver.org/1-9/sutton_tribble.html> (accessed 15 December 2015).

Symons, John and Paco Calvo, 'Computing with Bodies: Morphology, Function, and Computational Theory', in *Brain Theory: Essays in Critical Neurophilosophy*, ed. Charles T. Wolfe (London: Palgrave MacMillan, 2014), pp. 91–106.

Vaucanson, Jacques, *Le Mécanisme du Flûteur automate, présenté à Messieurs de l'Académie Royale des Sciences* (Paris: Conservatoire National des Arts et Métiers, [1738] 1985).

Walker Bynum, Caroline, 'Why All the Fuss about the Body? A Medievalist's Perspective', *Critical Inquiry*, Vol. 22 (1995), pp. 1–33.

Wheeler, Michael, 'Mind, Things and Materiality', in *The Cognitive Life of Things: Recasting the Boundaries of the Mind*, ed. Lambros Malafouris and Colin Renfrew (Cambridge: McDonald Institute for Archaeological Research Publications, 2010).

Wolfe, Charles T., 'Le mécanique face au vivant', in *L'automate: modèle, machine, merveille*, ed. Aurélia Gaillard, Bernard Roukhomovsky and Sophie Roux (Bordeaux: Presses Universitaires de Bordeaux, 2012), pp. 115–38.

Wolfe, Charles T., 'The Allure of the Flesh and the Vitality of Materialism: Aporias of Embodiment', in *Aesthetics of the Flesh*, ed. Felix Ensslin and Charlotte Klink (Berlin: Sternberg Press, 2014), pp. 83–106.

Wolfe, Charles T., 'Teleomechanism Redux? Functional Physiology and Hybrid Models of Life in Early Modern Natural Philosophy', in *Gesnerus* (Special Issue: *Teleology and Mechanism in Early Modern Medicine*), Vol. 71, Issue 2 (2014), pp. 290–307.

Wolfe, Charles T., 'Was Canguilhem a Biochauvinist? Goldstein, Canguilhem and the Project of "Biophilosophy"', in *Medicine and Society, New Continental Perspectives*, ed. Darian Meacham (Dordrecht: Springer, 2015), pp. 197–212.

Wolfe, Charles T. and Motoichi Terada, 'The Animal Economy as Object and Program in Montpellier Vitalism', *Science in Context*, Vol. 21, Issue 4 (2008), pp. 537–79.

CHAPTER 12

Generative Futures: On Affirmative Ethics

Rosi Braidotti

Introduction: After Foucault

One of Foucault's greatest insights was that the bio-political cuts both ways. Philosophically, this produced a multi-faceted vision of power as both restrictive or coercive (*potestas*) and empowering or productive (*potential*) that is intended as a critique of and an alternative to the binary dialectical schemes and their resolutely negative vision of power. The core conceptual issue is precisely that of the negative and its relation to the constitution of ethical and political subjectivity.[1] The main political aim, on the other hand, is to focus on bio-power as a way of highlighting the complex and contradictory mechanisms of care and control, empowerment and entrapment that lie at the core of the political economy of liberal democracies and their 'welfare' states.

This multi-layered vision of bio-power resulted in a second bifurcation, this time in relation to political activism and the very notion of politics. Foucault and Deleuze posited a crucial distinction between politics (*LA politique*) and the political (*LE politique*).[2] Politics focuses on the protocols of management of civil society and its institutions, the political on the transformative experimentations with new arts of resistance and existence. In other words, politics is made of progressive emancipatory measures predicated on chronological continuity, whereas the political is driven by transformative collective actions that require the non-linear time of critical praxis.

Foucault's bio-political approach combines the analysis of the web-like circulation of discursive practices with attention to the material grounding of such production, notably the social and institutional structures that sustain it. What primarily interests Foucault, however, is the practice of subject-formation as a discursive and material process of negotiation with the multi-layered and multi-directional structures of power. Becoming-a-subject (*assujetissement*) in and through *potestas* and *potentia* consists of a number of heterogeneous and potentially contradictory steps and elements. The idea of the bio-political functions within this process as an analytical tool that Foucault employs to examine the combination of care and control, discipline and punishment, stimulation and regulation, through which liberal democracies dispense their allegedly benevolent political economy to the citizens. The subject as an embodied and embedded relational entity is framed by these complex discursive and material mechanisms of care and control. Foucauldian discourse analysis rests on the assumption that these complex discursive and material mechanisms join forces to co-construct specific polemical targets or bio-political functions that are then deemed as essential to well-functioning subjects at different moments in history. Thus, an 'object' of scholarly enquiry does not precede the process that constitutes it. It rather emerges from the enmeshment with institutional, legal, political, economic and cultural conditions. The critical approach consequently consists in asking genealogical questions. For Foucault they are: what is this 'Man' whose humanistic era is now proving to be over? How do the classical humanities disciplines relate to the 'death of Man'? And what comes after this discursive death?

Genealogical questions are both critical and generative; they construct the thinker as a geologist, a genealogist, a clinician and a critical subject. For instance, Foucault focused on issues such as medical and mental health and hygiene, demography, biology and sexuality as the main objects of both discursive production and material control. In so doing Foucault redefines the meaning of critique itself by dis-engaging it from the Kantian tradition of unitary and transcendental consciousness.

This approach has important methodological implications. Foucault goes through different experimental phases of thought: from the archeology of knowledge, to a more systematic genealogical approach that grounds discursive practice upon the analysis of the conditions of possibility for their emergence. This produces a new cartographic method in critical thought that aims precisely to turn critique into an account – in equal parts genealogical as well as political – of the gradation and scale of the subjects' inscription into the web-like workings of power and knowledge, entrapment and empowerment, *potestas* and *potentia*. This internally contradictory set of negotiations is for Foucault the core of the bio-political management of both human and non-human life in advanced democracies.

It is worth stressing therefore that Foucault's bio-political work was essentially an analysis of the political economy of liberal democracies and the notion of moralised liberal individualism they produce. Foucault's analysis exposes the instrumental use made of human capacities and bodily abilities in a political system that celebrates and over-values individualism but does not prioritise social justice between individuals. Foucault strikes a radical note by connecting these bio-political systems of both individualisation and control to the universalist, humanistic idea of the 'Man of reason'.[3] This humanistic interpretation of the human subject defines 'Man' as the unitary, mortal being endowed with self-regulating rationality and universal moral goodness. The hermeneutics of suspicion towards individualism, which Foucault helps to develop, therefore also calls to task humanism, as elucidated in *The Order of Things*.[4] This double-cutting edge results in the critique of unitary subjectivity on the one hand, and the suspension of belief in the unfulfilled promises of the Enlightenment-based vision of the human as rational consciousness on the other. The humanistic Man of reason was pulled down from his pedestal, situated both geo-politically and historically within the Eurocentric tradition and made accountable for his deeds.

Feminist thinkers like Luce Irigaray – a contemporary of Foucault's – pointed out that the allegedly universal deal of 'Man' as a humanist symbol is, in fact, very much a male of the species.[5] Moreover, he is white, European, handsome and able-bodied. Feminist critiques of patriarchal posturing through abstract masculinity[6] and triumphant whiteness[7] argued that this Humanist universalism is objectionable on both epistemological and political grounds. Anti-colonial thinkers adopted a similar but distinct critical stance by questioning the implicit assumption of whiteness in the humanist ideal of 'Man'. Re-grounding the lofty humanist claims in the violent history of colonialism, anti-racist and post-colonial thinkers explicitly questioned the relevance of this ideal, in view of the obvious contradictions imposed by its Eurocentric assumptions. They also held Europeans accountable for the uses and abuses of this ideal by looking at colonial history and the violent domination of other cultures. These radical critiques of the Enlightenment ideal constitute the core of critical theory around and after Foucault and a necessary supplement to his discourse analysis.

So far, so good, but the essential problem with Foucault's bio-political project is that it is unfinished and incomplete, which means that his brilliant intuitions were not fully worked out and consequently have left a mixed legacy. In contemporary scholarship, the notion of the bio-political has been stretched beyond measure. Take, for instance, the flagrant contradictions between the two main contemporary schools of bio-political thought. First, the thanato-political reading of Foucault proposed by Agamben[8] presents the bio-political management of the living in terms of a 'bare life' that can be left to die.[9] The cartography of subjectivity produced here is entirely negative and power is exclusively bound to *potestas* (that is, a regulatory control system of domination and exclusion). Foucault's emphasis on the materiality of the subjects becomes in Agamben's work the source of a fundamental vulnerability to techno-industrial exploitation that exposes the self to the abuses of political regimes. Agamben translates this vulnerability into a full-scale indictment of the project of modernity as a whole.

Second, there is a neo-Kantian school of scholars that focuses on a variety of modes of 'bio-political citizenship', as an instance of liberal governmentality.[10] This line of thinking takes Foucault's work on the technologies of the self mostly as a relational ethics – rather than a complex grid of power relations. It argues that this ethics can assist in the emancipatory process of resisting the instrumental aspects of bio-political management of Life in advanced capitalism. In addition to these two well-established traditions of post-Foucauldian thought, multiple other renditions of the term bio-politics are circulating today. This proliferation of contradictory theoretical developments around the same term leads me to conclude that the bio-political has lost its critical edge and become another 'buzz' word.[11]

In my assessment, the first theoretical drawback of Foucault's unfinished anatomy of the bio-political is that it produces the analytics of a system of governmentality at the apex of its evolution and hence on the edge of implosion. Considering the fast rate of bio-political transformations propelled by contemporary technologies and the challenges they throw to the sovereign status of the human, Foucault's work has been criticised, notably by Haraway, for relying on an outdated vision of these very same technologies.[12] Haraway suggests that Foucault's bio-power provides the cartography of a world that no longer exists, in so far as the bio-political has mutated into the informatics of domination. Some of the most significant developments in bio-political analysis emerge from other strands of critical theory, notably feminist and queer,[13] environmentalist[14] and race theorists,[15] who have addressed the shifting status of embodiment and difference in advanced capitalism in a manner that reflects the complexities of global power relations.

The second major drawback of the Foucauldian method is that it remains firmly inscribed in an anthropocentric tradition. In the rest of this essay I want to explore this aspect further and develop an alternative case for vital neo-Spinozist materialism inspired by Deleuze and

Guattari's philosophy. My argument is that it offers a more adequate paradigm and a sharper methodology to deal with the multiple unfoldings of contemporary power formations. Deleuzian analyses based on the radical immanence of vital matter empower us to explore the political economy of advanced capitalism in ways that move beyond the anatomies of bio-political control. Let me expand on this.

From Bio to *Zoe*-politics, Or: Vital Materialism

Deleuze takes off from the same qualitative distinction between politics and the political as Foucault did, but he goes further into developing an alternative and well-structured political ontology. This is based on an ethological coding of political passions that emphasises the difference between the reactive and centralised, that is: majoritarian, and agonistic character of politics (*LA politique*) and the active or minor/minoritarian, nomadic, character of the political (*LE politique*). Politics is the management of what there is (Machiavellians – pretending to be pragmatists – would say 'the art of what is possible'), whereas the political is about qualitative transformations. This ethical distinction is replicated at the level of time and the forms of relational affectivity that different temporalities may engender. Thus, politics is postulated on *Chronos* – the linear time of institutional deployment of norms and protocols. It is a reactive and majority-bound enterprise that is often made of flat repetitions and reversals that may alter the balance and the agents but leave basically untouched the structure of power. The political, on the other hand, is postulated on the axis of *Aion* – the non-linear time of becoming and of affirmative critical practice. It is minoritarian and it aims at the counter-actualisation of alternative states of affairs in relation to the present. An example of the latter is nomad thought, as a *zoe*-centred[16] form of material vitalism[17] that sets the desire for transformations in the sense of becoming-ethical/nomadic at the centre of the theoretical but also the political agenda.

In other words, contemporary neo-Spinozist monism goes beyond Foucault's idea of the bio-political in that it implies a notion of subjectivity as vital and self-organising matter, that is to say an embedded form of 'matter-realism'.[18] This vital materialism[19] is intrinsically connected to the post-human definition of Life as *zoe*, or dynamic and generative post-anthropocentric force.[20] Vitalist materialism and its monistic political ontology engender a transversal and trans-species relational ethics that entails significant changes in the status and structure of what counts as human. The central discrepancy between Foucault's notion of bio-power and contemporary posthuman political structures, therefore, has to do with the displacement of anthropocentrism.

This point is extremely important in the light of another crucial element for a cartography of the present, namely the technologically mediated structure of advanced capitalism, also known as 'cognitive capitalism'.[21] This is built on the convergence between different and previously differentiated branches of technology, notably biotechnologies and information technologies. More specifically, what the neoliberal market forces are after and what they financially invest in, is the informational codes of living matter itself – bio-genetic and computational codes. The opportunistic political economy of bio-genetic capitalism has also turned Life/*zoe* – that is to say human and non-human intelligent matter – into yet another commodity for trade and profit. Advanced capitalism both invests and profits from the scientific and economic control and the commodification of all that lives – both human and non-human organisms. I have argued that this political economy displaces the centrality of Anthropos and produces a paradoxical and rather opportunistic form of post-anthropocentrism on the part of market forces that happily trade on all that lives. Data banks of bio-genetic, neural and media information about individuals are the true capital today.[22]

In the same frame of reference, Patricia Clough provides an impressive list of the concrete techniques employed to test and monitor the capacities of affective or 'bio-mediated' bodies: DNA testing, brain fingerprinting, neural imaging, body heat detection and iris or hand recognition.[23] These also double up as the contemporary forms of electronic surveillance: big data and i-clouds functioning at a speed that our brains cannot match. They go well beyond the sites of confinement that Foucault analysed in the political economy of the nineteenth and early twentieth-century techniques of discipline and punishment. This is indeed an instance of what Haraway calls 'the informatics of domination'.[24]

Furthermore, the capitalisation of living matter across all species produces a new political economy based on 'the politics of life itself'.[25] It introduces discursive and material political techniques of both individual self-management and population control of a very different order from the administration of demographics, which preoccupied Foucault's work on bio-political governmentality. Today, we are undertaking 'risk analyses' not only of individual propensities but also of entire sections of the population in the world risk society.[26] Moreover, we are subjecting such data to the imperatives of a market economy that turns informational codes into capital, in a manner that Melinda Cooper calls 'Life as surplus'.[27]

This high degree of technological mediation, however, does not mean that traditional patterns of exploitation and oppression are resolved, far from it. Multiple forms of new 'clinical labour'[28] are still preying upon the corporeal forces and abilities of contemporary embodied subjects. They are marked off for a new range of 'risk analyses' and also left open to profit-indexed labour market practices that include a significant amount of exploitation. Think, for instance, of the digital hubs that stand side-by-side to e-waste dumping sites[29] and the slums that house the e-proletarians who have to disassemble our dead electronic devices. Another example is the call-centre labourers in off-shore information hubs, who provide material and immaterial services on a 24/7 basis, for sub-standard wages. The mechanisms of capture of these new bio-labourers, or the digital proletariat, still penalise the sexualised and racialised 'others' in emerging or declining economies. Think, for example, of the global chain of care[30] and other, more extreme cases of bodily commerce, in sexuality and sex-work, reproduction and surrogacy, medical and health practices, organ 'donation' and transplants and others. This combination of high-tech advances and low-life survival is one of the most problematic political aspects of advanced capitalism, namely its necro-political face. Because these flagrant internal contradictions in the system of advanced capitalism are not accidental but structural, they benefit from being analysed within a vital materialist political philosophy. This allows for a mind–body and nature–culture continuum to be brought to bear on the analysis of the power relations of advanced capitalism.

Necro-politics

The death-bound, destructive aspects of advanced capitalism come under sharper scrutiny in a vital materialist perspective that concentrates on the complexity of Life as *zoe*. We saw in the previous section that 'advanced' capitalism is a misnomer in many respects, in that technological advances co-exist with brutal power relations: not only is basic access to advanced technologies unevenly distributed worldwide (with only one-third of the world's households actually having electricity), but so is access to the benefits of bio-genetic capitalism.

Such a system is not only inherently discriminatory, but also racist at some basic level of the term, as Foucault first understood. The new interconnections between forms of political governance and the farming, retrieving and evaluating of genetic predispositions or risk factors, constitute a technique that Foucault defined as racism. It configures – and actually engenders as 'raced' – entire populations in a hierarchical scale, not determined this time by pigmentation, but by other bio-genetic characteristics or dispositions. Because the aim of this economic and political exercise is to estimate a given (human and non-human) population's chance of survival or of extinction, the bio-political management of the living is inherently linked to death. As it operates transversally across many species, it is not restricted to human life (*bios*) but also encompasses non-human life (*zoe*). Thus, we are confronted by a number of schizoid features: the uncritical and instrumental post-anthropocentrism created by bio-genetic capitalism opens up some kind of posthuman perspectives, but its high necro-political charge mostly makes it into an inhuman(e) system. This encourages a confusion between posthuman and inhuman actions – including violence, devastation and extinction – that does not do justice to the complexity of the issues involved in a monistic understanding of Life, or critical materialist vitalism. This means that equal attention needs to be given to the novelty of the posthuman predicament, but also to the perpetuation of traditional forms of marginalisation and oppression within it. I have argued that structural injustices,[31] far from being eliminated by the conditions of advanced bio-genetic capitalism and its opportunistic margins of posthuman access, are currently intensified and exacerbated.

This opportunistic dimension of the contemporary political economy and management of posthuman 'Life' in cognitive capitalism can be exposed by looking, for instance, at the public debates on the availability of pharmaceutical drugs against human immunodeficiency virus (HIV), or large-scale vaccines against malaria. A whole under-class of genetically over-exposed and socially under-insured disposable bodies is engendered in this process, both within the Western world and in the emerging global economies.[32] Another example is the high levels of forced mobility, due to war, climate change and poverty and how they create patterns of global migration that are structural, not incidental: the global city and the refugee camps are two sides of the same coin.[33] This political economy of forced eviction and under-paid inclusion in the technological revolution calls for a population control that goes beyond Foucault's analysis of the bio-political. It does not function by techniques of discipline and control, but rather by necro-political neglect, bio-genetic farming of data, commercialisation of informational data and by 'bio-piracy'.[34] As Mark Halsey put it: 'Where once the sole objective was to control the insane, the young, the feminine, the vagrant and the deviant, the objective in recent times has been to arrest the nonhuman, the inorganic, the inert – in short, the so-called "natural world".[35] This is posthuman *zoe*-politics, not bio-political governmentality.

The next element that needs to be added to a cartography of contemporary power is mediation. The political management of embodied subjects nowadays can no longer be understood within the visual economy of bio-politics in Foucault's sense of the term.[36] The representation of embodied subjects is not visual in the sense of being scopic, as in the post-Platonic notion of the simulacrum. Nor is it specular, as in the psychoanalytic mode of redefining vision within a dialectical scheme of oppositional recognition of self and/as other. The representation of embodied subjects has been replaced by simulation and has become schizoid, or internally disjointed: on the one hand it enhances the vital capacities of the body. On the other it also tends to be spectral, structurally connected to practices of dying and culturally fascinated by corpses, vampires and zombies.[37] The contemporary social management of 'Life itself' has immersed carnal matters – bodies and their derivatives – in a logic of boundless circulation, that is multilayered. Bio-mediated bodies are caught in the commodification of all that lives, but are also suspended somewhere beyond the life and death cycle, becoming an image of themselves. The bio-genetic economy has consequently become forensic in its relationship to the body as virtual corpse and in the quest to control a life that cannot be contained within anthropomorphic parameters. But is it also spectral in being haunted by self-representation or the doubling up of bodies as images. It follows that contemporary embodied subjects have to be accounted for in terms of their surplus value as bio-genetic containers on the one hand, and as visual commodities circulating in a global media circuit of cash flow on the other. They are therefore doubly mediated by both bio-genetic and informational codes. In conclusion, the central insight of Foucault's political anatomy remains valid: bio-power also involves the management of dying and the question of the governance of life contains that of extinction as well. But in order to deploy the full ethical and political potential of this brilliant insight, however, we do need to move conceptually in a different direction.

Monism and Complexity

A monistic philosophy of critical vital materialism is my preferred alternative to the opportunistically posthuman management of 'Life' that marks the axiomatic mode of governance of advanced capitalism. An axiomatic system[38] refuses to provide definitions of the terms it works with, but prefers to order certain domains into existence with the addition or subtraction of certain norms or commands, their objects being treated as purely functional. Axioms operate by emptying flows of their specific meaning in their coded context and thus by decoding them. This produces both deformations of received meanings and the despotic imposition of groundless variations on them. Think, for instance, of the uses and abuses of the term 'freedom' by authoritarian and nationalistic political parties in Europe.[39] The aim of this political economy, operating on a larger scale through processes of overcoding pre-existent regimes of signs, is to decode and subject them to a centralising hierarchical machine that turns activity into labour, territories into land and surplus value into profit.[40] The political process of nomadic becoming, on the other hand, encourages flows or diagrams of subversion, without the insertion of axioms. Nomadic neo-materialist thought rejects the ways in which capitalism captures and arrests qualitative flows of becoming. It defines political praxis as the ethically driven collective construction of alternative models of actualising what bodies are capable of becoming.

Within a monistic materialist and vitalist universe, phenomena and subject-formations are approached as actualisations of different ethical forces and differential modes of becoming. The univocity of being, theorised by Deleuze, means that we have to deal – relationally – with one matter, which is intelligent, embedded, embodied and affective. In order to account for the actualisation of transversal subject formations, also known as 'assemblages',[41] we require a subtler analysis of differential variations in the process of subjectivation. These differences are not quantitative but rather qualitative; they are ethological variations, not normative judgements: they have to do with relational forces and degrees of intensity.

As I have often argued, the nomadic subject is a materially embodied and historically embedded entity, in that it is a bound instantiation of a common and ever-shifting matter.[42] Each singular self is an actualised and temporarily bound expression of the on-going process of becoming. According to the monistic vision, matter is intelligent and self-organising; specific forms of individuation are carved out of this vital material. In the specific case of human individuation that is anthropomorphic, monism implies the 'embrainment' of the body as well as the embodiment of the mind.[43] Neo-Spinozist vital materialism defies the oppositional character of dialectical thought and posits a pacifist ontology of mutual specification as the motor of processes of individuation[44] and auto-poietic self-styling. Working within a Spinozist framework,

I propose an affirmative emphasis on critical vitalism as a relentlessly generative force.[45] This requires an interrogation of the shifting inter-relations between human and non-human entities. *Zoe*/Posthuman monistic vitalism stresses a constitutive sense of ontological pacifism and a sense of entanglement in a web of immanent and ever-shifting relations in perpetual becoming. Georges Bataille's agnostic spirituality is of great inspiration for posthuman nomadic thought, in that it leads to a non-theistic form of naturalism that rejects all transcendental mystifications[46] and honours what Bryden calls 'a dynamism of the void'.[47] The idea that we are all 'part of nature', as Lloyd put it,[48] generates not only vital monism, but also alternative visions of how matter and mind interact and join forces to co-create affirmative becomings. Intimacy with the world speaks of our ability to re-collect it and re-connect to it and hence of our capacity to find our 'homes' within it, in the pursuit of nomadic sustainable relations.[49] Relational nomadic subjects engage in transversal connections with – Haraway speaks of 'becoming-with'[50] – multiple human and non-human others. Such webs of connections and negotiations define belonging not as attachment to static identity lines but as dynamic transversal moves across eco-sophically inter-connected categories. Relationality consists of a deep sense of negotiations with the multiple ecologies – social, environmental and psychic,[51] that constitute us. A sense of familiarity with the world flows from the simple fact that we are the products of such ecological interconnections and notably of the nature–culture continuum[52] that marks our era.

Deleuze's monistic idea of the radical immanence of embodied brain and embrained matter is a vitalist anti-theology. It provides the conceptual grounds to assert a non-unitary ethical subject immersed in the intelligent and self-organising structure of Life itself. It therefore infuses affect and endurance at the heart of the embodied and embedded materialism of the subject and of matter itself as a nature–culture continuum. The proposed methodology is not social constructivism, but rather neo-Spinozist expressionism.[53]

Moreover, monistic neo-materialism is a practice of affirmation, not of negativity and this commitment to positive passions constitutes not only its core ethical value, but also its political force. Neo-Spinozist monism places a different emphasis on the affective elements of human subjectivity under advanced capitalism and on the process of political subject-formation. Rejecting the Lacanian conceptual structure and terminology, vital neo-materialist thinkers stress the generative importance of affects and connect them to a positive view of desire as plenitude, not as Lack.[54] The unconscious drives, instead of being played back upon a sort of negative filter linked to the 'black box' of desire as Lack with its corollary of negative passions like envy, resentment and perennial frustration, are approached affirmatively. Affects are the autonomous visceral elements of our allegedly rational belief system.[55] What they express is the profoundly relational nature of human subjectivity and its constitutive drive for the freedom of expression of its powers (*potentia*).

By way of contrast, the Hegelian-Marxist school of dialectics of consciousness equates critical political subjectivity with negative, oppositional or 'unhappy' consciousness. Such reactive vision of the subject banks on negativity and even requires it, because it builds on the assumption that the critical position consists in analysing negative social and discursive conditions, in order the better to overthrow them. In other words, it is the same conditions that construct the negative moment – for instance, the experience of oppression, marginality, injury or trauma – and also the possibility of overturning them. The same analytic premises provide both the damages and the possibility of positive resistance, counter-action or transcendence.[56] The 'wounded attachments'[57] that trigger and at the same time are engendered by this process of vulnerability and resistance constitute the paradoxical core of oppositional consciousness.

As an alternative, Deleuze and Guattari construct a non-Hegelian, monistic and vital-materialist account of the genesis of political subjectivity that foregrounds the relational, negotiation-driven and affirmative elements of this process. The political is sustained by a relational affirmative ethics that aims to cultivate collectively and produce the conditions of its own expression. The ethics of affirmation frame and generate the political: it is an auto-poietic praxis based on a positive definition of the subject as an ontologically positive and process-driven 'di-vidual'. A subject's ethical core is clearly not their moral intentionality, as much as the effects of power (as repressive – *potestas* – and positive – *potentia*) – their actions – are likely to have upon the world. It is a process of engendering empowering modes of becoming.[58]

Here is the punchline of contemporary *zoe*/posthuman neo-spinozist materialist politics: affirmative ethics defines our politics. Given that the ethical good is equated with radical relationality aiming at affirmative empowerment, the ethical ideal is to increase one's ability to enter into modes of relation with multiple others. Oppositional consciousness as a reactive mode is replaced by affirmative praxis and political subjectivity is redefined as a process or assemblage that actualises this ethical propensity. This position aspires to the creation of affirmative alternatives by working through the negative instances so as to collectively transform them into affirmative practices. The drive towards affirmation is a key feature of neo-spinozist nomadic political subjects.

This view of subjectivity does not condition the emergence of the subject on negation but on creative affirmation – not on loss but on vital generative forces. The rejection of the dialectical scheme implies also a shift of temporal gears. It means that the conditions for political and ethical agency are not dependent on the current state of the terrain: they are not oppositional and thus not tied to the present by negation. Instead, they are projected across time as affirmative praxis, geared to creating empowering relations aimed at possible futures. Ethical relations create possible worlds by mobilising resources that have been left untapped in the present, including our desires and imagination. They are the driving forces that concretise in actual, material relations and can thus constitute a network, web or rhizome of interconnection with others. This qualitative shift engages our collective imaginings[59] and desire[60] – in response to world-historical structural transformations.

Deleuze's eco-sophy of radical immanence and intensive transformative subjects is an affirmative answer to the unsustainable logic and internal contradictions of advanced capitalism. The vital materialist body is in fact an eco-logical unit. This *bios-zoe-technos*-body is marked by the interdependence with its environment, through a structure of mutual flows and data-transfer that is best configured by the notion of viral contamination, or intensive inter-connectedness. This ecology of belonging is complex and multi-layered. This environmentally bound intensive subject is a collective entity, an embodied affective and intelligent entity that captures, processes and transforms energies and forces. Being environmentally bound and territorially based, a rhizomatic embodied entity is immersed in fields of constant flows and transformations. Philosophy therefore needs to create forms of ethical and political agency that reflect this high degree of complexity.

Affirmative Politics

Affirmative politics is my answer to these challenges and contradictions. It indicates the process of transmuting negative passions into productive and sustainable praxis, which does not deny the reality of horrors, violence and destruction of our times but proposes a different way of dealing with them. What is positive in the ethics of affirmation is the belief that negative affects can be transformed. This implies a dynamic view of all affects, even the traumas that freeze us in pain, horror or mourning. The slightly de-personalising effect of the negative or traumatic event involves a loss of ego boundaries, which is the source of both pain and potentially energetic reactions. Multi-locality and multi-directional memory[61] are the affirmative translation of this negative sense of loss. Let me illustrate this controversial point with an example drawn from diasporic subjects.

Following Glissant,[62] 'becoming-nomadic' marks the process of positive transformation of the pain of loss into the active production of multiple forms of belonging and complex allegiances. Every event contains within it the potential for being overcome and overtaken – its negative charge can be transposed. The moment of the actualisation is also the moment of its neutralisation. The ethical subject is the one with the ability to grasp the freedom to depersonalise the event and transform its negative charge. Affirmative ethics puts the motion back into e-motion and the active back into activism, introducing movement, process, becoming. This shift makes all the difference to the patterns of repetition of negative emotions. It also reopens the debate on secularity, in that it actually promotes an act of faith in our collective capacity to endure and to transform.

What is negative about negative affects is not a normative value judgement but rather the effect of arrest, blockage and rigidification, that comes as a result of a blow, a shock, an act of violence, betrayal, a trauma or just intense boredom. Negative passions do not merely destroy the self, but also harm the self's capacity to relate to others – both human and non-human others, and thus to grow in and through others. Negative affects diminish our capacity to express the high levels of inter-dependence, the vital reliance on others that is the key to both a non-unitary vision of the subject and to affirmative ethics. Again, the vitalist notion of Life as *zoe* is important here because it stresses that the Life I inhabit is not mine, it does not bear my name – it is a generative force of becoming, of individuation and differentiation: a-personal, indifferent and generative. What is negated by negative passions is the power of life itself – its *potentia* – as the dynamic force, vital flows of connections and becoming. And this is why they should neither be encouraged nor be rewarded for lingering around them too long. Negative passions are black holes.

This is an antithesis of the Kantian moral imperative to avoid pain, or to view pain as the obstacle to moral behaviour. It displaces the grounds on which Kantian negotiations of limits can take place. The imperative not to do on to others what you would not want done to you is not rejected as much as enlarged. In affirmative ethics, the harm you do to others is immediately reflected on the harm you do to yourself, in terms of loss of *potentia*, positivity, capacity to relate and hence freedom. Affirmative ethics is not about the avoidance of pain, but rather about transcending the resignation and passivity that ensue from being hurt, lost and dispossessed. One has to become ethical, as opposed to applying moral rules and protocols as a form of self-protection: one has to endure. Endurance is the Spinozist code word for this process. Endurance has a spatial side to do with the space of the body as an enfleshed field of actualisation of passions or forces. It evolves affectivity and joy, as in the capacity for being affected by these forces, to the point of pain or extreme pleasure. Endurance points to the struggle to sustain the pain without being annihilated by it and hence opens up to a temporal dimension, about duration in time.

Affirmative ethics is based on the praxis of enduring by constructing positivity, thus propelling new social conditions and relations into being, out of injury and pain. It actively constructs energy by transforming the negative charge of these experiences, even in intimate relationships where the dialectics of domination is at work.[63] We need to actively and collectively work towards a refusal of horror and violence – the inhuman aspects of our present – and to turn it into the construction of affirmative alternatives. Such an approach aims to bring affirmation to bear on undoing existing arrangements, so as to actualise productive alternatives.

As critical thinkers we are always trying to be worthy of the times, to interact with them, in order to resist them, that is to say differ from them. It is a form of *amor fati*, a way of living up to the intensities of life, so as to be worthy of all that happens to us – to live out our shared capacity to affect and to be affected. Beyond negative dialectics, we need to disengage the process of subject formation from negativity to attach it to affirmative otherness. This involves a change of conceptual references: reciprocity is no longer defined dialectically as the struggle for recognition, but rather auto-poietically as mutual definition or specification. Violence is bypassed by the ontological pacifism of a system based on monistic vital materialism and on the processes of differing that rest upon it.

Amor fati is not passive fatalism, but a pragmatic and liable engagement with the present in order to collectively construct conditions that transform and empower our capacity to act ethically and produce social horizons of hope, or sustainable futures. The ethical cultivation of positivity moreover does not exclude, either logically or practically, situations of antagonism or conflict. If we follow the Spinozist rule and de-psychologise the discussion about affirmation and negativity, to cast it instead in terms of an ethics or an ethology of forces, it follows that some of the relations we are likely to establish with others may well be of the antagonistic kind. What matters – and this is the shift of perspective introduced by affirmative ethics – is to resist the habit of inscribing antagonistic relations in a logic of dialectical negativity. The transcendence of dialectics, in other words, has to be enacted in the inner structure of relations – of the inter-personal as well as the non-human kind. Antagonism need not be inscribed in the lethal logic of the dialectical struggle of consciousness. This habit of thought needs to be resisted and re-coded away from the necessity to establish negativity as the pre-condition for the process of subject-formation.

In other words, the 'worthiness' of an event – that that ethically compels us to engage with it, is not its intrinsic or explicit value according to given standards of moral or political evaluation, but rather the extent to which it contributes to conditions of becoming. It is a vital force to move beyond the negative. Protevi argues that in this nomadic view, the political is the non-reactive and the non-habitual response of reactive engagement with the events of one's life that can reshape one's becoming[64] – a sort of creative dis-organisation of the negative that aims at keeping life immanent, non-unitary and non-reified according to dominant codes and hegemonic traditions of both life and thought.

My ethical stance is that there is no logical necessity to link political subjectivity to oppositional consciousness and reduce them both to violence and negativity. Political activism can be all the more effective if it disengages the process of consciousness – raising from negativity and connects it instead to creative affirmation and the actualisation of virtual potentials. In this process, theoretically based and politically infused cartographies are of crucial importance not only to ground the practice of material and discursive critique, but also as ethical connectors across multiple human and non-human actors. If it is indeed the case that the collective construction of affirmative values requires the confrontation with but also transformation of the negativity of the present, cartographies fulfil a clinical as well as an ethical function. Because these transformative practices are by definition not contained in the present conditions and cannot emerge from them, they have to be brought about or generated creatively by a qualitative leap of the collective praxis and of our ethical imagination. We simply do not know what such a collectively sustained, ethically affirmative body-as-transversal-assemblage can actually do. To find out, we have to experiment with its intensities.

Notes

1. Noys, *Persistence of the Negative*.
2. Foucault and Deleuze, 'Intellectuals and Power'.
3. Lloyd, *Spinoza and the Ethics*.
4. Foucault, *The Order of Things*.
5. Irigaray, *Speculum of the Other Woman*; *This Sex Which Is Not One*.
6. Hartsock, 'The Feminist Standpoint'.
7. hooks, *Ain't I a Woman*; Ware, *Beyond the Pale*.
8. Agamben, *Homo Sacer*.
9. Agamben's notion of 'bare life' has been taken to task by feminist scholars for his denial of feminist politics of natality and his fixation on mortality. See: Cooper, 'The Silent Scream'; Colebrook, 'Agamben: Aesthetics, Potentilality, Life'; Braidotti, *The Posthuman*.
10. See: Rose, *Politics of Life Itself*; Rabinow, *Anthropos Today*; Esposito, *Bios: Biopolitics and Philosophy*.
11. Lemke, *Biopolitics: An Advanced Introduction*.
12. Haraway, *Modest_Witness@Second_Millennium*.
13. See Grewal et al., eds, *Scattered Hegemonies*; Braidotti, *Metamorphoses*; Butler, *Undoing Gender*; Grosz, *The Nick of Time*.
14. Shiva, *Biopiracy*.
15. See: Hall and du Gay, *Questions of Cultural Identity*; Gilroy, *Against Race*.
16. Braidotti, *Transpositions*.
17. Deleuze and Guattari, *A Thousand Plateaus*.
18. Fraser et al., eds, *Inventive Life*.
19. See: Braidotti, *Transpositions*; Bennett, *Vibrant Matter*.
20. See: Braidotti, *Patterns of Dissonance*; Braidotti, *Metamorphoses*; Braidotti, *Nomadic Subjects*; Braidotti, *Nomadic Theory*.
21. Moulier Boutang, *Cognitive Capitalism*.
22. Braidotti, *The Posthuman*.
23. Clough, 'The Affective Turn'.
24. Haraway, *Simians, Cyborgs and Women*, p. 161.
25. Rose, 'The Politics of Life Itself'.
26. Beck, *World Risk Society*.
27. Cooper, *Life as Surplus*.
28. Cooper and Waldby, *Clinical Labour*.
29. Gabrys, *Digital Rubbish*.
30. Hochschild, 'Global Care Chains', p. 131.
31. Richest sixty-two people as wealthy as half of world's population. In: <http://www.theguardian.com/business/2016/jan/18/richest-62-billionaires-wealthy-half-world-population-combined> (accessed 12 April 2016).
32. Duffield, 'Global Civil War'.
33. See: Sassen, *The Global City*; Sassen, *Expulsions*; Diken, 'From Refugee Camps to Gated Communities'.
34. Shiva, *Biopiracy*.

35. Halsey, *Deleuze and Environmental Damage*, p. 15.
36. Foucault, *History of Sexuality. Vol. I.*
37. See: Braidotti, *Metamorphoses*; Braidotti, *Nomadic Theory*.
38. Toscano, 'Axiomatic'.
39. Both Berlusconi in Italy and Geert Wilders in the Netherlands have called their movements 'Freedom Party'.
40. Protevi and Patton, eds, *Between Derrida and Deleuze*.
41. Deleuze and Guattari, *A Thousand Plateaus*.
42. Braidotti *Nomadic Subjects*; Braidotti *Nomadic Theory*.
43. Marks, *Deleuze: Vitalism and Multiplicity*.
44. Simondon, *Being and Technology*.
45. This includes: Deleuze and Guattari, *Anti-Oedipus*; Guattari, *Chaosmosis*; Glissant, *Poetics of Relation*; Balibar, *Politics and the Other Scene*; Hardt and Negri, *Empire*.
46. See: Bataille, *The Accursed Share*; Braidotti, 'In Spite of the Times'.
47. Bryden, ed., *Deleuze and Religion*, p. 5.
48. See: Lloyd, *Part of Nature*; Lloyd, *Spinoza and the Ethics*.
49. Braidotti, *Transpositions*.
50. Haraway, *When Species Meet*.
51. Guattari, *The Three Ecologies*.
52. Haraway, *Modest_Witness@Second_Millennium*.
53. Braidotti, *Transpositions*; Braidotti, *The Posthuman*.
54. Braidotti, *Transpositions*.
55. Connolly, *Why am I not a Secularist?*
56. Foucault, *Discipline and Punish*.
57. Brown, 'Wounded Attachments'.
58. See: Deleuze, *Spinoza et le problème de l'expression*; Braidotti, *Transpositions*.
59. Gatens and Lloyd, *Collective Imaginings*.
60. Braidotti, *Nomadic Theory*.
61. Rothberg, *Multidirectional Memory*.
62. Glissant, *Poetics of Relation*.
63. Benjamin, *The Bonds of Love*.
64. Protevi and Patton, eds, *Between Derrida and Deleuze*.

Bibliography

Agamben, Giorgio, *Homo Sacer: Sovereign Power and Bare Life* (Stanford: Stanford University Press, 1998).
Balibar, Etienne, *Politics and the Other Scene* (London: Verso, 2002).
Bataille, Georges, *The Accursed Share* (New York: Zone Books, 1988).
Beck, Ulrich, *World Risk Society* (Cambridge: Polity Press, 1999).
Benjamin, Jessica, *The Bonds of Love: Psychoanalysis, Feminism and the Problem of Domination* (New York: Pantheon Books, 1988).

Bennett, Jane, *Vibrant Matter* (Durham, NC: Duke University Press, 2010).
Braidotti, Rosi, *Patterns of Dissonance* (Cambridge: Polity Press, 1991).
Braidotti, Rosi, *Metamorphoses: Towards a Materialist Theory of Becoming* (Cambridge: Polity Press, 2002).
Braidotti, Rosi, *Transpositions: On Nomadic Ethics* (Cambridge: Polity Press, 2006).
Braidotti, Rosi, 'In Spite of the Times: The Postsecular Turn in Feminism', *Theory, Culture & Society*, Vol. 25, Issue 6 (2008), pp. 1–24.
Braidotti, Rosi, *Nomadic Subjects: Embodiment and Sexual Difference in Contemporary Feminist Theory*, 2nd edition (New York: Columbia University Press, 2011).
Braidotti, Rosi, *Nomadic Theory: The Portable Rosi Braidotti* (New York: Columbia University Press, 2011).
Braidotti, Rosi, *The Posthuman* (Cambridge: Polity Press, 2013).
Brown, Wendy, 'Wounded Attachments', *Political Theory*, Vol. 21, Issue 3 (1993), pp. 390–410.
Bryden, Mary, ed., *Deleuze and Religion* (New York and London: Routledge, 2001).
Butler, Judith, *Undoing Gender* (New York and London: Routledge, 2004).
Clough, Patricia, 'The Affective Turn: Political Economy, Biomedia and Bodies', *Theory, Culture & Society*, Vol. 25, Issue 1 (2008), pp. 1–22.
Colebrook, Claire, 'Agamben: Aesthetics, Potentiality, Life', *South Atlantic Quarterly*, Vol. 107, Issue 1 (2009), pp. 107–20.
Connolly, William, *Why am I not a Secularist?* (Minneapolis: University of Minnesota Press, 1999).
Cooper, Melinda, *Life as Surplus: Biotechnology & Capitalism in the Neoliberal Era* (Seattle: University of Washington Press, 2008).
Cooper, Melinda, 'The Silent Scream: Agamben, Deleuze and the Politics of the Unborn', in *Deleuze and Law: Forensic*, ed. Rosi Braidotti, Claire Colebrook and Patrick Hanafin (Hampshire and New York: Palgrave Macmillan, 2009), pp. 142–62.
Cooper, Melinda and Catherine Waldby, *Clinical Labour: Tissue Donors and Research Subjects in the Global Bio-Economy* (Durham, NC: Duke University Press, 2014).
Deleuze, Gilles, *Spinoza et le problème de l'expression* (Paris: Minuit, 1968).
Deleuze, Gilles and Felix Guattari, *Anti-Oedipus: Capitalism and Schizophrenia* (New York: Viking Press; Richard Seaver, 1977).
Deleuze, Gilles and Felix Guattari, *A Thousand Plateaus: Capitalism and Schizophrenia* (Minneapolis: University of Minnesota Press, 1987).
Diken, Bulent, 'From Refugee Camps to Gated Communities: Biopolitics and the End of the City', *Citizenship Studies*, Vol. 8, Issue 1 (2004), pp. 83–106.
Duffield, Mark, 'Global Civil War: The Non-insured, International Containment and Post-interventionary Society', *Journal of Refugee Studies*, Vol. 21 (2008), pp. 145–65.

Elliott, Larry, 'Richest 62 People as Wealthy as Half of World's Population, Says Oxfam', *The Guardian*, 18 January 2016 <http://www.theguardian.com/business/2016/jan/18/richest-62-billionaires-wealthy-half-world-population-combined (accessed 12 April 2016).

Esposito, Roberto, *Bios: Biopolitics and Philosophy* (Minneapolis: University of Minnesota Press, 2008).

Foucault, Michel, *The Order of Things* (New York: Pantheon Books, 1970).

Foucault, Michel, *Discipline and Punish* (New York: Pantheon Books, 1977).

Foucault, Michel, *The History of Sexuality. Vol. I* (New York: Pantheon Books, 1978).

Foucault, Michel and Gilles Deleuze, 'Intellectuals and Power', in *Language, Counter-Memory and Practice*, ed. D. F. Bouchard (Ithaca: Cornell University Press, 1977), pp. 205–17.

Fraser, Mariam, Saraha Kember and Celia Lury, eds, *Inventive Life: Approaches to the New Vitalism* (London: Sage, 2006).

Gabrys, Jennifer, *Digital Rubbish: A Natural History of Electronics* (Ann Arbor: University of Michigan Press, 2011).

Gatens, Moira and Genevieve Lloyd, *Collective Imaginings: Spinoza, Past and Present* (New York: Routledge, 1999).

Gilroy, Paul, *Against Race: Imagining Political Culture beyond the Color Line* (Cambridge, MA; London: Harvard University Press, 2000).

Glissant, Edouard, *Poetics of Relation*, trans. Betsy Wing (Ann Arbor: University of Michigan Press, [1990] 1997).

Grewal, Inderpal and Caren Kaplan, eds, *Scattered Hegemonies: Postmodernity and Transnational Feminist Practices* (Minneapolis: University of Minnesota Press, 1994).

Grosz, Elizabeth, *The Nick of Time* (Durham, NC: Duke University Press, 2004).

Guattari, Felix, *Chaosmosis: An Ethico-Aesthetic Paradigm* (Sydney: Power Publications, 1995).

Guattari, Felix, *The Three Ecologies* (London: The Athlone Press, 2000).

Hall, Stuart and Paul du Gay, *Questions of Cultural Identity* (London: SAGE, 1996).

Halsey, Mark, *Deleuze and Environmental Damage* (London: Ashgate Publishing, 2006).

Haraway, Donna, *Simians, Cyborgs, and Women: When Species Meet* (New York: Routledge, 1991).

Haraway, Donna, *Modest_Witness@Second_Millennium. FemaleMan©_Meets_OncoMouseTM* (London; New York: Routledge, 1997).

Haraway, Donna, *When Species Meet* (Minneapolis: University of Minnesota Press, 2007).

Hardt, Michael and Antonio Negri, *Empire* (Cambridge, MA: Harvard University Press, 2000).

Hartsock, Nancy, 'The Feminist Standpoint: Developing the Ground for a Specifically Feminist Historical Materialism', in *Feminism and Methodology*, ed. Sandra Harding (London: Open University Press, 1987), pp. 35–54.

Hochschild, Arlie R., 'Global Care Chains and Emotional Surplus Value', in *On The Edge: Living with Global Capitalism*, ed. W. Hutton and A. Giddens (London: Jonathan Cape, 2000), pp. 130–46.
hooks, bell, *Ain't I a Woman* (Boston: South End Press, 1981).
Irigaray, Luce, *Speculum of the Other Woman* (Ithaca, NY: Cornell University Press, 1985).
Irigaray, Luce, *This Sex Which Is Not One* (Ithaca, NY: Cornell University Press, 1985).
Lemke, Thomas, *Biopolitics: An Advanced Introduction* (New York: New York University Press, 2011).
Lloyd, Genevieve, *The Man of Reason: Male and Female in Western Philosophy* (London: Methuen, 1984).
Lloyd, Genevieve, *Part of Nature: Self-knowledge in Spinoza's Ethics* (Ithaca, NY; London: Cornell University Press, 1994).
Lloyd, Genevieve, *Spinoza and the Ethics* (London; New York: Routledge, 1996).
Marks, John, *Gilles Deleuze: Vitalism and Multiplicity* (London: Pluto Press, 1998).
Moulier Boutang, Yann, *Cognitive Capitalism* (Cambridge: Polity Press, 2012).
Noys, Benjamin, *The Persistence of the Negative: A Critique of Contemporary Continental Theory* (Edinburgh: Edinburgh University Press, 2010).
Protevi, John and Paul Patton, eds, *Between Derrida and Deleuze* (London: Continuum, 2003).
Rabinow, Paul, *Anthropos Today* (Princeton: Princeton University Press, 2003).
Rose, Nicholas, 'The Politics of Life Itself', *Theory, Culture, and Society*, Vol. 18, Issue 6 (December 2001), pp. 1–30.
Rose, Nicholas, *The Politics of Life Itself: Biomedicine, Power and Subjectivity in the Twentieth-first Century* (Princeton: Princeton University Press, 2007).
Rothberg, Michael, *Multidirectional Memory: Remembering the Holocaust in the Age of Decolonization* (Stanford: Stanford University Press, 2009).
Sassen, Saskia, *The Global City* (Princeton: Princeton University Press, 1992).
Sassen, Saskia, *Expulsions* (Cambridge, MA: Harvard University Press, 2014).
Shiva, Vandana, *Biopiracy: The Plunder of Nature and Knowledge* (Boston: South End Press, 1997).
Simondon, Gilbert, *Being and Technology*, ed. Arne De Boever, Alex Murray, Jon Roffe and Ashley Woodward (Edinburgh: Edinburgh University Press, 2012).
Toscano, Alberto, 'Axiomatic', in *The Deleuze Dictionary*, ed. Adrian Parr (Edinburgh: Edinburgh University Press, 2005).
Ware, Vron, *Beyond the Pale: White Women, Racism and History* (London: Verso, 1992).

Notes on Contributors

Andrej Radman has been teaching design and theory courses at Delft University of Technology, Faculty of Architecture since 2004. A graduate of the Zagreb School of Architecture in Croatia, he is a licensed architect and recipient of the Croatian Architects Association Annual Award for Housing Architecture in 2002. Radman received his Master's and Doctoral Degrees from TU Delft and joined the Architecture Theory Section as Assistant Professor in 2008. He is a member of the editorial board of the architecture theory journal *Footprint*.

Heidi Sohn is Associate Professor of Architecture Theory at the Architecture Department of the Faculty of Architecture, TU Delft. She received her doctoral title in Architecture Theory from the Faculty of Architecture, TU Delft in 2006. Since 2007 she has been academic coordinator of the Architecture Theory Section. Her main areas of investigation include genealogical enquiries of the postmodern theoretical landscape from the 1970s to the present, as well as diverse geopolitical and politico-economic expressions typical of late capitalist urbanisation.

Rosi Braidotti (BA Hons Australian National University, 1978; PhD, Université de Paris, Panthéon-Sorbonne, 1981; Honorary Degrees Helsinki, 2007 and Linkoping, 2013) is Fellow of the Australian Academy of the Humanities (FAHA), 2009; Member of the Academia Europaea (MAE), 2014; Knighthood in the order of the Netherlands Lion, 2005; and is Distinguished University Professor and founding Director of the Centre for the Humanities at Utrecht University. Her latest books are: *The Posthuman* (2013), *Nomadic Subjects* (2011) and *Nomadic Theory: The Portable Rosi Braidotti* (2011). www.rosibraidotti.com

Jenny Dankelman has a degree in Mathematics, System and Control Engineering from the University of Groningen and is professor in Minimally Invasive Technology at the Delft University of Technology since 2001. Her work focuses on the design of novel medical instruments, medical haptics, training and simulation systems, and patient safety. See www.misit.nl. She cooperates with surgeons and interventionalists of, among others, Leiden University Medical Center (UMC), Erasmus Medical Center (MC) Rotterdam, Reinier de Graaf Gasthius (RdGG) Hospital Delft and Academic Medical Center (AMC) Amsterdam. Application areas are minimally invasive surgical, needle and endovascular interventions.

Christian Girard is a licensed architect and a theoretician living in Paris. He did a Master's in Urban planning at Ecole Nationale des Ponts et Chaussées (ENPC), received a Doctorate in Philosophy (Université Paris I Sorbonne) and holds an Habilitation à Diriger des Recherches from Université Paris 8. After serving as Professeur at Ecole d'Architecture Paris-Villemin (1993–9) where he was Chair (1996–8), he was a founding member in 2000 of the Ecole N.S. d'Architecture Paris-Malaquais, where he is Professor and heads the Digital Knowledge Department.

Arie Graafland was a visiting researcher to Nanjing University, faculty of Architecture, working on a publication on urbanism (*Cities in Transition*, NAi/010 Publishers, 2015). He was a visiting research professor at the University of Hong Kong in 2014–15, and was the Deutscher Akademischer Austauschdienst (German Academic Exchange Service) (DAAD) professor at Anhalt University, teaching at the Dessau Institute of Architecture (Anhalt Hochschule Anhalt/Anhalt University of Applied Sciences HS)/Dessau International Architecture Graduate School (Anhalt HS) (DIA). in Dessau. He was professor in Architecture Theory at the Faculty of Architecture, TU Delft until 2011 (emeritus). He has lectured internationally and published extensively in these areas.

Keith Evan Green is Professor of Design and Mechanical Engineering at Cornell University. His research focuses on addressing problems and opportunities of an increasingly digital society by developing and testing meticulous, artfully designed, cyber-physical (robotic) artefacts at larger scale. His abstract draws partly from his book, *Architectural Robotics: Ecosystems of Bits, Bytes and Biology* (MIT Press, 2016).

Stavros Kousoulas studied Architecture at the National Technical University of Athens and at TU Delft. Since 2012 he has been involved in several academic activities at the Theory Section of the Faculty of Architecture of TU Delft as a guest researcher and lecturer. Currently, he is a PhD candidate at IUAV Venice participating in the Villard d' Honnecourt International Research Doctorate. His doctoral research focuses primarily on morphogenetic processes framed within assemblage theory. He has published and lectured in Europe and abroad. He is a member of the editorial board of Footprint since 2014.

Katharina D. Martin read Aesthetics and Media Art in Germany and the Netherlands. She is Lecturer for Philosophy and Moving Image Theory and

Moving Image Programme Coordinator at the ArtEZ University of the Arts, Arnhem, Netherlands. Her research gravitates around the concept of milieu as methodological instrument for the analyses of digital technology. Martin is associated member in the research project 'Form, Code, Milieu' in the Cluster of Excellence 'Image Knowledge Gestaltung' at the Humboldt University of Berlin.

Kas Oosterhuis studied architecture at the Delft Technical University. In 1987–8 he taught as Unit Master at the AA in London and worked and lived in the former studio of Theo van Doesburg in Paris together with visual artist Ilona Lénárd for one year. Since 2000 Oosterhuis runs the chair of digital architecture / Hyperbody at the TU Delft. In his book *Towards a New Kind of Building* (2010) Oosterhuis shares his thoughts on parametric design, non-standard and interactive architecture.

Rachel Prentice is an Associate Professor in the Department of Science & Technology Studies at Cornell University. Her work focuses on questions about bodies, training, knowledge and the senses in biomedical education and animal studies. In 2013, she published *Bodies in Formation: An Ethnography of Anatomy and Surgery Education* (Durham: Duke University Press, 2013).

Peg Rawes is Professor in Architecture and Philosophy and Programme Director of the Masters in Architectural History at the Bartlett School of Architecture, University College London (UCL). Recent publications include: co-author with Beth Lord, *Equal By Design* (Lone Star Productions, 2016); co-editor of *Poetic Biopolitics* (London: IB Tauris, 2016); editor of *Relational Architectural Ecologies* (London: Routledge, 2013); and author of 'Humane and Inhumane Ratios' in The Architecture Lobby's *Asymmetric Labors* (2016) and 'Spinoza's Geometric and Ecological Ratios', in *The Politics of Parametricism: Digital Technologies in Architecture*, edited by Manuel Shvartzberg and Matthew Poole (London: Bloomsbury Academic, 2015).

Chris L. Smith is the Associate Professor in Architectural Design and Technê in the Faculty of Architecture at the University of Sydney. His research is concerned with the interdisciplinary nexus of philosophy, biology and architectural theory. Chris has published on the political philosophy of Gilles Deleuze and Félix Guattari; technologies of the body; and contemporary architectural theory. His recent work, *Architecture in the Space of Flows* was a book co-edited with Professor Andrew Ballantyne (London and New York: Routledge, 2013). Presently, Chris is concentrating upon an Australian Research Council project focused on the architectural expression of scientific ideals in bio-medical laboratories and a book currently in review titled *Bare Architecture: A Schizoanalysis*.

Sjoerd van Tuinen is Assistant Professor in Philosophy at Erasmus University Rotterdam and holds a PhD from Ghent University. He is editor of several books, including *Deleuze Compendium* (Boom, 2009), *Deleuze and The Fold: A Critical Reader* (Palgrave Macmillan, 2010), *De nieuwe Franse filosofie* (Boom, 2011), *Speculative Art Histories* (Edinburgh University Press, forthcoming) and *The Polemics of Ressentiment* (Bloomsbury, forthcoming) and has authored *Sloterdijk: Binnenstebuiten denken* (Kampen: Klement, 2004).

Charles T. Wolfe is a Research Fellow in the Department of Philosophy and Moral Sciences, Ghent University. He works primarily in history and philosophy of the early modern life sciences, with a focus on materialism and vitalism. He is the author of *Materialism: A Historico-Philosophical Introduction* (Springer, 2016), and has edited volumes including *Monsters and Philosophy* (College Publications, 2005), *The Body as Object and Instrument of Knowledge* (Springer, 2010, with O. Gal), *Vitalism and the Scientific Image* (Springer, 2013, with S. Normandin) and *Brain Theory* (Palgrave MacMillan, 2014).

Index

abstraction, 9, 39, 44–5, 133–4
  abstract machines, 5–6, 11, 128, 162
  see also destratification
access, 62–5, 67–70, 72–4; *see also* correlationism
activism, 87, 288, 301, 303
actual(isation), 4, 9–10, 64, 136, 261, 292, 297, 301–3
aesthetics, 6, 32, 34, 37, 53n, 67, 69, 75, 95, 117n50, 162, 203n, 233, 281
  aesthetic concept, 10
  aesthetic invention, 107, 110
  aesthetic object, 25
  aesthetic paradigm, 7
  aesthetic theory, 68, 71, 140n25
  see also ethico-aesthetic
affect, 8, 10, 13n, 26, 33, 37, 42, 45, 76n, 84–5, 269, 281, 292–3, 297–9, 302; *see also* Deleuze; Spinoza
affordance, 24n, 31, 33; *see also* Gibson
Agamben, Giorgio, 244, 291
agencement, 28; *see also* assemblage
AI *see* artificial intelligence
Aion, 292
Alberti, Leon Battista, 65, 68–70, 72
Allen, Stan, 23, 36, 39
allostasis, 195–6, 259
Amnesty International, 25
AMO (research office of OMA), 47
amor fati, 302
anaesthesia, 189, 194, 207–8, 245–6
anthropology, 26, 30, 39, 107
  embedded anthropology, 21, 44, 50n1
anthropomorphism, 129, 143, 296
anthropotechnics, 244, 248–50, 253, 260
anthropo-urgy, 248
apparatus, 28, 41, 61–3, 189, 192, 209, 214, 219n14
Apple, 161, 163
archaeology, 61, 188; *see also* genealogy
architecture (T), 2–5, 7, 9–10, 13n14, 15n46, 21–4, 29–30, 33–40, 42–9, 59–115, 143, 158, 160–4, 168–83, 234, 247

architect, 2, 24, 30–1, 36–7, 39–44, 46–7, 67–71, 81, 87, 90–1, 95, 101, 103, 107, 109, 112–13, 124–6, 128–9, 131–3, 136–7, 162–3, 179
architectonics, 12
architectural body, 34
architectural critique, 175
architectural discourse, 42, 62, 64
architectural form, 33, 35, 37, 53n, 65, 139n, 166n; *see also* form
architectural history, 112, 126, 129–30, 133
architectural production, 66, 75
architectural robotics *see* robotics
Architectural Robotics Lab, 143
architectural theory, 37, 42, 48, 50n, 62–3, 66, 73
automated architecture *see* digital architecture
digital architecture, 42, 129, 131, 135–6
starchitecture, 37, 162–4
xenoarchitecture *see* Chu
see also architectural design; architectural education; autonomy turn; Baroque; built environment; envelopes; Gothic; plan; projective city; typology
Aristotle, 66, 272
art, 2, 7, 9, 17n65, 39, 42, 44, 46, 53n63, 65, 71, 82, 107, 109–10, 113, 130, 162, 194, 249–50, 253, 255, 261
arthritis, 41, 190, 212–13, 217
arthroscopy, 40, 211, 214, 216, 222
artificial intelligence (AI), 125, 138n1, 163
ascesis, 249
a-signifying, 14n21, 101, 109, 117n50, 187; *see also* semiotics
assemblage, 4, 7, 9, 11, 28, 72, 134, 214, 281, 297, 299, 303
Assemblage journal, 134, 140n20
Assistive Robotic Table (ART), 145–50, 152–4, 156–7, 160, 162
asymmetry, 250

augmented reality, 136, 194, 201; *see also* virtual reality
automata, 129, 270–1, 273, 275–7
automation, 123–4, 128–9, 134–5, 137–8
autonomy, 10, 46–8, 85, 125–6, 136–7
autopoiesis, 32; *see also* heteropoiesis; open system; Varela

Barad, Karen, 62–3, 192
bare life, 260, 291
Baroque, 129–30, 134
Barthes, Roland, 101, 108–9, 112–13
Bataille, Georges, 298
beauty, 10, 25, 67, 72, 103, 247; *see also* charis; grace; Spuybroek
Benedict, Ruth, 30
Bennett, Jane, 45
Bentley System's Generative Components, 36
Bergson, Henri, 257
bifurcation
    of architecture, 67–8, 73
    of nature, 64, 68, 74, 247, 253
    see also Whitehead
biogenetics, 47
bio-piracy, 295
biopolitics, 80–2, 84, 87–9, 251, 256, 260; *see also* Foucault; Spinoza
bio-power, 288, 291, 293, 296
bios, 84, 295, 300
bioscience, 26
biosphere, 44, 252
Birth of the Clinic, The, 188; *see also* Foucault
black hole, 7, 257, 301
body, 3, 10–11, 22–6, 28–9, 31–41, 45–8, 65, 67–8, 126, 128, 131–3, 169–75, 187–95, 198–201, 205–18, 221, 226, 229, 243, 253–4, 270, 272–80, 283n20, 293, 296–7, 300, 302–3
    bio-technical body, 162
    body-brain, 257
    body-centrism, 280, 284n39
    body drift, 26; *see also* Kroker
    body map, 25, 33, 36, 188–90, 192, 194, 198; *see also* Damasio
    body multiple, 11, 24; *see also* Mol
    body plan, 63, 174–5
    body without organs, 3, 9; *see also* Deleuze and Guattari
    cross-body movements, 156–7
    habit-body, 206–7
    Hyperbody, 182
    mind and body, 25, 32, 37, 40–1, 85, 131, 243, 259–61, 277, 280, 284, 294, 297

    social body, 11, 251
    see also corporeality; embodiment
Borelli, Giovanni Alfonso, 270, 275
both-and *see* Venturi
Bourdieu, Pierre, 207, 248
Braidotti, Rosi, 9, 28–9, 32, 47, 50n4, 81–3, 87
brain, 7, 25, 41, 55, 103, 115, 132, 137, 138n13, 142, 161, 165n24, 198, 222, 243–4, 254–62, 264n36, 280–1, 284, 293; *see also* neurosciences
Building Information Management (BIM), 44
built environment, 75, 90–1, 95, 98, 103, 126, 152, 161, 168–9, 172, 179–80, 186, 189, 191

CAD (computer-aided design), 132
CAD–CAM (computer-aided design–computer-aided manufacturing), 132
Canguilhem, Georges, 2, 275, 284n38
capitalism, 44, 46, 86, 244, 255, 297
    advanced, 291–6, 298, 300
    bio-genetic, 295
    cognitive capitalism *see* advanced capitalism
    neurocapitalism, 258
care, 11, 21–4, 27, 30, 56, 80–7, 95, 123–6, 131, 137–8, 143–50, 152, 222, 244–5, 255, 262, 288–9, 294
    care of the self, 81–7, 95; *see also* Foucault
Cartesian coordinate systems, 231–2
cartography, 3–4, 12, 23, 38, 50n4, 125, 133, 227, 233, 235, 243, 289, 291, 293, 296, 303; *see also* mapping
catastrophe theory, 44; *see also* chaos theory
Centre for Contemporary Architecture, 182
Champalimaud Centre for the Unknown, 101–7
chaos theory, 27; *see also* catastrophe theory; complexity theory
charis, 10; *see also* beauty; Spuybroek
Chronos, 292
Chu, Karl S., 46–7
Clark, Andy, 278, 280
clinic, 21, 24, 32, 112, 146, 188
clinical, 1–2, 6, 12n2, 26–8, 35, 40, 101–3, 109–11, 115n1, 143, 146–7, 149–50, 187, 189, 195, 227, 233, 243–4, 250, 253, 257, 294, 303
Clough, Patricia, 293
CNC (computer numerical machine), 181–2
coding, 128, 132, 134, 292–3, 296–7, 302–3
collective enunciation, 10

complexity, 134, 172
computation, 27, 35, 40–2, 46–8, 125, 128–38; see also programming
computational turn see digital turn
conatus, 84
conceptual persona, 65, 72
concrete rules, 6, 11, 162; see also stratification
context, 12, 23, 28, 33, 90–1, 110, 112, 196
control society, 82
convergence, 161–3, 293
corporeal space, 9, 187, 189–90, 192–3, 195, 198
corporeality, 9, 68, 273
correlationism, 61, 63–4; see also access
craft, 48, 131–3, 205–7, 217
critique, 1–2, 6, 12n2, 27–8, 39, 47, 86–7, 101, 109–10, 123, 243, 250, 253, 257, 288–92, 295–6, 298–9, 302–3
culture, 10, 16n58, 23, 27–8, 30, 42, 46–8, 61, 128, 136, 161, 192, 217, 247–50, 253, 290
  digital culture, 35
  Greek culture, 10
  high-technology culture, 26, 37
  material culture, 27
  network culture, 260
  surgical culture, 218
  see also nature and culture
cybernetics, 6, 32, 34–5, 44, 133, 192, 256–7
  cyber-sociability, 44
  cyberspace, 23, 136
  cyberPLAYce, 163–4
  cyber-physical system, 163–4
  post-cyber, 32

Da Vinci system, 226, 229
Damasio, Antonio, 25, 32
data, 23, 36, 41, 96n4, 133–4, 136, 144, 154–5, 170–1, 182–3, 188, 190, 194, 230, 236, 293, 295, 300
  big data, 125, 134, 293
degrees of freedom (DOF), 221, 224–5, 228, 231
DeLanda, Manuel, 134
Deleuze, Gilles, 1–3, 5, 7–10, 44–5, 65, 72, 101–2, 109–11, 113, 115, 137, 160–2, 196–8, 248–9, 252–3, 256–7, 269, 288, 291–2, 297–300; see also affect; body without organs; fold; health; plastic principle; semiotics; symptom; A Thousand Plateaus

Deleuze and Guattari, 10, 45, 50n5, 102, 109, 113, 115, 137, 160–2, 196–8, 256–7, 269, 291–2, 299; see also A Thousand Plateaus
dematerialisation see abstraction
Derrida, Jacques, 260, 305
Descartes, René, 76n16, 209, 219n14, 270, 273–5
design, 34, 47, 53n63, 80–95, 143–51, 1 55–8, 172, 174–5, 235, 243, 246, 249, 255, 262
  architectural design, 10, 21–4, 30–1, 33–7, 39–42, 47–9, 61, 66–7, 70–2, 101–2, 125–6, 130–2, 136–8, 162–4, 179–82
  digital design, 29, 130–5, 181–2
  urban design, 61–2, 133–4
desiring-production see production
destratification, 8–9, 108; see also abstraction
determination, 3–5, 11, 32, 37, 72–3, 84–5, 254, 256
diagnostics, 1–2, 13n14, 28, 38, 189, 191–4, 198, 201, 221
diagram, 5, 132, 139n14
dialectics, 123, 243, 259, 288, 296–7, 299–300, 302–3
dialysis, 21–2, 24, 30–3, 246
Diderot, Denis, 271–3, 275
differential, 9, 62–3, 86, 123, 297
digital (Φ), 36, 39, 130, 132–3, 134, 137, 152, 154, 163–4, 174, 191, 194, 200–1, 235, 294
  digital culture, 28, 35
  digital operating room assistant (DORA), 222, 235–7
  digital techniques, 22, 33, 40–2, 45, 50n1, 129–30, 139n15, 144, 175, 190
  digital turn, 6, 34, 41–2
  see also digital architecture; digital design
disciplinarity, 7, 22–4, 33–4, 61, 75, 123–5, 137, 143, 222, 230
dissimilarity, 84–6, 89
dividual, 299
DLR (Deutsches Zentrum für Luft- und Raumfahrt), 41
DOF see degrees of freedom
DORA see digital operating room assistant
dramatisation, 74–5
drawing, 22–3, 36–7, 39, 45, 47, 126, 130, 135, 235
Dreyfus, Hubert, 278
duration, 73, 302

education, 82–3, 103
　architectural education, 22–3, 33–4, 37–41, 46, 48–9, 50n1, 181–2
　medical education, 32–3, 48, 188, 190–1
　see also learning; pedagogy; surgical education
ecology, 16n51, 21, 32, 38–9, 50n4, 84, 164, 185, 197–8, 201, 252, 298, 300
　ecology of practices, 61–3
　see also diffractive methodology; material-discursive practices; umwelt; *Three Ecologies*
ecosophy, 87
edge overlay, 195
elasticity, 243–68
embodied cognition, 32, 205, 277–8, 296
embodiment, 3, 23–4, 26, 28, 37, 39–40, 62–3, 68, 107, 125–6, 128, 132, 135, 137, 171, 182–3, 205, 207, 209–10, 212–18, 252, 257, 269–70, 274–80, 289, 291, 294, 296–8, 300
emergence see event
emotion, 182
empiricism, 4, 28, 73, 248, 252, 281
enactivism, 278
endoscopy, 40–1, 191, 194, 199–200, 222, 224, 227, 229
energetics, 7
enfant terrible, 250
Engels, Friedrich, 271, 273, 275
enlightenment, 24, 81–3, 86, 266n83, 290
entailment, 12
envelopes, 39
epistemology, 29, 68, 74, 125, 132, 137, 187–8, 192, 196, 254, 290
　episteme (Foucault), 61–3, 188
　epistemology of simulation, 139n15
　folk epistemology, 24
essence, 63, 66, 73, 260
eternal return, 252, 259
ethics, 6, 26, 45, 258
　affirmative ethics, 299–303
　ethical ratio, 82
　ethico-aesthetic, 6
　Ethics, The see Spinoza
　housing ethics, 83
　relational ethics, 291–2
ethology, 12, 303
etiology, 21, 109–10, 260
e-topia, 163
event, 4, 7, 11, 23, 37, 45, 73, 137, 196, 218, 245, 251, 257, 265n81, 301, 303

evolution, 5, 7, 29, 32, 129, 132–3, 136, 168–70, 175, 178, 248, 291
experience, 13n9, 23–5, 27, 33, 49, 72–5, 85, 112, 114, 190–1, 205–6, 211–12, 214–18, 234, 256, 258, 262, 279, 302
　bodily experience, 44–5, 277–8, 283n28
　cognitive experience, 254
　emotional experience, 22
　eyes-on-experience, 200
　patient-centered experience, 31
　phenomenological experience, 74, 219n14
　sensory experience, 217
　subjective experience, 68, 74–5, 192
　see also sentience; touch

Facebook, 163, 178
feelings, 25, 193, 248; see also Damasio
fiction, 126, 134, 199
　philosophical fiction, 68
　science-fiction, 123, 129
firmitas, 66–7, 75, 160; see also Vitruvius
flickering signifier, 42, 45
floating signifier, 101–2, 107–10, 112, 115
flow (F), 4–6, 39, 44, 160–1, 171, 173, 178, 297, 300–1
　workflow, 222, 235
fold, 31–2, 44, 110, 244, 253–4, 262
form(alism), 9, 15n44, 33, 35, 37, 39, 43, 53n63, 65–6, 68–72, 75, 84, 109, 111–12, 115, 126, 129–30, 133, 161–3, 198, 200, 244, 251, 253, 255–7, 259, 272
　(de)formation, 71, 75, 251, 297
　form(ation), 71, 75, 84, 112, 243, 251, 297
　see also architectural form
Foucault, Michel, 11, 13n9, 61–2, 81–4, 87, 188–9, 209, 247, 288–96; see also *The Birth of the Clinic*; *le regard*; *The Order of Things*; panopticon; 'Society Must Be Defended'; technologies of the self
Fuller, Buckminster, 133

Gattungswesen, 247
Geerz, Clifford, 30
Gehry Technologies, 36
Gelassenheit, 246, 262
gender, 26, 46, 216, 281n3, 283n28
genealogy, 47, 50n4, 62–5, 71, 250, 289; see also archaeology
generator of diversity (GOD), 26
generic, 2
genetics, 2, 21, 23, 46–7, 63, 73, 115, 175, 243, 261, 277, 293–6; see also morphogenesis

Gibson, James Jerome, 31, 33, 124
gift economy, 10
global positioning system (GPS), 225
Goethe, Johann Wolfgang von, 2, 243
Golub, Edward, 26
Good, Byron, 24–8, 30, 32, 38, 48–9
Gothic, 161
GPS *see* global positioning system
grace, 70, 152, 246–7, 279; *see also* beauty; Three Graces
ground (T), 6, 11, 25, 62–3, 66, 82, 217, 260, 262, 288–90, 301; *see also* subterranean trends
Guattari, Félix, 3–8, 10, 32, 45, 48, 81–3, 87, 102, 109, 113, 115, 137, 160–2, 165–6, 196–8, 256–7, 269, 281, 292, 299; see also *Schizoanalytic Cartographies*; *A Thousand Plateaus*; *Three Ecologies*

habit, 35, 95, 206–7, 217, 244, 248, 251–3, 255–6, 258, 260–2, 270, 303
habit-body *see* body
habitus, 207, 261
hand, 36, 41, 126, 131–2, 147, 171, 189, 192, 194, 198, 205, 207–12, 214, 216–17, 221, 224, 227–31, 234, 261, 272, 293
haptic, 131, 214, 221–2, 224, 226, 229–30, 234
Haraway, Donna, 21, 23, 26–7, 29, 32, 37, 50n5, 82, 205, 270, 281n3, 291, 293, 298
Hayles, Katherine, 23, 25, 27, 29, 32, 34, 42, 137
health, 1–3, 22, 83, 87–8, 110, 189–90, 199, 221, 223, 233, 289, 294
  great health, 2
  see also Deleuze; Nietzsche
Heidegger, Martin, 219n14, 244, 246, 262, 281
Helmholtz, Hermann von, 271, 273, 275
heteropoiesis, 1
heuristics, 12, 137, 270, 275
HIV (human immunodeficiency virus), 295
Hollein, Hans, 126–7
home+, 143–50, 152, 158, 160, 162
homeostasis, 7, 195–6, 253, 259; *see also* symmetry
homo faber, 247
housing, 34, 80–1, 83–91, 95, 126, 175
  Housing and Planning Bill, 80
  Housing Standards Consultation, 95
  social housing, 80–1, 85–91, 95
  see also Starter Homes Design

humanism, 243–4, 249, 251–2, 260, 290; *see also* posthuman
humanities, 22, 24, 27–8, 47, 107, 289
human-machine interface *see* interface
hylomorphism, 64

icon, 26–7, 108–9
ideation, 49
Illich, Ivan, 151, 162–4
illness, 2–3, 24–5, 27, 49, 71, 190, 192–3, 198–9
immanence, 3, 6–7, 9, 11, 66, 74, 243, 246, 252, 256–7, 260–2, 292, 298, 300, 303
immunology, 26–7, 31, 243–4, 250–3, 256–8, 260–2
  immunological orchestra, 26
  see also Haraway; Sloterdijk
individuation, 2, 10, 74–5, 87, 297, 301
information
  information and communications technology (ICT), 174
  information technology, 46, 161, 174, 293
  information theory, 125, 133
instrument, 22, 24, 41, 169, 187, 194–5, 198–201, 208, 210, 212–16, 218, 221–9, 231–2, 234–6, 272
instumentalism, 32, 34, 43, 290–1, 295
interface, 43, 131, 171, 214, 262, 275
  human-machine interface, 21–4, 29, 124, 132, 151, 162–3, 182
intra-action, 62–3
Irigaray, Luce, 290

James, William, 248

Kant, Immanuel, 2, 75, 76n16, 289, 301
kinaesthetics, 40–1, 210, 215
kinematics, 152–5
Klein bottle, 111
Koolhaas, Rem, 46–8, 139n14; *see also* OMA
Kraftwerk, 129
Kwinter, Sanford, 9, 16n58, 64

La Mettrie, Julien Offray de, 269, 272–3, 275–6, 281
landscape, 23, 34–5, 115
laparoscopy, 40, 211, 231, 234
Latour, Bruno, 28–9, 31, 38, 50n5, 244
Le Corbusier, 48, 65, 70–2, 75, 86, 128

learning, 12n2, 39–40, 48–9, 205, 208–9, 230, 255–6, 260, 265n64
  cognitive approach to learning, 40
  kinaesthetic approach to learning, 205, 208
  knowledge-based learning, 230
  problem-based learning, 24, 33, 48
  rules-based learning, 39, 230
  skill-based learning, 230
  see also surgical education
Leibniz, Gottfried Wilhelm von, 129
Lenin, Vladimir, 258
Lévi-Strauss, Claude, 101, 107–11, 115, 251, 264
libido, 9
Library-Cubed, 164
lineamenta, 68–9, 75; see also Alberti
Lynn, Greg, 182

machine
  abstract machine, 5–6, 11, 69, 128, 162
  affective machine, 269, 281
  machine age, 124
  machinic phylum (Φ), 4, 63
magnetic resonance imaging (MRI), 22, 190, 277
majoritarian, 292
Malabou, Catherine, 243–4, 253–60
man and nature, 22, 24
mana, 101, 107–8, 111–12, 115
Marx, Karl, 50n5, 244, 247–8, 271, 299
Massumi, Brian, 3, 10, 62, 71–2, 109
materia, 68–9
materialism, 32, 42, 136, 271, 277, 280, 291–2, 296–8, 302; see also new materialism
materiality, 3, 14, 14n19, 22–3, 38–9, 42, 53n82, 62, 66–8, 111, 128, 130, 136, 278, 291
matters of concern, 31, 38
Maya, 36
mechanism, 269–71, 273–80
  iatromechanism, 270, 273
  teleomechanism, 269, 273
mechatronics, 41, 125, 135
medicine (F), 187–237; see also medical education
Merleau-Ponty, Maurice, 206–7, 212, 218, 273, 278–9; see also habit; habit-body
metamodelling, 6
Microsoft, 154, 163
milieu, 4, 12, 26, 67, 187, 193, 195–9, 201, 206, 248, 254
minoritarian, 292
  minor tradition, 9
  see also vortical model

MISIT (minimally invasive surgery and interventional technology) group, 222, 227
mnemotechnics, 249
Mobius strip, 111
modernity, 29, 64, 66, 71, 244, 246–51, 291
  modernism, 71, 130
  see also rupture
Mol, Annemarie, 22–3, 25–6, 28–9, 32, 38, 41
Monism, 28, 292, 295–9, 302
morphogenesis, 2, 9
multiplicity, 26, 125, 201, 252

N-1, 9
NASA (National Aeronautics and Space Administration), 37
nature and culture, 7, 247–8, 294, 298; see also man and nature
near absolute material architecture, 137
necro-politics, 294
neurosciences, 33, 103, 125, 196, 243, 254; see also brain
new materialism, 6, 271, 280
Nietzsche, Friedrich, 2, 5, 243, 249, 251–2
nomadism, 292, 297–8, 300–1, 303
nomos, 6, 10
non-, 74, 102, 114
non-linguistic sign, 101
non-organic vitality see zoe
norm(ative), 2, 6, 26, 36, 217
nosology, 188–9, 195
NSA Nonstandard Architecture, 182
NURB (non-uniform rational Basis spline), 36

OctArm, 153–4
OMA (Office for Metropolitan Architecture), 47, 139; see also AMO; Koolhaas
ONL, 175–8, 180, 182
ontology, 62, 64, 68, 134, 252, 276, 281, 292
  mechanistic, 270, 274–6
  ontology of the event, 73
  political, 292
  relational, 84, 86
operable man, 245–6, 252
operating room (OR), 2, 209, 212, 217–18, 221–2, 234–7, 246
Oppositions journal, 134
Order of Things, The, 61; see also Foucault

pain, 25, 64, 189, 193, 223, 225, 249, 300–2
Pallasmaa, Juhani, 132
panopticon, 247
parametricism, 16

particle-signs, 108
passive syntheses, 249
pathology, 30, 38, 189, 193, 205–6, 216
pedagogy (U), 6, 33; *see also* education; learning
'people who are missing' (Deleuze), 110
philosophy (U), 243–308
phylum *see* machinic phylum
plan, 47, 70, 91, 103
  body plan, 63, 174–5
plane of consistency, 6, 12
plasticity, 72, 243, 251–62
  plastic principle, 5, 72
polar coordinate systems, 228, 232; *see also* Cartesian coordinate systems
positron emission topography (PET), 23
posthumanism, 28, 32, 48, 293, 295–6, 298–9
postmodernism, 26, 46, 134, 278
potential, 288, 301, 303
potestas, 288–9, 291, 299
power of the false, 252
practice
  hermeneutic practice, 38; *see also* representation
  material practice, 9, 39–40, 42, 46
  material-discursive practices, 63
  see also Allen; design practice
pragmatism, 109
Prentice, Rachel, 25–6, 29, 32–4, 39–42, 190–3
production
  desiring-production, 1
  image production, 36–7
  knowledge production, 61, 187, 191, 193
  production of variation, 6, 14
  see also synthesis
projective city, 44–5
Prometheanism, 244
proprioception, 40–1, 213, 216, 229
prosthetics, 162, 248, 253
Protevi, John, 303

queer, 291

ratio *see* reason
rationalism, 4, 85
  digital rationalism, 130
  geometric ratio, 81, 83, 85, 89, 95, 130
  humane ratio, 85
Ravaisson, Félix, 248, 261
realism, 36, 123, 192, 247, 292
reason, 45, 66–8, 80–7, 89, 243, 290
regard, le, 209

representation, 23, 30, 32–3, 36–7, 39–43, 47, 68, 74–5, 107, 129, 133, 190, 200, 205–6, 209, 217, 222, 227, 233–5, 277; *see also* visualisation
RFID (radio-frequency identification), 235
rheostasis, 196
rhizome, 6, 9, 300
Robin Hood Gardens, 87, 90
robotics (Φ), 123–83
  architectural robotics, 138, 143, 147, 160–1
  continuum robotics, 147, 152–4, 158, 160
  robot, 41, 123, 126, 129, 133, 143, 145, 152–4, 158, 172–4, 233
  robotic condition, 168
rupture, 69–71, 74, 257

schizoanalysis, 9
Schizoanalytic Cartographies, 4–5, 8
semiotics, 7, 47, 101; *see also* a-signifying; floating signifier; non-linguistic sign; particle-sign; sign
sentience, 64, 74
sign, 108–9, 113, 115
Simondon, Gilbert, 133
simulation, 15n47, 129, 132–8, 144, 201, 210, 234, 296
singularity, 7–8, 135
skill, 26, 33, 42, 72, 128, 190, 194, 205–7, 209–10, 212, 216–17, 221, 230
Sloterdijk, Peter, 31, 243–8, 250, 252–3, 255–6, 258, 261–2
smart
  SMART (simple, minimal dimensions, application based, reliable and transparent to use) approach, 226–9
  smart city, 163
  smart design, 179
Smithson, Alison and Peter, 87, 90; *see also* Robin Hood Gardens
social constructivism, 192, 298
social technology of belonging, 62
'Society Must Be Defended', 11
socius, 32, 44; *see also* cyber-sociability
spatium, 3
speculation, 42, 162
sphereology, 31, 44, 48, 72, 252, 261; *see also* Sloterdijk
Spinoza, Benedict de, 2, 25, 67, 80–6, 95, 278
Spuybroek, Lars, 10, 107
Starter Homes Design, 95
Stengers, Isabelle, 61–2
stratification, 6, 8, 10; *see also* destratification; individuation

subjectivation, 7–8, 12, 28–9, 63–4, 82, 84, 87, 196–7, 199, 218, 244, 246, 249–52, 258, 262, 269–70
subterranean trends, 44
surgery, 25, 71, 124, 189–91, 193–5, 199, 245, 253
  minimally invasive (key-hole) surgery, 40–1, 44, 206, 209–10, 212, 214–17, 221–7, 229–31, 236–7
  surgical education, 205–18
symmetry, 6, 8, 27, 66, 175, 269; *see also* asymmetry
symptom(atology), 1–3, 11, 13n, 28, 49, 71, 109–10, 187–8, 193, 198–9, 201
synthesis, 2, 10, 26, 252, 262
swarm, 170, 172, 182, 275

tableau, 188–9, 198
technology (F)
  biotechnology, 125, 135, 293
  nanotechnology, 135
  technologies of the self, 81
  see also digital; mechatronics
technoscience, 23, 29, 32
tectonics, 29, 35
teleomechanism *see* mechanism
territory (T) existential, 4, 6–9, 12, 160–1, 300
therapeutics, 1, 13n14, 110, 155–6
Thousand Plateaus, A, 6, 11, 102, 113, 160; *see also* Deleuze and Guattari
Three Ecologies, 48, 87, 298
Three Graces, 10–11
tomography, 190, 194
Tools for Conviviality see Illich, Ivan
touch, 37, 40, 131, 140, 152, 189, 205, 209, 212, 229
transductive relation, 262
transversality, 7, 32, 48, 86–7, 192, 195, 297–8, 303

TU Delft, 1, 33–4, 124, 182, 227
two cultures, 28
typology, 66, 152

Uexküll, Jacob von, 14n23, 196–7, 201
umwelt (Uexküll), 4, 196–7
uncanny valley, 222, 233–5, 237
unconscious, 9, 126, 189, 251, 255, 258, 298
universes of reference (U), 4–5
Urbild, 197
utilitas, 66–8, 75; *see also* Vitruvius

Varela, Francisco, 32, 278
Venturi, Robert and Denise Scott Brown, 161–2
venustas, 66–7, 75; *see also* Vitruvius
virtual(isation), 4, 6–7, 9–10, 35, 64–6, 75, 136, 189, 194, 200, 208–10, 218, 229, 303
virtual reality (VR), 35, 42, 136, 200, 230, 233–4; *see also* augmented reality
Visible Human Project, 41–3, 45, 190
visualisation, 35, 37–8, 40, 48, 146, 195, 201, 227; *see also* representation
Vitruvius, 65–70, 72, 166n26; *see also firmitas; utilitas; venustas*
vortical model, 160–1

'We are the Robots' *see* Kraftwerk
welfare, 80, 82–3, 88, 288
What You See Is What You Get (WYSIWYG), 135
Whitehead, Alfred North, 64, 67, 73–4, 247
World Wide Web, 163

X-ray, 191–5, 198–9, 226–7

zoe, 28, 32, 81, 84, 86–7, 292–6, 298–301
zone of indiscernibility, 111

EU representative:
Easy Access System Europe
Mustamäe tee 50, 10621 Tallinn, Estonia
Gpsr.requests@easproject.com

www.ingramcontent.com/pod-product-compliance
Lightning Source LLC
Chambersburg PA
CBHW080923300426
44115CB00018B/2924